381

An Ocean in Common

For Gordon Hamilton,
— the investment
paid off!

thanks

[signature]

An Ocean in Common

AMERICAN NAVAL OFFICERS, SCIENTISTS, AND

THE OCEAN ENVIRONMENT

Gary E. Weir

TEXAS A&M UNIVERSITY PRESS ■ COLLEGE STATION

The paper used in this book meets the minimum requirements of the American National
Standard for Permanence of Paper for Printed Library Materials, Z39.48-1984.
Binding materials have been chosen for durability.
∞

Library of Congress Cataloging-in-Publication Data

Weir, Gary E.
 An ocean in common : American naval officers, scientists, and the ocean environment /
Gary E. Weir.
  p.cm. — (Texas A&M University military history series ; 72)
Includes bibliographical references and index.
 ISBN 1-58544-114-7
 1. Naval art and science—United States—History—20th century. 2. United States.
Navy—Officers—History—20th century. 3. Marine sciences—United States—History—
20th century. 4. Naval research—United States—History—20th century. I. Title.
II. Series
 V55.U55 W45 2001
 359'.07'0973—dc21   00-011707

FOR PROFESSOR JOHN HOWARD MORROW, JR.

Patient Mentor
and
Valued Friend

CONTENTS

ILLUSTRATIONS

PREFACE

As the twentieth century opened, it seemed unlikely that the strictly hierarchical, uniformed, war-fighting world of the naval officer would regularly intersect with the ocean scientist's theory, university laboratories, and devotion to freedom of inquiry. After 1918, however, this intersection did occur—and more completely than either culture ever expected. After World War I, the interests of scientists and American naval officers converged dramatically to produce a professional relationship of immense consequence to both the national defense and our understanding of the world ocean. This association laid the foundation for long-range ocean communication, modern submarine warfare, ocean-bottom drilling, popular and compelling ocean exploration, and the formation of oceanography as a discrete discipline. Necessity, not natural inclination, gave birth to the productive dialogue that emerged from this encounter. As it happened, neither culture could satisfy its most fundamental purpose without the other.

The world ocean defined both communities. Without the ocean, words like *ship, navy,* and *oceanography* have no meaning. The water covering 70 percent of the Earth's surface has always given life to these terms and governed their cultural context. Thus, this study focuses on the compelling motives and carefully engineered means that brought civilian scientists and naval officers together to achieve a more comprehensive understanding of the ocean environment. The level of understanding sought went beyond the traditional map and chart functions of the Navy Hydrographic Office to address the ocean in all of its dynamic complexity. While learning to communicate with each other, the ocean science community and the navy also had to confront the advent and implications of oceanography. New, powerful, and multidisciplinary, this approach to the ocean environment suggested a potential for discovery and understanding that would challenge both the traditional scientific disciplines and the customary practice of naval warfare. This study offers a historical perspective on cross-cultural communication, naval patronage, and warfare at sea.

Obliged to work in concert at first by war and then by economic adversity, civilian science and the naval community overcame near incommensurability to develop a fruitful professional dialogue. Both the demands of war and the desire to contribute to scientific understanding resulted in solutions heterogeneously engineered to accommodate the needs of both communities. In nearly every case, the process of arriving at a potent dialogue rested less with institutions and agencies than with individuals. The latter brought action, insight, leadership, and critical cultural translation skills to a situation demanding effective communication.

On the eve of World War I, naval warfare remained the officer's art. Historically, scientific and environmental considerations had a greater effect on tactics and ship handling than general strategy. While some applied ocean sciences and engineering techniques certainly proved important, they primarily enhanced an officer's ability to fight a traditional battle on a larger scale, on the surface, and in a slightly more sophisticated manner. Besides needing to understand winds and currents, naval warfare did not make demands that would prompt officers to look to science and the environment for combat solutions. While most officers and seamen certainly viewed the environment as the ultimate challenge, they derived personal solutions from experience at sea, reliable charts, and skill. For them, little else could adequately address weather, waves, and rugged coastal geography.

Theirs was a professional naval culture, transmitted to the novice by the masters at the U.S. Naval Academy and the Naval War College. This culture had its own language, mythology, validation system, and strict fundamental expectations. It existed to permit the United States to wage war at sea, a style of conflict this particular community and its European antecedents had virtually defined. On the fundamental aspects of war at sea, naval officers did not usually seek professional advice outside their own community of practitioners.

In 1914, the effectiveness of the German U-boat changed everything. Suddenly, naval warfare had a potent third dimension that extended below the keel. The submersible presented a truly unusual strategic and tactical threat without conventional solution. Where was it and what would it do? Naval officers now faced the prospect of having to look to the past. They drew on the efforts of HMS *Challenger;* the discoveries of Lt. Matthew Fontaine Maury, the navy's premier nineteenth-century ocean scientist; the exploits of the Wilkes Expedition; and their university-based contemporaries for significant strategic and tactical answers drawn from ocean and atmospheric science.

Interwar marine science had a minimum number of patrons, mostly in universities or foundations, yet there was an increasing need for direct access to the ocean well beyond the beach or shoreline. Furthermore, driven by early leaders like Frank Lillie, Thomas Wayland Vaughan, and Henry Bryant Bigelow, many within the ocean science community sought a more inclusive, multidisciplinary approach to the ocean. Oceanography, the proposed means of achieving the desired comprehensive quality, competed with the strong, traditional disciplines of biology, chemistry, geology, and physics for professional and financial support as well as recognition within the academic community. This cultural and methodological struggle affected the evolution of the new discipline and its relationship with the navy well into the last half of the twentieth century.

Motives for commitment to the ocean sciences varied considerably in Europe and America. Their profound dependence on the ocean for both food and climate prediction drew Scandinavians and northern Europeans to oceanography for a more comprehensive picture of the ocean environment. In the United States, a concerted effort at fisheries development by Spencer Fullerton Baird and his successors found only modest support and patronage at the turn of the century. As a direct consequence of the war at sea between 1914 and 1918, ocean physics, chemistry, and geology rather than marine biology took the lead in American ocean studies, giving the foundation of oceanography on the western shore of the Atlantic Ocean a very different emphasis.

Unlike the fisheries or its recognized academic competitors, infant interwar American oceanography had an unusual ally. When physicist Harvey Cornelius Hayes suggested in 1923 that the navy might build a constructive postwar image for itself that isolationist America would find comforting, officers relegated to watching torch-wielding workers scrap their battleships began to listen. Hayes, a former Swarthmore College physics professor and wartime anti-submarine acoustics researcher, suggested that the navy should begin serious work in oceanography. Like the army's Corps of Engineers, naval oceanography could provide wealth and bounty to millions in peacetime while increasing the navy's knowledge of its own combat environment.

While Hayes genuinely sought to serve the navy and improve its peacetime image, he did not neglect his own interests. A commitment to oceanography and its positive political effect might finally induce Congress to commit authorized money to open the new Naval Research Laboratory. This facility would constitute the best possible home for Hayes's acoustics research, then based at the navy's Engineering Experiment Station in Annapolis.

The navy attempted to act on the Hayes initiative in 1924, only to find the president and his budget managers less than enthusiastic. While budgetary realities restrained naval ambition, the priorities set by those excited by naval oceanography influenced the plans and programs of the interwar Navy Hydrographic Office. The scientific community had a wealth of expertise, and the navy had modest financial resources to commit as well as the ability to go to sea anywhere in the world ocean. The Hayes initiative led to a very productive interwar partnership between ocean scientists and those responsible for the navy's charts, maps, and hydrographic programs. Born of mutual poverty and limited patronage, this partnership formed a personal foundation for the ambitious and successful programs directed by federal agencies both after 1940 and in the first two decades after World War II.

In addressing these developments, this study comes to terms with a variety of critical matters. The issue of cultural engagement required careful reflection on a variety of personalities in the navy and civilian science communities. Individual attitudes illustrated an interesting mix of paralyzing parochialism and creative open dialogue. Fortunately, but not inevitably, professional and personal relationships significantly magnified the effect of the latter, fostering communication and collaboration across invisible but very real cultural boundaries.

The problem of disciplinary identity and the professional acceptance of oceanography, especially as it emerged from World War II, evoked other pivotal questions. The largely informal relationship that governed interwar cooperation between ocean scientists and the navy after 1923 demonstrated the dual application of oceanography to both civilian purposes and the national defense. During the interwar period, the small numbers of people involved, the limited yet mutually serving ambitions, and the simple nature of the relationship made it relatively easy to direct and control. After 1945, advocates of the new multidisciplinary science welcomed the opportunities presented by abundant postwar naval patronage and the potential for quick, exponential growth, but they worried that the navy would turn oceanography into an ill-defined, Pentagon problem-solving tool. Many advocates of oceanography also questioned the emphasis on the physics, chemistry, and geology of the ocean driven by particular naval interest to the possible detriment of disciplinary coherence and fields like marine biology and biological oceanography. Scientists welcomed naval funding, but they also wanted to use their newly acquired wealth and visibility to fashion oceanography into an accepted, university-based discipline, with programs, courses, teachers, and students. The navy regularly acted as a partner in this effort, but at times its programs and funding habits presented obstacles.

Patronage habits changed fundamentally and dramatically between 1919 and 1961. From the near absence of navy money for oceanography after the demise of the Hayes initiative, naval funding supported the ocean sciences through World War II and culminated in the antithesis of the 1924 executive branch budgetary policy. With the creation of the Office of Naval Research (ONR) in 1946, the navy established the most liberal, effective, and flexible federal patronage agency in history. With several ONR divisions supporting oceanography, the new discipline grew at an extraordinary rate, multiplying both the scientific possibilities and important professional concerns. With oceanography finally defining itself as an independent discipline in the postwar years, generous funding and support could pose as many problems as possibilities. Who would define the new science, its practitioners, or the levels of funding flowing from its generous naval patron?

Even if the navy made money available, did science and scientists belong on warships? Only during World War II did oceanography begin to assume the role of effective ally and shipboard tool for the navy. The conversion process that led combat officers to accept ocean scientists as part of a war-fighting team proved difficult and required careful cultural translation.

In the end, the relationship between the navy and the civilian scientific community—an association viewed as natural by many contemporary scholars—was not natural. Even with the advent of the submarine, naval officers found it difficult to make a place for ocean scientists in war at sea. While their research projects and odd instruments offered valuable basic knowledge and a more acceptable peacetime image, a war-fighting partnership remained problematic. Until World War II suggested alternatives, oceanographers and admirals did not coexist as professionals under fire or in strategic and tactical debates. After 1945, circumstances changed rapidly and fundamentally. Roughly fifty years after Pearl Harbor, an officer with a degree in oceanography rose to flag rank and employed his broad knowledge of ocean currents as a matter of routine in tracing the origins of free-floating Iranian mines during the Persian Gulf conflict. Oceanography had finally qualified as a permanent resident on the bridge of American fighting ships. This is the history of that significant change in relationship, role, and perspective. It is based upon a cultural dialogue as significant for humanity's appreciation of the world ocean as for fighting a war.

This study divides easily into three parts. The first five chapters cover the period 1919–40. They trace the development of the interwar relationship between the Hydrographic Office and the major civilian scientific institutions, as well as the extent of ocean exploration by civilian science and the navy. After these five, the first of two interpretative chapters occurs. Called *interpolations,* these essays at the end of each major section provide an opportunity to step back and view the factual and interpretive aspects of the preceding material in broad context. This technique permits the resolution of particular issues of causation, relative significance, or general interpretation requiring a horizon well beyond one or two chapters. The second section of the study—chapters 6 through 10—covers the most critical event in the development of American oceanography: World War II. The second interpolation is devoted entirely to the significance of the war. The early Cold War dominates the third and last section of the study, examining the years 1946–61 in chapters 10 through 17. Chapter 18 places the Cold War perspective in the context of the emerging National Oceanographic Program and the environmental policies of the Kennedy administration.

The choice of some terminology used in examining these issues and personalities also deserves explanation. To this day, it is sometimes difficult to draw a distinction between those scientists who are, for example, physicists working on ocean questions and those who are physical oceanographers. The emphasis is different and the latter's view of the ocean as a dynamic whole requiring methods derived from multiple disciplines to serve this "physical" aspect of oceanography certainly sets that particular practitioner apart. For the longest time, a similar debate raged over the difference between marine biology and biological oceanography, fueled in part by the views of Gordon Riley, then at Yale University. Since the process of defining oceanography extends throughout the period of time covered by this study, I decided to employ the words *oceanography* and *oceanographer* when a practitioner gave it that name or claimed that title. If I suspected from the context that oceanography more accurately described the process at hand, then I used that term. Otherwise, I employed terms like *ocean science, ocean scientist,* or *ocean science community.* In this way, the vocabulary assures that the focus remains on the ocean as the subject, but otherwise acknowledges that roles, approaches, and categories for the period remained in a state of flux or transition because oceanography was still taking shape relative to other sciences examining the ocean.

The term *dialogue* also appears rather extensively and serves a critical role. While implying understanding and effective communication, a dialogue also addresses the problem of bridging strict professional and cultural boundaries, demonstrating that potent motives and forces can oblige seemingly absolute and exclusive cultural distinctions to yield. This study demonstrates that while ocean scientists and naval officers have never escaped the prison of their particular cultural and linguistic frameworks, the imperative of war, both hot and cold, called forth translation skills that fashioned effective cultural bridges. These links formed the basis for a permanent and remarkably fruitful relationship. In the context of this study, *dialogue* suggests that an individual or community of practitioners can reach beyond a personal cultural domain to achieve a state or relationship that will permit effective communication, understanding, and professional cooperation. The essence of this study is an analysis of the evolving dialogue between the civilian oceanographic community and American naval officers.

In a study of this scope some possible subjects must take a back seat to the central focus. Unfortunately, I was not able to treat the contributions of the Royal Canadian Navy and the Canadian oceanographic community to the desired extent. To give that subject proper treatment would easily require another book. The Canadians, as this study often demonstrates, qualify as one of the U.S. Navy's best and most competent scientific and operational allies. Their contributions, especially in the northern latitudes, in the polar regions, and in the Gulf Stream proved vital to oceanographic knowledge acquired for both scientific and national defense purposes.

ACKNOWLEDGMENTS

My interest in navies and oceanography grew from an abiding professional interest in submarines and undersea warfare. Having spent many years examining the naval-industrial relationship that created and sustained American submarines for nearly a century, it is only natural that I developed a desire to appreciate the extent of naval understanding of the undersea operating environment. After all, it is hard to establish a rationale for a navy without reference to the ocean.

This project first emerged from a scholarly paper offered to a conference on the Undersea Dimension of Maritime Strategy sponsored by the Centre for Foreign Policy Studies at Dalhousie University in Halifax, Nova Scotia, in June, 1991. The research for that paper raised fundamental questions about the navy's oceanographic perspective that I could not answer. After completing a project then under way on the naval-industrial-scientific relationships governing American submarine design and construction between 1941 and 1961, I decided to return to those unanswered questions. In 1993, the U.S. Navy's Director of Naval History, Dean Allard, approved my proposal for a history of the navy's role as a participant and patron in oceanographic research. After Allard's retirement, his successor, William Dudley, generously supported the project through to the completion of the manuscript in 1998.

In the course of my research and writing, a great many people and institutions made vital contributions to the success of this project. Since oceanographers live in delightful but rather far-flung locations, supplementary financial support proved vital. The ONR—represented by Eric Hartwig, Gordon Hamilton, and Mel Briscoe—offered more financial and moral support than I had a right to expect. In his support of not only this project but also the Maury Workshop series on the history of oceanography instituted by the Naval Historical Center (NHC) and the Naval Research Laboratory, Briscoe demonstrated the ONR's continuing and enlightened interest in using the past and present to shape the future of science in the navy and in the civilian sector.

The oceanographer of the navy also provided enthusiastic support during the tenure of Rear Adms. Geoffrey Chesborough, Paul Tobin, and Jerry Ellis. Admiral Tobin even agreed in retirement to act as one of the naval reviewers for the final manuscript. The oceanographer's public affairs officer, Gail Cleere, also provided critical information and assistance with an exhibit at the Navy Memorial Museum, which was inspired by the present project.

Various scientific institutions permitted me to use their resources in very special ways to further my inquiry. The Scripps Institution of Oceanography in La Jolla, California, gave me time as a visiting scholar to perform a series of oral history interviews intended to explore deep submergence and the story of *Trieste*. Those interviews were made possible by the historical insight and financial generosity I have always found at the Navy Undersea Museum, directed by William Galvani. The Marine Biological Laboratory at Woods Hole, Massachusetts, awarded a Frank Brown Memorial Fellowship that permitted free access to a study desk in their incomparable library. Finally, the Woods Hole Oceanographic Institution's Marine Policy Center granted me continuing Guest Investigator status that enormously furthered this project. I want to thank the late James Broadus, head of marine policy, and his successor, Andrew Solow, as well as Robert Gagosian, the institution director, for their generosity and support.

All the moral support and financial aid in the world will not offer the promise of a scholarly result if the archival resources and archivists working with them are unavailable. I have benefited greatly in a personal and professional sense from archivists who went beyond the call of duty. Merely mentioning their names to the informed will grant my work further credibility: Deborah Day at Scripps; Janice Goldblum at the National Academy of Sciences; Michael Walker, Kathy Lloyd, John Hodges, and Dr. Regina Akers of the U.S. Navy's Operational Archive; William Dunkle, Steven Gegg, and Margot Garritt at the Woods Hole Oceanographic Institution (WHOI); Barry Zerby, Cary Conn, A. J. Daverede, and Michael Waasche at the National Archives; and Ralph Elder of the Center for American History at the University of Texas, Austin. Archival and technical advice from WHOI's John Portius, Tom Kleindienst, and Garfield Arthur provided insight into undersea photography, imaging technology, and the processing of images that helped both to illustrate this volume and to inform its content.

I must also acknowledge the patience and generosity of many fellow historians who have helped this project evolve in various ways. My friend and colleague in the Contemporary History Branch of the NHC, Robert Schneller, carefully examined parts of this work and his criticisms prompted many improvements. Three very able colleagues outside the NHC agreed to give the entire study a close reading. Ronald Rainger, Eric Mills, and David van Keuren survived the experience, remained enthusiastic, and provided the kind of well-considered criticism that characterizes the historical profession at its best. I am fortunate to have such colleagues. I would also like to thank Rear Adm. Paul

Tobin and Capt. Don Walsh, USN (Ret.), for providing a "sanity check" from the perspective of professional naval officers deeply involved in some of the events and practices described in this study.

Finally, I wish to make mention of a few special contributions. I want to thank Bill Dunkle of WHOI not only for carefully preserving the records and artifacts of his institution with such devotion and care, but also for the trust he placed in me when I walked through his door for the first time. He was then and still is a breath of fresh air. He offered honesty and expected the same in return. Such men are rare. I want to acknowledge WHOI archivist Steve Gegg for his friendship, professionalism, and foresight. Shipmates do not come much better than he. I also need to thank the late WHOI oceanographer Allyn Vine for offering friendship, infecting me with a small measure of his all-consuming curiosity, and displaying his constant energetic pursuit of scientific and historical knowledge. He regularly demonstrated the infinite fascination the ocean inspires.

If this study has any merit at all, it is because Prof. John Howard Morrow, Jr., formerly of the University of Tennessee, Knoxville, took the time to teach me well. The dedication that appears in this book is long overdue. I hope his present institution, the University of Georgia, knows how fortunate its students are in having him.

My family keeps me sane and smiling. Thus, I must mention how much I cherish the love and friendship of my daughter, Lili. She is sixteen, becoming wonderful, and the best reason in the world for turning off the word processor. My last words here are for my wife. Just being with Catherine is still the most delightful thing in life.

An Ocean in Common

Selling Bellevue, 1914–24

■ If it is necessary to depart from research work devoted entirely to military purposes to secure the interest of a member of the Naval Consulting Board, there is no hope of "selling Bellevue" to Congress or the nation unless such a departure is definitely made.

—PHYSICIST HARVEY C. HAYES, U.S. Naval Engineering Experiment Station, Annapolis, Maryland, 19 February 1923

THROUGHOUT THE HISTORY OF THE U.S. NAVY, SURVIVING the peace has proved as difficult as winning the war. At the conclusion of the worldwide carnage of 1914–18, most Americans turned their back on the prospect of armed conflict and foreign entanglements. In this political environment, the Navy Department searched for ways to demonstrate its peacetime utility and continuing financial need to a war-weary public and skeptical Congress. After displaying its utility during the war, oceanography figured prominently in a political survival strategy adopted by the navy during the early 1920s.[1]

Oceanography began taking form during the fifty years before the Great War, emphasizing an international, multidisciplinary effort to understand the character of the ocean. In our own time, oceanographers traditionally trace the study of the marine environment back to 21 December 1872, when the British research vessel HMS *Challenger* departed on its three and one-half year voyage of exploration. Under the direction of the University of Edinburgh's professor of natural history, Charles Wyville Thomson, this expedition formulated the goals, systematic methods, and careful analysis that later became the standard for

3

oceanographic research around the world. At the same time, his ambitious endeavor provided a historical foundation and working model for later efforts to
apply both an integrated, multidisciplinary approach and modern scientific analysis to the search for a more comprehensive picture of the marine environment.[2]

In discussing the scientific importance of the *Challenger* Expedition, the
role of the Royal Navy often falls by the wayside. The HMS *Challenger* was a
warship, as was the HMS *Beagle,* the vessel that provided Charles Darwin with
his seminal experiences at sea. Scientific exploration often took its place next to
politics, trade, and the economy in the effort to make far-flung lands familiar and
profitable for the British Empire. It came as no surprise that emerging naval
powers like Germany and the United States did the same in the new century.

Naturalist to Analyst

Before the turn of the century, most investigators studied the ocean within
the context of independent and distinct traditional disciplines. Usually trained
in biology, geology, or chemistry, the late-nineteenth-century naturalist on a
mission to discover, describe, classify, and catalogue marine life determined the
nature of research. This approach filled many voids in mankind's knowledge of
marine life, ocean chemistry, coastline configuration, and ocean-bottom geology and topography.

While naturalists certainly cooperated informally in their effort to understand the ocean, traditional distinctions between disciplines remained the rule.
In spite of significant advances in particular fields, these circumstances bridled
the potential of oceanography. Individual interests, traditions, and scientific
training led investigators to appreciate the ocean in its particulars rather than
more comprehensively as a system.[3]

After *Challenger,* it took nearly thirty years for strategy and methodology
to move beyond the particular and descriptive approach. Although understanding the ocean would always require these skills, early in the twentieth century a more inventive and inclusive approach with an energy born of political
and economic necessity cast regularly gathered information in a new light and
offered a more comprehensive understanding of the ocean.

This more dynamic, analytical approach took hold largely in response to
the need of northwestern Europe for well-managed fisheries. Scandinavia,
Great Britain, and Germany embraced oceanography because of its practical
application to an industry vital to the well-being of their people. Modern fishing techniques threatened to reduce drastically the annual catch, an economic
pillar of the region. A science preoccupied with studying marine life, currents,
the chemistry of seawater, and other major factors in fostering and sustaining
life in the sea became an important tool in the effort to insure the physical and

material welfare of several countries. In 1902 Britain, Germany, Denmark, Sweden, Finland, Norway, Holland, and Russia became charter members of the International Council for the Exploration of the Sea (ICES). Efforts like this confirmed the importance of the ocean sciences and ensured official and private support for research. Thus, while Spencer Fullerton Baird presided at the birth of the American Fish Commission in 1871, having lobbied hard for continued congressional funding, oceanography flourished in Europe, funded by the fisheries and fueled by economic need.[4]

Reflecting on these issues and events nearly thirty years later, the first director of the Woods Hole Oceanographic Institution, Harvard's Henry Bryant Bigelow, suggested that oceanography's comprehensiveness and utility emerged early as its greatest asset. In Europe, practical applications and economic value attracted talented people, stimulated new perspectives, and prompted generous public sponsorship. In a report published in 1931 under the auspices of the Committee on Oceanography of the National Academy of Sciences (NAS), Bigelow wrote, that "a period of general oceanographic stagnation might have succeeded to the preceding peak of activity [in the late nineteenth century] (this did, in fact, happen in America), had there not arisen new schools, centering their attention on the biologic economy of the inhabitants of the ocean as related to their physical-chemical environment, on mathematical analysis of the internal dynamics of the sea water, and on the geologic bearing of submarine topography and sedimentation, rather than on . . . surveys of one or another feature of the sea."[5]

Although the Europeans were the first to apply the new approach, Bigelow became its earliest American spokesman and reflected on its consequences and methodology. Did the Scandinavians, for example, realize at the turn of the century the cultural bridges that practitioners had yet to build between discrete scientific disciplines to achieve the comprehensive understanding that oceanography sought? In his 1931 essay titled "A Developing Viewpoint in Oceanography," Bigelow displayed the contrast between old and new and a clear appreciation that this transformation constituted a revolution within science. In *Oceanography,* he wrote,

> the keynote must be physical, chemical, and biological unity, not diversity, for everything that takes place in the sea within the realm of any one of these artificially divorced sciences impinges upon all the rest of them. In a word, until new vistas develop, we believe that our ventures in oceanography will be most profitable if we regard the sea as dynamic, not as something static, and if we focus our attention on the cycle of life and energy as a whole in the sea, instead of confining our individual outlook to one or another restricted phase, whether it

be biologic, physical, chemical, or geologic. This applies to every oceanographer; every one of us, if he is to draw the veil backward at all, must think and work in several disciplines. He must be either something of a jack-of-all-trades or so closely in tune with colleagues working in other disciplines that all can pull together.[6]

Decades before Bigelow put pen to paper, converting civilian scientists to the value of a multidisciplinary endeavor seemed daunting. Oceanography required marine biologists, seawater chemists, ocean-floor geologists, and those interested in the physical behavior of the ocean to build bridges between the traditional discrete disciplines that governed their professional lives. University faculties in marine biology, for example, defined the subject, approved the accepted methodologies, set guidelines for published and oral dialogue, and controlled the system for reward and censure. Oceanography challenged this closed world.

As that process began for the United States in the early twentieth century, the prospect of naval involvement presented additional daunting challenges. Whereas ocean scientists might find common denominators in their training, methods, and goals, naval officers came from a different world. While they depended completely upon the ocean, the institution that regulated their existence had limited contact with the scientific community and saw few reasons to change. With rare exceptions, the practitioners of naval warfare seemed to share few if any of the goals of science and basic research. The central problem in the growth of naval oceanography emerged from this difference. Did their interests and independent missions intersect? If they did, only a determined effort to establish a dialogue, to understand the same professional language, and to share the same determination to secure critical scientific knowledge and an effective national defense might open a productive and interesting future. All of these challenges remained unaddressed at the turn of the century, and the determination to resolve them had yet to emerge.[7]

Undersea Warfare as a Catalyst

World War I and the U-boat menace provided the catalyst that accelerated American naval oceanographic studies, dramatically altered scientific practice, and profoundly affected the selection of new subjects for investigation. Wartime projects under the aegis of the Naval Consulting Board (NCB) and the NAS's National Research Council (NRC) drew scientists from every specialty out of their normal academic or industrial environment to address the critical needs of the operational forces.[8] Antisubmarine warfare and prosubmarine inquiries provided considerable incentive, and added new avenues of investigation to the study of the ocean depths which some scientists continued to pursue after the war ended.[9] In the course of this work, oceanography came of age in America and revealed itself to the U.S. Navy.

Between 1914 and 1918, oceanographic antisubmarine research, as opposed to prosubmarine investigations, dominated the attention of the Allied scientists who were asked to generate an effective way of neutralizing the German submarine threat. In the United States, the organization of this effort took place along two parallel lines: one directed by the navy, and the other by the civilian scientific community at the request of the Navy Department.

In the navy, the primary effort to draft scientists into the war effort took form with the NCB, created in July, 1915.[10] Secretary of the Navy Josephus Daniels initially established this civilian scientific board and placed it under the direction of the famous inventor Thomas Alva Edison to evaluate suggestions and inventions offered to improve the navy's performance should America become involved in the war.

Throughout its existence, the NCB remained an advisory body to the secretary of the navy. It could encourage the research and development of systems such as the magnetic submarine detector invented by physicist Vannevar Bush.[11] However, having no research and development money of its own, the board and its committees could only act as advocates, and urged Secretary Daniels to support promising developments in the private sector.[12]

When the United States became a belligerent in 1917, Daniels expanded the NCB's powers and it instituted special committees to explore difficult wartime problems. As early as 26 October 1915, the secretary ordered the navy's Bureau of Construction and Repair (BuC&R) to investigate a means of detecting submerged submarines from a surface ship. One week after President Woodrow Wilson severed relations with Germany on 3 February 1917, the NCB created the Committee on Special Problems to coordinate naval and civilian efforts to detect and destroy U-boats—including efforts initiated by the naval bureaus and sponsored by the NRC. This committee, chaired by board member Lawrence Addicks, divided the problem of antisubmarine warfare (ASW) into its component parts for consideration by subcommittees. These subdivisions of the Addicks Committee explored all available antisubmarine tactics and techniques, including underwater sound, nets, magnetic and electrical means, underwater searchlights and visibility, and air attack.[13]

To focus naval resources on the best areas of inquiry, Edison's staff invited experts in ASW-related fields nominated by the NCB to gather in New York at the Engineering Society Building on 3 March 1917. These specialists concluded that underwater sound and echo ranging offered the most promising avenue of exploration for ASW scientists in the war effort. Physics and physical oceanography thus immediately became vital to the national war effort.

One month later, the NCB recommended that Daniels divert $10,000 of the money earmarked for the development of the Naval Research Laboratory (NRL) to the Committee on Special Problems. The U-boat threat had become so important that the board voted unanimously to place research on submarine detection above the creation of the long-desired NRL.[14]

After the New York conference, the Submarine Signal Company of Boston, a specialist in underwater sound, won the support of the NCB's Submarine Detection by Sound Subcommittee for its promising work on submarine detection.[15] This firm incorporated a powerful oscillator developed by Reginald A. Fessenden into a practical device for detecting icebergs, and demonstrated the possibility of determining ocean depth by means of echo ranging. When the company's first U-boat detection device failed to impress the Navy Department, the NCB encouraged cooperative research by Submarine Signal, General Electric, and Western Electric at the latter's facility in Nahant, Massachusetts. Armed with complete state-of-the-art knowledge, the firms at Nahant continued to explore various methods of submarine detection, including echo ranging and promising hydrophone listening devices. The NRC followed a similar course, supporting the creation of the Naval Experimental Station in New London and recruiting, among others, Robert Millikan of the University of Chicago and Wisconsin's Max Mason to apply their skills to the ASW problem. Mason provided the creative genius behind several generations of the navy's "M" or multiple tube, passive submarine sensors. This apparatus focused the sound drawn from the water to ascertain its source. To determine the direction from which the sound came, the operator needed only to seek the maximum output in his earphones by turning a dial.

In June, 1917, the NRC furthered the process of cooperation and education on the U-boat detection problem by arranging an international conference. The council brought to Washington British experts like Sir Ernest Rutherford, and his French military counterparts, Majors Fabry and Abraham and Captain Dupray—all three of whom were trained in the pioneering underwater sound techniques of Paul Langevin and Constantin Chilowsky.[16]

In addition, the NAS collected scientific intelligence from around the world through its Research Information Service, established during the war by the NRC in Washington, D.C., Rome, London, and Paris. Information from participating scientists kept the NRC and the navy abreast of the latest work done on underwater sound and echo ranging.[17]

Shortly before the armistice, American naval representatives journeyed to Paris for a conference on "supersonics" or underwater echo ranging. The Americans met with their French and British counterparts from 19–22 October 1918 and received more complete information about Langevin's progress in piezo-electric research as well as one of the underwater sound transmission devices the French had designed to apply the theories developed by Chilowsky and Langevin.[18]

Reports on the conference were prepared by both the American associate scientific attaché in Paris, Karl T. Compton, and one of the leading scientists in the American effort to build an operational supersonic device, Prof. J. H. Morecroft of Columbia University.[19] They not only described in great detail the performance of the Langevin device, but also demonstrated a heightened appreciation of the properties of the ocean that affect undersea sound transmission. In the course of American experiments in underwater signaling, Compton "noticed, as have all those who have been engaged in listening under water, great irregularities in transmission due certainly to the influence of the water medium." He went on to discuss the viscosity of the water, its temperature, the presence of marine life and debris, and the affect of bubbles on sound transmission.[20] Oceanography was quickly becoming indispensable to modern ASW.

In the short period of time America participated in the conflict, scientific research helped keep the U-boats at bay. When the advent of convoys in 1917 required some capability for detecting U-boats, U.S. industry manufactured three thousand SC hydrophones with their characteristic rotating T bar and stethoscope listening set. Although primitive, the detectors protruding from the bottom of American and British submarine chasers forced German submarine commanders to take greater care in approaching convoys. In many instances developments took longer to reach the operational forces. Vannevar Bush's device for detecting a U-boat as it broke a magnetic field was barely installed in British minesweepers for testing before the worldwide struggle ended. These and other wartime experiences identified science as an important partner in modern naval warfare. As historian A. Hunter Dupree observed many years later, effective weapons, doctrine, and seamanship now required a scientific supplement. Dupree saw "the very approach to the problem as one that could be solved only by massed and coordinated scientific resources demonstrated clearly that a new era of warfare had arrived and that science had an essential place in it."[21]

What sort of warfare lay in the future? In 1919 very few Americans wanted to address that question. Peace, not war, was uppermost in their minds.

Tight Budgets and Arms Control

Departing from earlier opinion, President Warren G. Harding sought to identify with this postwar American mood. On inauguration day in 1921, he offered to place the very ambitious 1916 warship construction program on the table at an international conference to reduce naval armaments. When Secretary of State Charles Evans Hughes formally proposed the meeting, the sheer cost of war and a possible naval arms race—especially between the United States, Great Britain, and Japan—provoked wide public support for his proposal. A naval building competition similar to that between Britain and Imperial Germany before the Great War, was perceived by the voting public as well as by many in Congress as destabilizing, a waste of resources, and a threat to national security. Hughes also argued that a policy of conciliation combined with a willingness to negotiate would diffuse international tensions with Japan over the Anglo-American presence in the western Pacific, and with Britain over naval supremacy.

The conference that began in Washington on 12 November 1921 halted the substantial American construction program authorized five years before and established a fixed ratio of relative battleship strength between Britain, the United States, Japan, France, and Italy. These limits and other restrictions accepted by the participants signing the treaties at the Washington Naval Conference laid the foundation for interwar American naval policy. The Congress not only accepted the limits set by the conference in 1922, but for the next twelve years refused to appropriate funds for new construction to meet the minimum force levels permitted by the agreements.[22] As these events unfolded, the navy struggled both to meet its operational commitments and to convince the public and the Congress of its peacetime value.

Limited resources also plagued postwar civilian oceanography. In 1919 the NRC appointed its first Committee on Oceanography. Henry B. Bigelow of Harvard's Museum of Comparative Zoology convinced the council's Division of Biology and Agriculture to encourage a cooperative survey of ocean life. Henry F. Moore of the U.S. Bureau of Fisheries served as chairman of a three-man committee that included Bigelow and Alfred G. Mayor from the Carnegie Institution's marine laboratory. Unfortunately, the paucity of public and private funds terminated the project and the NRC dissolved the committee before it accomplished anything substantial.[23]

Under the leadership of Rear Adm. Edward Simpson, the U.S. Navy Hydrographic Office (Hydro) avoided the committee's fate and flourished in spite of tight budgets and naval arms limitations. Hydro aggressively pursued old and new tasks just as the NRC committee folded its tent. Attached to the Bureau of Navigation (BuNav) since 1866, Hydro traced its history to the creation of the Depot of Charts and Instruments in December, 1830. In its various forms, the Hydrographic Office boasted affiliation with the Naval Observatory dating from 1854, and participation in Lt. Charles Wilkes's U.S. Exploring Expedition of 1838–42. It also provided a stimulating home for the pioneering oceanographic work of Comdr. Matthew Fontaine Maury before his entry into Confederate service in 1861.[24]

No sooner did Simpson's command resume its work on charts and maps after the Great War than it stepped upon the world stage in response to an invitation from Great Britain.[25] In 1919, the Admiralty proposed an international hydrographic conference. From London, the American naval attaché, Rear Adm. Harry S. Knapp, described the purpose of the conference very precisely in the standard staccato language of a dispatch: "Object conference consider advisability of all maritime nations adoption similar method in preparation construction and production of charts and all hydrographic publications; instituting prompt system mutual exchange Hydrographic information between all countries; provision opportunity for consultation and discussion of Hydrographic subjects generally by Hydrographic experts of the world."[26]

At a time when America began disengaging from European involvement, this invitation to remain committed in matters relating to the ocean and navigation offered important advantages as well as the possibility of unwelcome international complications. Simpson successfully argued that Hydro should participate and received permission to make the journey to London for the June, 1919, meeting along with his colleague from the Hydrographic Office, George W. Littlehales. Navy Secretary Josephus Daniels applied to the House Foreign Affairs Committee for the authority to let the navy's delegation attend the conference. This procedure was set down in the Deficiency Act of 4 March 1913. However, while waiting for congressional approval, the secretary issued the travel orders and told the commander of U.S. naval forces in Europe to treat American attendance at the conference as an intradepartmental matter. Daniels justified the early dispatch of Simpson and Littlehales by arguing that they could collect valuable technical information at the invitation of the British Admiralty. Congress gave its approval after the fact. Along with scientists from many other nations meeting in London between 24 June and 16 July 1919, the two representatives from Hydro participated in the creation of the International Hydrographic Bureau (IHB) and, in the early postwar years, kept the United States involved internationally in ocean surveying and marine science.[27]

Those scientists and naval officers representing the twenty-seven nations belonging to the IHB envisioned a mission for the new organization that went beyond the standardization of terms and techniques in chart production. Admiral Simpson agreed to take on administrative responsibility as part of a three-man provisional committee empowered by the London conference to give form to the bureau. Although its creators vowed to avoid any interference in matters of international policy, the bureau had from the outset a mandate to act as an international advisory body. It promoted both the creation of hydrographic offices in countries lacking such expert institutions, and productive relationships among existing offices. The IHB also lobbied with member governments to grant financial and material support to those engaged in research of general interest to member countries.[28]

Although Simpson left Hydro and surrendered his position on the provisional committee in June, 1920, his French and British colleagues continued preparations and sent a final copy of the bureau's statutes to the prospective member nations on 1 January 1922.

Other nations responded to the bureau's request for ratification and formal adherence more quickly than did theUnited States. One of its statutes proclaimed that the new bureau would function under the auspices of the League of Nations. Although the League was President Wilson's brainchild, the Senate decided against U.S. membership as part of its repudiation of the Versailles peace settlement. After Wilson left office, President Harding's State Department made sure that the bureau would not entangle the United States in the kind of international commitments the Congress avoided in rejecting the formula composed by the Big Four at war's end. Throughout the early years of the bureau's existence, both civil and naval authorities in the United States repeatedly made it clear that they expected neither an extension of the League's influence into bureau technical matters nor any disadvantage or prejudice to result from Washington's refusal to join the League. Early in the game, the administration decided that the bureau's value did not justify the risk of international political complications. After receiving further clarification and adequate assurances, the State Department informed IHB headquarters in Monaco of America's participation.[29]

Hydro got off to a slow start in its bureau participation, but it quickly made up for lost time. Simpson's departure and the briefly delayed American covenant with the IHB caused the first election of bureau officers to pass without candidates from the United States. Once Congress, the State Department, and the navy determined the proper level of participation and financial commitment to Monaco's satisfaction, steps were taken to propose an American candidate for a seat on the bureau's board of directors due to open in 1923.

Reviewing a short list of possibilities, the navy, in cooperation with a Commerce Department concerned about international fisheries management, nominated Rear Adm. Albert P. Niblack for consideration by IHB members. Niblack became part of the provisional organizing committee in June, 1920, after Simpson's departure, and participated in drawing up the bureau statutes later that summer. After witnessing the composition and passage of those regulations while serving as commander of U.S. naval forces in Europe, the admiral urged the Navy and State Departments to ensure American commitment to the IHB. He retired in 1923 as a vice admiral and then accepted the joint nomination from the navy and the Department of Commerce to fill the vacancy on the bureau's board of directors created by the death of J. Renaud, head of the French navy's Hydrographic Service. He won election to the board in 1924 and then to the IHB presidency before his death in Monte Carlo in 1929.[30]

Bunnies, Birds, and the SDF

Hydro's international activities with the IHB and practical wartime experience attracted the attention of many scientists in search of logistical and financial support for their ocean research. However, the navy had no money to give in a time of contracting budgets and naval arms limitation. But those studying the ocean, like Prof. William H. Hobbs, a geologist at the University of Michigan, had other ideas. Would the navy permit scientists and officers on board its operational vessels to collect scientific data in the course of a regular deployment? Hobbs not only addressed this question to Adm. Robert E. Coontz, the chief of naval operations (CNO), in May, 1921, but he also wondered if the Navy Department would permit the use of a ship's sophisticated instrumentation to accumulate scientific information while the crew carried out its assigned task for the fleet.[31] Going to sea was essential for those studying oceanography, and the navy had ships traveling to all parts of the world. Furthermore, navy physicist Harvey Hayes had developed a sonic depth finder (SDF) based upon his research into active sonar during the war. This device projected the sound generated by a Fessenden oscillator toward the bottom of the ocean and used the time it took for the echo to return as an indication of depth. Hobbs and others in the civilian science community quickly recognized the scientific possibilities of the SDF. Might the professor's summer research trip to the Celebes Sea in Indonesia between Borneo and New Guinea not provide the best possible chance to test the SDF? If a naval vessel proved unavailable, Hobbs suggested the regularly scheduled army transport ships running between the West Coast and Manila via Honolulu as the perfect alternative.

Barely five days after his request, Admiral Coontz advised Hobbs that he was granting Hayes permission to make the trip, but added that he could not commit the War Department and army transport vessels to the project. In the end, the army's cooperation proved unnecessary. Professor Hobbs made the trip by army transport, but Acting Secretary of the Navy Charles E. McVay found Hobbs and Hayes a naval vessel already in the vicinity of the Philippines for their SDF tests and scientific work. After carrying out oceanographic investigations in the western Pacific with the permission of the Japanese navy, Hobbs met Hayes and the ship in Manila.[32]

Their experiments, as well as earlier evaluations conducted by the navy at the Engineering Experiment Station (EES) in Annapolis, Maryland, confirmed the importance of the SDF for both the navy and the scientific community. At the NRC's annual meeting in April, 1922, Harvard geologist William M. Davis suggested more extensive testing of the SDF, including alterations to the device permitting determination of the ocean-bottom slope as well as the depth at any given point.[33] To the universal acclaim of the scientific community, Hayes used his invention to make the first complete bottom profile of any ocean during the June, 1922, transatlantic crossing of the destroyer *Stewart* (DD-224) from Newport, Rhode Island, to Gibraltar. With Hayes on board, *Stewart,* under the command of Lt. Comdr. Norman R. Van De Veer, made nine hundred soundings of the ocean bottom to depths greater than three thousand feet. The news of this accomplishment went through the scientific community like a bolt of lightning. As historian Susan Schlee observed, "The results were indeed spectacular. The *Challenger* in her entire three-and-a half-year voyage had taken less than three hundred soundings in depths exceeding 1,000 fathoms, and in the same years the Coast Survey considered it a good field season when 100 or so deep soundings were collected."[34]

The navy's new instrument gave scientists their first comprehensive look at the configuration of the ocean floor in all its irregularity. Sound at last began to reveal what years of work with rope and wire sounding lines had only suggested. Civilian science quickly concluded that the number and range of naval vessels as well as the SDF's revolutionary potential made the U.S. Navy an indispensable partner in ocean exploration.

Scientists committed to oceanography eagerly explored the possibilities of a partnership with the navy. Although NRC officials appreciated the uncertainties faced by the Navy Department in the wake of the Washington Naval Conference, Yale's Herbert E. Gregory proposed a project that helped define the navy's role in the investigations deemed most important by American scientists. Gregory served as director of the Bernice Pauahi Bishop Museum in Honolulu, an institution that had become a center for the scientific exploration of the Pacific Ocean. While serving as volunteer chairman of the Committee on Pacific Investigations of the NRC's Division of Foreign Relations, Gregory proposed that the division sponsor a project to probe the agricultural, marine, and mineral resources of the Pacific region "to serve as an intelligent guide for economic, social, and political development." The NRC found the proposal appealing and Albert L. Barrows, secretary of the Division of Foreign Relations, communicated with George Littlehales of Hydro, asking him to cast Gregory's ideas into a form suitable for presentation to the secretary of the navy. Gregory and the NRC wanted the secretary to commit naval vessels, instruments, and personnel to exploration in the Pacific. About a month before the *Stewart* performed its revolutionary SDF profile of the Atlantic, Gregory's committee asked E. W. Nelson of the U.S. Biological Survey to approach officials of the War and Navy Departments in search of support for Gregory's planned investigations of the Pacific.[35]

In his letter, Nelson told Assistant Secretary of the Navy Theodore Roosevelt, Jr., that Gregory was worried about birds and rabbits. Although this combination seemed to have little to do with ocean science of interest to the navy, conditions created by these animals offered an excellent opportunity for hydrographic and oceanographic studies. The Hawaiian Islands Bird Reservation, a chain of small islands jutting into the Pacific northwesterly from Honolulu, had long suffered from a major environmental imbalance. At the turn of the century, small domestic rabbits were introduced on Lysinski Island, one of the smaller links in the chain and a breeding ground for a wide variety of rare birds. The rabbits quickly multiplied and consumed virtually all of the island's vegetation, leaving a desert behind them. An earlier expedition reduced the number of rabbits on the islands to a manageable number, permitting the foliage to grow and the birds to return. Gregory wanted to investigate the current size and effect of the rabbit population while conducting a thorough scientific survey of the islands and their surrounding waters. This would also provide money and support for mapping work and a program of offshore SDF soundings that would reveal much about the ocean-bottom topography of the region.[36]

Roosevelt's considerable enthusiasm for the project spoke volumes about postwar problems and planning within the navy. Given popular public reaction to the horrors of the Great War, the navy desperately needed a new, constructive peacetime role to improve its image and increase its political strength. The number of American warships had fallen drastically with no sign of recovery in the foreseeable future. The public and Congress did not care to hear about the need for increased appropriations for implements of war. With naval strategic planning emphasizing more than ever the primary importance of the Pacific Ocean, perhaps a scientific venture near Hawaii would help reshape the navy's public image while furnishing important information about the waters near one of the navy's most important Pacific bases, Pearl Harbor. Besides, George Littlehales of Hydro had called the attention of the scientific community and the navy to the importance of Pacific oceanography as far back as 1916 in a paper he presented to the NAS.[37] Gregory thus found the assistant secretary very enthusiastic and receptive when they spoke a few days after Nelson's letter reached the Navy Department.[38] Roosevelt felt strongly that committing a destroyer or minesweeper for two months to support an expedition to the Hawaiian bird islands might produce considerable dividends for the navy.

Gregory's proposed expedition received considerable support from others, both within the navy and the federal government. Admiral Simpson, the former head of Hydro who then was serving as commandant of the Fourteenth Naval District at Pearl Harbor, reacted very positively to the bird island proposal when Gregory informed him of the project by letter in July, 1922. The admiral thought that a minesweeper, ironically of the Bird Class, would prove very suitable if the Navy Department would assign one of these ships to the expedition. Reporting on the project to the CNO, Simpson described meeting

with Gregory and making him fully aware of the shortage of available naval vessels. However, he diplomatically reminded Admiral Coontz that both Roosevelt and Secretary of Agriculture Henry A. Wallace supported the project and
had suggested that "an expedition of that sort might, in addition to their work,
prove of value as presenting opportunities for Naval or hydrographic investigation, as the charts in that vicinity were based on old and uncertain data, and
that while scientists were ashore the vessel would have the opportunity to do a
certain amount of surveying work."[39]

It did not take long for the Navy Department to declare its support. In
May, 1922, Assistant Secretary Roosevelt told E. W. Nelson that the navy would
definitely help the bird island project. In expressing his support nine months
later, Henry Wallace suggested that the expedition might prove even more
valuable to both his department and the navy if the venture included surveys of
Wake, Midway, and Johnston Islands. Secretary of the Navy Edwin Denby
approved the enterprise on 5 February 1923 and delegated responsibility for
the ship and naval personnel to the commandant of the Fourteenth Naval
District.[40]

Hayes, Gregory, and Australia

Shortly after Denby announced his decision, Harvey Hayes wrote a memorandum to his supervisor, Capt. John Halligan, Jr., the officer in charge of the
EES. This correspondence best marks the beginning of an official naval commitment to modern oceanographic research. Hayes suggested that activities
like testing the SDF with Davis in the Philippines or supporting the Hawaiian
Bird Island Expedition provided a key to the navy's future. He had become
frustrated with congressional reluctance to provide regular and adequate funding for the NRL and voiced concern about the adverse effect that this might
have on his underwater sound work and other projects destined to move to the
laboratory. At the end of the memorandum's second paragraph the author asserted that "the Bellevue Station [NRL] will never be definitely and adequately
supported by Congress until its members are made to realize the importance
of military research, as such, or until their interest in the station is aroused
through the successful application of the results of these researches for other
than military purposes."[41]

Professor Hayes reasoned that any field of naval research designed to persuade Congress must meet certain criteria. For example, the new endeavor
should complement and not interfere with the NRL's main mission of conducting applied research in support of naval operations. Furthermore, the work
would have to fall exclusively within the navy's sphere of influence and should
complement established naval policies. Most important, Hayes wanted to gen-

erate, on a regular basis and at minimum expense, data valuable enough to attract the attention of civilian scientists and the press. In his 19 February 1923
memo to Halligan, Hayes concluded "that these researches should be undertaken in the field of oceanography."[42] The multifaceted nature of this science
might attract the attention of many diverse and talented investigators to the
study of the navy's natural environment. Along with biologists, geologists,
chemists, and physicists, the navy would contribute to human knowledge in a
way that could potentially affect the economic welfare of millions.

He suggested further that the navy mount an oceanographic expedition
covering a precisely defined area of the Pacific Ocean. Hydro and the Bureau of
Fisheries would contribute a large portion of the necessary equipment and the
expedition could turn to the Bureau of Engineering (BuEng) for the radio
equipment and the SDF. If the Navy Department carefully courted and selected
civilian scientists and institutions, the reputations and achievements of those
involved would soon make the navy's project the center of scientific attention.
To sustain this credibility and identify it with the navy, the NRL would publish
the results of the expedition in a laboratory contributions series. If properly organized, Hayes believed that this project would draw financial grants and gifts
galore while at the same time placing the peacetime navy and the NRL in the
limelight.[43]

As the Hayes memorandum made its way through channels, it gathered
positive endorsements from all quarters. The BuNav applauded Hayes's initiative, and Capt. Frederic B. Bassett, hydrographer of the navy, enthusiastically
supported the proposal, citing eight precedents for navy-supported oceanographic research. Assistant Secretary Roosevelt noted that the existing operating force plan made Hayes's recommendations impossible, but insisted that the
CNO allow for such a project in a revised operations plan in the near future.[44]

Hayes's proposals struck a chord within both the navy and the scientific
community that generated an instant response. Two cooperative projects began almost simultaneously within days of the Hayes memorandum. The Navy
Department elected to participate in the 1923 Pan-Pacific Science Congress by
agreeing to showcase Hayes's SDF for the scientists in attendance and Gregory's Hawaiian Bird Island Expedition finally got under way.

The excitement caused by the potential of the depth finder and Hayes's
memorandum to Halligan prompted many within the scientific community to
recommend immediate use of the SDF for oceanographic exploration. Ocean-
bottom profiles similar to those taken by the *Stewart* in the Atlantic Ocean
seemed the natural course. Professor Davis, now emeritus at Harvard University, suggested naval participation in the Second Pan-Pacific Science Congress
scheduled for Sydney and Melbourne, Australia, in August, 1923. Secretary
Denby had received a notice from the State Department on 22 January that

Great Britain had asked the navy to participate. Davis proposed that a naval vessel equipped with an SDF conduct a series of bottom profiles while en route to the conference. He predicted that "If successfully carried through there can be no doubt that the achievement would be the outstanding feature of the Congress. It would be a handsome and generous testimony on the part of our Navy to the importance of the Congress and it would give a great impulse to the exploration of the oceans. It would receive the recognition that it would deserve."

Nevin M. Fenneman, chairman of NRC's Division of Geology and Geography echoed Davis's opinions in a letter to Denby on 30 April. Fenneman observed that, in the event of naval participation, he "might be pardoned for a certain amount of pride in the traditional value of the American Navy as an ally to scientific research."

To the acclaim of the congress organizers, Secretary Denby agreed to allow the navy to participate. After initially rejecting the old light cruiser *Denver* (CL-16) and investigating the availability of funds for the enterprise, he sent the newly commissioned light cruiser *Milwaukee* (CL-5), with Capt. William C. Asserson in command, to undertake the Australian adventure. Captain Frederic B. Bassett, the new Hydro chief, instructed Asserson to make a series of ocean-bottom profiles en route to illustrate a presentation Asserson would make to the congress on the SDF and its operation. Bassett assured him that upon arrival he could count on the assistance of Alfred Brooks of the Department of the Interior and Professor Hobbs.

The SDF presentation was one of the highlights of the meeting. When he read his paper on 27 August 1923, Captain Asserson illustrated his comments with a chart showing a line of soundings and bottom profiles from the Columbia River in the American Northwest to Sydney via Honolulu, Samoa, and the Fiji Islands. The delegates showed considerable interest in his illustrations and afterward came aboard *Milwaukee* to pepper him with questions about the SDF. If the navy and Harvey Hayes wanted public attention, the *Milwaukee*'s summer voyage to Australia gave them a first taste of public acclaim and provided a very satisfactory experience.[45]

While these events unfolded, the navy launched Professor Gregory's expedition. In the spring of 1923, Admiral Simpson of the Fourteenth Naval District at Pearl Harbor placed Lt. Comdr. Samuel W. King in command of the naval contingent assigned to the Hawaiian bird island project. The minesweeper *Tanager* (AM-5) served the explorers and surveyors from April through August, 1923. While shore parties surveyed island contours—including those of Johnston, Wake, and Midway—the ship conducted hydrographic work that included depth soundings in the vicinity of more than a half-dozen reefs and islands. Other naval vessels transported staff and supplies from the mainland to the Hawaiian Islands in support of the Gregory project.

The success of this expedition inspired a second, more ambitious round of Pacific exploration. Professor Gregory, with the support of his fellows on the NRC's Committee on Pacific Investigations, approached the new Fourteenth Naval District commandant, Rear Adm. John McDonald, with three tasks for the summer of 1924.[46] First, Gregory wanted a ship for a thorough reexamination of the Hawaiian Islands Bird Reservation confirming the restoration of the ecological balance. Then he proposed a comprehensive scientific survey of Howland, Baker, Jarvis, and Washington Islands, which lay between eight hundred and fifteen hundred miles to the south and southwest of the Hawaiian chain. These islands were American property, but scientifically very little was known about them. In a letter penned on 10 March to Albert Barrows, secretary of the NRC's Division of Foreign Relations, Gregory noted that some participants at the Washington disarmament conference had disputed American jurisdiction over these islands two years earlier. Additional information about this real estate and the surrounding waters might prove politically useful in the future. The last of Gregory's three proposed summer activities involved a six-month investigation of the Tuamotu Island group southeast of Hawaii.[47]

Secretary of the Navy Curtis Wilbur, who took office in March, 1924, found the resources to support all of the proposed research save for the six-month trip to the Tuamotu group. Gregory's appeal received hearty recommendations from Admiral McDonald and a variety of federal and private scientific organizations, including the NRC, the Scripps Institution for Biological Research, the Coast and Geodetic Survey, and the Smithsonian Institution. Early in his endorsement, the admiral called attention to the fact that "The desirability of the Navy participating in and rendering every possible assistance to such scientific research work has been repeatedly recognized by the Department. The Navy's past participation in such work is an asset of no mean proportions to the naval service. In speeches, publications, and annual reports of the Navy to Congress, it is always found of value to the Navy to be able to cite such important contributions to national peaceful pursuits." Obviously, Harvey Hayes was not the only one within the navy thinking in these terms.[48]

Admiral McDonald furnished two minesweepers with orders to take soundings and perform surveys in addition to transporting scientists and equipment. The minesweeper *Whippoorwill* (AM-35) thus joined *Tanager,* which had been supporting the Bird Island project since the previous summer. Lieutenant Commander King again took command of the naval contingent. In his report to the NRC on the summer's toil, E. W. Nelson of the Agriculture Department's Bureau of Biological Survey applauded the work of the naval officers and men attached to the minesweepers under King's direction. He concluded that profitable ventures of this sort should not happen by chance or only on occasion, and suggested a more permanent relationship. "Would it not be possible," Nelson queried, "through the National Academy of Sciences and the National Research Council, to bring about a permanent policy of cooperation in scientific research by the Navy Department and thus promote a rapid increase of our knowledge of oceanic areas?"[49]

The journey toward creating a dialogue between two cultures, scientific and naval, had only just begun.

The Hayes Initiative Bears Fruit, 1923–25

■ Moreover it seems to me that the Navy can thus do a great deal to justify its existence in the minds of people who are calling for a further reduction of armament.

—PROF. ELLSWORTH HUNTINGTON, Geologist, Yale University, 17 March 1924

I N HIS FEBRUARY, 1923, MEMORANDUM, HARVEY HAYES DEM-onstrated political sensitivity, loyalty to the navy, personal ambition, and a talent for weaving these disparate threads into a practical and appealing policy proposal. The importance of his proposition lay not in its suggestion that the navy commit itself to science, but in its new perspective on the future and its author's call for a broader and deeper collaboration between science and the navy. With a tradition that included Lt. Charles Wilkes's leadership of the 1838–41 U.S. Exploring Expedition and Lt. Matthew Fontaine Maury's groundbreaking work on physical oceanography, the navy and the ocean sciences were old friends.[1]

A variety of ambitions drove the former Swathmore physicist. Saving the NRL and nurturing its scientific programs emerged asthe primary public motive. If the NRL received adequate funding, then his own research on underwater sound and sonar would flourish in the navy's most advanced scientific facility. While this certainly provided a personal catalyst, a more productive naval-scientific relationship emerged from the memorandum as the author's central ambition. It suggested the need to go beyond the necessary and useful work of developing charts, instruments, and aids to navigation at the Hydrographic Office and the Naval Observatory. All of Hayes's proposals pointed toward a long-term, mutually beneficial partnership in basic research between the navy and civilian professionals dedicated to the perennial accumulation and analysis of oceanographic data.

Hayes's proposals meant a shift in the navy's emphasis and a higher priority for basic research. He wanted the navy, along with the civilian scientific community, to seek fundamental insights into the geology, chemistry, and physical attributes of the ocean environment for the benefit of naval activities and the general public. The activities Hayes proposed offered the navy a practical benefit as well. If it stood on the cutting edge of oceanographic knowledge, the navy would surely enhance combat readiness and easily support Hydro's vital chart-making services.

This commitment to science would drive the navy's research at the NRL and the Hydrographic Office, while the shift in priorities would provide an appreciation of the navy's natural province, which exceeded basic navigational skills. The result would enhance Hydro's traditional activity and provide an abundance of oceanographic knowledge to benefit the navy and the nation.

To implement his ideas, Hayes suggested mounting an expedition not unlike that led by Charles Wilkes nearly a hundred years earlier. He envisioned a naval vessel manned by a navy crew with accommodations for approximately fifty government and civilian scientists. Specialists in the various disciplines of ocean science would assist the navy in selecting the region of investigation and the academic and research institutions sponsoring the participants would help defray the trip's expenses. Hayes insisted upon naval direction in every phase of preparation so that maximum publicity and popular goodwill might accrue to the navy for conducting the expedition and for whatever successes it produced. Although prominent academic institutions and the NAS would certainly assist in the selection of participating scientists, the final invitation to join the expedition would come from the secretary of the navy.

Hayes clearly sought to surpass past expeditions by arming American scientists and naval ships with the latest instrumentation developed by the navy. He noted that even the great British *Challenger* Expedition only provided a limited amount of information on the contour of the ocean bottom. With the SDF developed by Hayes and his EES team, navy and civilian scientists could explore, for the first time, the topography of the ocean floor. The exciting possibility of examining the American continental borders and the Pacific rim, with their intriguing volcanic and seismic activity, suddenly presented itself. He also noted the abundance of marine life in these same regions.

Hayes clearly saw that the initiative he proposed went beyond the navy and any single expedition. In an age of naval armaments treaties, force reductions, and budget cutbacks, the navy needed the financial and professional support of the academic community and private research institutions. It would also have to draw on the resources of the Bureau of Fisheries for instrumentation to supplement equipment provided by the BuEng and the NRL. Furthermore, Hayes believed that once begun, the program would take on a life of its own: "I am of the opinion that if the proposed research work is once started by the Navy that it will continue indefinitely, and, if this proves true there is no doubt but that the researches will suggest many improvements in the apparatus that will result in continuously making the work more effective. It is along these lines that the Bellevue Station will cooperate."

The navy's premier expert on underwater sound also realized that vanity and reputation would play a long-term role in sustaining the navy's oceanographic program. If the navy supplied ships, instruments, and commitment, a truly excellent series of official reports would emerge from the data accumulated and digested by naval, federal, and civilian specialists in all of the ocean sciences. In short order it could well become a mark of eminence to have contributed titles to the navy's oceanographic series. Human ambition was a valuable and perennial source of energy and resolve.[2]

Seeking Advice: Hobbs and the NRC

Professor Hayes could not lay sole claim to these ideas, nor in some cases could he take credit for presenting them first. Unlike scientists who suggested this kind of commitment in the past, Hayes fully comprehended the advantage of making his suggestion from within the navy. He fashioned his proposal as a practical naval response to a political and financial crisis. His timing, experience within the Navy Department, and the argument that oceanographic research would not only provide knowledge but practical solutions to pressing naval problems made his proposal very appealing to all parties, whatever their motives or interests.

Geologist William Hobbs of the University of Michigan shared Hayes's desire to launch an oceanographic program in cooperation with the navy. Together they tested the SDF in the western Pacific in 1921, and Hobbs had long supported cooperative research ventures. In December, 1923, he visited Navy Secretary Edwin Denby, and followed up the encounter with a 7 January 1924 letter offering specific proposals for cooperative research. Three years earlier, Hobbs had suggested an oceanographic expedition in cooperation with the navy and he informed the secretary that the CNO during that period, Admiral Coontz, had looked favorably upon the proposal. In the intervening years, Hobbs lobbied the Geological Society of America and the NRC, and now saw in the Hayes recommendations an opportunity to launch the project with considerable support from the navy and the scientific community.

In his January letter, Hobbs suggested the Pacific and Caribbean as the focus of the joint oceanographic expedition and offered some ideas on organization and logistical support. The navy would supply the ship, officers, crew, scientific library, and a considerable amount of instrumentation—including the remarkable SDF. Research institutions and universities sponsoring the participants would pay salaries and the cost of food for the voyage. Hobbs insisted upon a comprehensive approach to the research, covering geology, ethnology, zoology, and botany as well as extraordinary efforts to collect the best data possible. The SDF would take bottom contour readings, scientists would measure waves, employ seismic instruments, and acquire an apparatus for rock-core drilling to secure deep samples from the Pacific coral reefs.[3]

In February, 1924, Secretary Denby instructed the Hydrographic Office, then under the direction of Captain Bassett, to seek NRC advice on the Hobbs and Hayes proposals. In his letter, Captain Bassett asked the NRC for an opinion on both the kind of oceanographic exploration the navy should pursue and the merit of Hobbs's proposal.[4] Gano Dunn, chairman of the NRC Executive Board, responded to Bassett's appeal and referred the questions to the Division of Foreign Relations's Committee on Pacific Investigations, chaired by Herbert Gregory, director of the Bishop Museum in Honolulu. Albert Barrows, secretary of the Division of Foreign Relations, provided the essential point of contact in Washington, D.C., for the activities of the various division committees.

Barrows immediately set to work collecting opinions from the committee members, including Gregory; John C. Merriam of the Carnegie Institution in Washington, D.C.; and Thomas Wayland Vaughan, incoming director of the Scripps Institution for Biological Research in La Jolla, California.[5] Hobbs had neglected to inform the NRC of his proposed venture, thus Dunn, Barrows, and committee members knew nothing of it before Bassett's communication.[6]

All of those contacted by Barrows quickly communicated their enthusiasm for cooperative oceanographic work with the navy, although a few had reservations about the particulars of Hobbs's proposal. Indeed, Walter T. Swingle of the U.S. Department of Agriculture commented in his reply that he had "long believed that it was a great mistake not to do a certain amount of scientific work under the auspices of the Navy Department." Barton W. Evermann of the California Academy of Sciences praised Hobbs's broad definition of the fundamental problems and the importance of cooperating with the navy. William Bowie of the U.S. Coast and Geodetic Survey in Washington, D.C., recommended strong direction by a single person or agency to provide central authority and a clear definition of the voyage's goals.

The parallel inquiries conducted by Bassett's second in command, Comdr.

Guy Davis, with Prof. Andrew Lawson, chairman of the NRC's Division of Geology and Geography, intimated that a major inaugural voyage might have its advantages.[7] On 29 February 1924, Davis wrote to Lawson, who taught mineralogy and geology at the University of California, asking for his views on cooperative oceanographic ventures.

Reporting on the request to the NRC's annual meeting in April, 1924, Lawson commented on the competence of his division to address the questions presented and the importance of the navy's appeal: "Any comprehensive plan of oceanography which the Navy may adopt will require the advice, if not the direction, of a disinterested body representing the various sciences concerned, and it would be difficult to find a more appropriate and competent body than the Division of Geology and Geography of the National Research Council. The proposal which the Navy has in contemplation is of the greatest importance for the extension of scientific knowledge in a domain where our knowledge is very scant."

Although Hydro received advice from a bevy of experienced scientists, Lawson's seven-page reply to Davis's inquiry was the most richly detailed. In it, Lawson made clear that he agreed with Hayes that the navy would perform a public service by pursuing oceanographic research while serving its own professional and financial needs in the process. He described an involved program of geological, physical, chemical, and biological studies requiring a staff of thirty-three scientists, observers, and assistants. To perform thorough and valuable scientific research, the navy and its associates in the endeavor would have to impose strict financial and geographic limits on the project. Lawson suggested a program of five years' duration to investigate the continental shelf off the West Coast and its relation to the continent and to the ocean bottom. The study would include mapping the region along with physical and biological investigations. He estimated that if provided with the proper laboratories, personnel, and equipment to the tune of $90,000, the navy and civilian scientists would emerge with scientifically worthwhile and publicly impressive results.[8]

As the responses solicited by Bassett and Davis accumulated, the difficulty of properly organizing and financing a single, comprehensive expedition emerged as the central concern. Leonhard Stejneger of the Smithsonian Institution informed Barrows and Dunn that although a large expedition sounded marvelous, he doubted its feasibility and long-term value. Instead, Stejneger suggested a series of smaller, more focused projects in conjunction with the navy. These would lend themselves to results of higher quality at affordable cost. He praised the work done on the famous voyage of the HMS *Challenger,* but insisted that general reconnaissance projects could not yield as much significant data on the immense Pacific Ocean as numerous, smaller, carefully organized expeditions. William E. Ritter replied that he heartily agreed, and the man who would soon succeed him as director of Scripps, Thomas Wayland Vaughan, went one step further. Vaughan enthusiastically supported collaboration between the federal government and civilian science and suggested that Secretary Denby or his successor should call a general conference to define the character and extent of any expeditions the navy chose to sponsor. Vaughan feared that a single major exploratory cruise would fall short, whereas a series of shorter, well-focused, and well-financed ventures might succeed beyond all expectation.

Barrows and Gano Dunn communicated the consensus of these opinions to Captain Bassett in their reply. The multiple expeditions and the strong central direction preferred by most of the NRC scientists more closely corresponded to some of Professor Hayes's suggestions than the plan put forth by William Hobbs. The NRC advice also provided the navy with a way to define more precisely the action it planned to take.[9]

After evaluating Lawson's proposals and Dunn's response to the Bassett letter, Secretary Denby and Assistant Secretary Theodore Roosevelt, Jr., concluded that the navy should pursue oceanographic studies in a way that would provide the broadest possible benefit. With the approval of incoming Navy Secretary Curtis Wilbur, Roosevelt took Vaughan's advice and began preparations to convene a federal Interagency Conference on Oceanography (ICO) in Washington on 1 July 1924. He wanted conference participants to suggest the most profitable application of naval and national resources for oceanographic exploration.

The usefulness of this type of venture was not lost on the fleet, but its officers had their own perspective on the role surveying ships might play. Rear Adm. William H. Standley, director of the CNO's War Plans Division, requested permission to send naval officers on any proposed oceanographic research voyages in the Pacific to gather intelligence on islands and ocean areas that would have strategic importance in a war with Japan. The motives for supporting the ICO proved as diverse as the potential benefits the exploration had to offer.[10]

The Interagency Conference on Oceanography

The ICO proposals determined the nature of the naval commitment to oceanographic research for the next two decades. On 2 June 1924, Acting Navy Secretary Roosevelt sent out invitations to prominent civilian scientists, scientific institutions, government agencies, and naval bureaus and activities. Following the suggestions made by Harvey Hayes, Roosevelt made sure that the navy maintained a high profile for the sake of political utility and public relations without losing sight of the genuine value of the scientific exploration under consideration. In the text of his invitation he said the navy would appeal to Congress for the necessary funds to finance the kind of exploration the ICO might recommend.[11]

Roosevelt received a very encouraging response to his invitation. Sixty-one scientists and officials from twelve different government agencies and private institutions announced that they would attend. Harvey Hayes attended under the aegis of the navy's Bureau of Engineering (BuEng). Other important participants included George Littlehales and Captain Bassett from Hydro, Comdr. H. S. Howard from the Bureau of Construction and Repair (BuC&R), Austin Clark from the Smithsonian, William Bowie and six colleagues representing the committees of the NAS and the NRC, and Harvard's Henry Bigelow, representing the Commerce Department's Bureau of Fisheries.[12]

The ICO opened at the Navy Department on 1 July with a keynote speech by Hydro's hydrographic engineer, George Littlehales. His short talk focused on the issues that those in the audience found most immediate. They had not come together to debate the worthiness of exploring the Earth's oceans and better exploiting its resources. Those ends, as well as their scientific curiosity, motivated all of them. Assistant Secretary Roosevelt had called them together to define more completely just how to accomplish their goals for the benefit of as many Americans as possible. Was it time for another general purpose voyage of exploration lasting several years on the HMS *Challenger* model? After reading the names of thirty-six expeditions undertaken by a dozen countries extending as far back as 1868, Littlehales remarked that general human knowledge of the world ocean had progressed significantly since the end of the nineteenth century. Did this progress warrant shifting from general studies to particular, more precisely defined investigations? Perhaps the scientific community should sponsor a series of cruises, each with a restricted a focus in a particular region. This could provide in-depth knowledge about marine life, water temperature and salinity, and, with the aid of the SDF, a view of ocean-bottom contours never seen. He concluded by emphasizing the interdisciplinary nature of oceanography and the broad understanding of the world ocean that lay in the future.[13]

Under the able direction of Hydro's Captain Bassett, the conference pursued the Hayes initiative well beyond its author's original intent. In their report to Secretary Wilbur, the ICO participants emphasized the great utility of the proposed investigations. Of the four primary objectives stated at the conference, improving humanity through discovery and exploration and preserving human life were third and fourth. Learning how to use the resources of the sea and improving communication through submarine cable and radio topped the list.

How should the scientific community explore? The consensus at the ICO counseled against another *Challenger*-like expedition and decided to devote themselves "to intensive study of selected regions and problems." In their specific recommendations, the conferees placed the Gulf of Mexico–Caribbean area first on its list. They requested at least a single vessel and crew from the navy and hoped for more to explore this body of water and the areas immediately adjacent in the North Atlantic and down to the Canal Zone. Thereafter, the work could expand into the Pacific with an initial emphasis on the northern portions of that ocean.

Universities, research institutions, and the federal government could together provide the balance of the expedition's requirements. A group of specialists assigned to the problem during the ICO estimated that instruments and equipment would cost $50,000, with the navy, other federal agencies, and private institutions sharing the burden. They envisioned a scientific staff of at least nine: an oceanographer, biologist, geologist, and a minimum of six technicians and assistants. The ICO committee demanded scientists of the highest caliber for the three senior positions. Their home institutions would absorb their basic expenses, save for subsistence costs for fieldwork and round-trip travel. The latter would run approximately $7,500,bringing the projected cost of the project to $57,500.

Although the conference placed the greatest emphasis on geology and geophysics, those problems given priority represented all of the major disciplines within the field of oceanography. The invention of the SDF created many opportunities to study the configuration of the ocean bottom. Those interested in the Earth's crust no longer suffered from blindness. Eager ICO participants wanted to study "changes in the size and shape of the bottoms of the seas, such as shifting of shorelines, warping of the margins of continents and submarine upheavals and dislocations." Greater understanding of the sediments that compose the ocean bottom as well as violent changes caused by earthquakes and volcanoes now appeared to scientists in the broader context of ocean-bottom profiles and some of the first regional maps of the ocean floor. To this the physicists added their desire to gain a greater appreciation of gravity variation in the Caribbean and Pacific areas. Other goals included the study of shallow and deepwater currents; the temperature, salinity, density, and general chemistry of seawater; and the interaction of the atmosphere with the ocean's surface.

Although Harvey Hayes often seemed to disappear in the context of the conference precipitated by his 1923 memo, the political, economic, and scientific value of oceanography's utility remained a theme central to the ICO, just as it was in Hayes's original proposal. Understanding the interaction between the ocean and the atmosphere would permit more accurate weather prediction. Knowledge of sedimentation, tides, and currents would ensure safer and easier navigation. Scientists foresaw the results of the research initiated by the conference influencing oil discoveries, radio, communication, and ship design.

In the spirit of the comments made by NRC scientists at the navy's request, the ICO did not deliberate without a commitment to the future. Although many ocean scientists would only intermittently take advantage of the programs begun by the navy, the conferees intended that the commitment, purpose, and programs should last and proliferate. Under Bassett's leadership, the ICO designated the U.S. National Museum as the permanent repository for the constant flow of specimens. For its part, the navy pledged to follow the example set by the *Challenger* scientists, who published a series of volumes based on their oceanographic data edited by C. Wyville Thomson and John Murray. The conference also projected the possibility and desirability of a SDF ocean bottom chart for the Gulf of Mexico–Caribbean area. In an era of dramatically reduced spending on the armed forces, the final conference report dared to emphasize this theme of optimism and permanence by asserting that "the recommendations embodied in this report are based upon the expectations that research in oceanography will take a permanent place among the activities of the Navy." The ICO report noted that the navy secretary should consider the appointment of a continuing advisory committee to serve as advocate for the project during the budget process and to fashion an efficient administrative system for the venture in its early stages. Captain Bassett, as head of Hydro, and his assistant, Lt. Comdr. George E. Brandt, received nominations and quick approval as chairman and secretary of the new committee. Other participants in the ICO who agreed to serve with Bassett and Brandt included Capt. R. O. Crisp of the U.S. Coast Guard, Lt. Col. C. A. Seoane of the army's Signal Corps, the NRC's David White and William Bowie, E. D. Ball of the Department of Agriculture, and George Littlehales, Austin Clark, and Henry Bigelow. Captain J. P. Ault of the Carnegie Institution also joined the committee. As the captain of the nonmagnetic Research Vessel (R/V) *Carnegie,* his advice would prove invaluable in the preparations for a research expedition. The ICO wanted the entire program named "Maury U.S. Naval Oceanographic Research," after the premier ocean scientist in the navy's history, Matthew Fontaine Maury.[14]

The most revealing aspect of the ICO emerged from the notes taken on the proceedings by Captain Bassett, the conference secretary general. Genuine excitement took hold of those in attendance. Federal agencies envisioned harvesting the sea with greater efficiency, making navigation safer, promoting communication and submarine cable projects, and countless other productive adventures. Ocean scientists, who perennially suffered from inadequate funding and the lack of suitable ships for research at sea, realized the potential for private and public gain in cooperative work with the navy and other federal agencies. In his effort to sustain the NRL and promote naval research, Professor

Hayes had unleashed a remarkable amount of pent-up enthusiasm, determination, and energy.

Development of the SDF also played a critical role. Usually the effect of an
instrument, no matter how useful, remains limited to the task at hand. It liber-
ates the scientist from repetitive chores, makes an awkward task easier, or helps
overcome physical obstacles. The SDF did all of this and more for those studying the ocean. It suddenly provided a way to examine the invisible, and in the
process provided material and psychological benefits. For centuries, scientists
had been forced to rely on rope, wire, grabbers, and weights to experience the
topography of the ocean's bottom. While it did not preclude the use of these
tools to attain certain research goals, the SDF provided a picture of the Earth's
contours that lay submerged and beyond our eyes for centuries. Although it
did not make the ocean transparent, to a remarkable degree it freed the scientific community from severe physical restraints, offered great promise, and
opened an array of research opportunities. Hayes's SDF turned on a sonic light
in a very dark room. It not only permitted science to see, but awakened scientists to new, stunning possibilities for oceanography.

Lieutenant Colonel Seoane's comments at the conference demonstrated
the new spirit generated by the advantages suddenly afforded by the SDF.
Seoane had worked for years on the problems involved in laying and maintaining submarine cables. Limited oceanographic knowledge forced planners
to incorporate in their length calculations an additional 10 percent for slack, or
the amount taken up by the often-irregular terrain of the ocean bottom. Seoane
said he could recall no cable being laid without at least 8 percent slack before
the advent of the SDF. Then he offered a current case of cable laying that employed only "4½ percent of slack. I doubt if any such saving as this has even
been accomplished before. That was due, first, to the Navy Department surveys
carried out, with the advantage which has been brought about by the sonic
depth finder, the United States Destroyers HULL and CORRY making this
survey. As an incident, it may be mentioned that there was brought out the fact
that this cable was laid over a mountain over 4500 feet in height—a submarine
mountain. With the sonic depth finder soundings can be taken at such frequent
intervals that a very definite knowledge of the ocean topography can be ascer-
tained."[15]

Captain Bassett's notes on Seoane's observations highlighted the advantages of using the SDF for cable laying. Slack cable came at a very high price,
and the SDF's sound profile of the bottom permitted very precise length calculation—not only for the cable itself, but also for the length of the cable-laying
voyage and all of the costly resources employed in the venture.

Bassett noted more than a score of presentations before the ICO that du-
plicated Seoane's testimonial and praised the possibilities of the SDF for un-

derstanding the geology and botany of the ocean bottom. Many of these suggestions and observations focused on the communities of fish frequenting the
American coastline. The SDF could provide data on the type of bottom topography that sustained these fish as a natural habitat and breeding ground,
knowledge that was vital to the fishing industry. Professor Hayes's creation
thus not only provided a remarkable view of the ocean floor, but also created
unanticipated opportunities and supported a general sense that the ocean environment and its resources could well turn into a very profitable new frontier.

The interim committee created by the ICO continued the conference work
beyond July, 1924, by promoting U.S. involvement in international scientific
meetings, by encouraging independent scientific investigation in pursuit of
conference goals, and by supporting the Maury research program. In October,
the committee resolved in a communication to Secretary Wilbur that the navy
should financially support the Carnegie Institution's effort to resume the observations of electrical and terrestrial magnetic phenomenon on board the R/V
Carnegie. Those experiments had been suspended in 1921.[16] The committee
also asked the secretary to request that President Calvin Coolidge institute
more liberal policy toward the participation of government scientists in international conferences. They hoped that representatives from Hydro and other
federal agencies might participate in the upcoming Third Pan-Pacific Science
Congress scheduled for Tokyo in October and November, 1926. American scientists and naval officers had made a significant contribution to the Second
Pan-Pacific Science Congress in 1923 when Captain Asserson of the USS *Milwaukee* first demonstrated the potential of the SDF beyond U.S. borders.[17] In
this spirit, Hydro kept the International Hydrographic Bureau constantly informed throughout its survey and map work in the Gulf of Mexico, the
Caribbean, and the Pacific. The bureau published information on available
charts and survey data in its circular letters to members.[18]

The ICO and its interim committee achieved mixed results. Encouragement from the navy and the offer of instrumentation from the NRL helped re-
vive the Carnegie Institution's R/V *Carnegie* research program. In 1926 the secretary did approve the participation of naval scientists, including George
Littlehales of Hydro, in the Pan-Pacific Science Congress in Tokyo. Unfortunately, the ICO's request for a supplemental appropriation of $152,000 in fiscal
year 1926 for initial research at sea in pursuit of conference goals did not meet
with the president's approval. H. J. Lord, director of the Bureau of the Budget,
informed Secretary Wilbur that Coolidge decided the "submission of these
estimates to the Congress at this time would be in conflict with his financial
program."[19]

Hydro's reaction to this setback demonstrated the determination that
emerged from the ICO and the devotion to the program and goals set by the

conference. Captain Bassett wrote to the secretary urging that "researches recommended by the Conference on Oceanography be begun as soon as possible using funds regularly appropriated for the Navy." He suggested that the secretary keep the USS *Rainbow* in commission to carry out the proposed research. Hydro felt that the value of the information obtained would more than offset the cost of operating the *Rainbow*. Bassett's own enthusiasm for the program emerged clearly in the correspondence as he reminded Wilbur of the broad base of support for joint oceanographic research in evidence at the ICO.[20]

Like Captain Bassett, many ICO veterans never considered forsaking the conference's goals. David White, chairman of the NRC's Division of Geology and Geography, noted the importance of the navy's effort in his bimonthly report to the NRC on 2 February 1925. President Coolidge's refusal to include the oceanographic proposal in the Emergency Appropriation Bill simply indicated to White that the project would have to wait until the next session of Congress. His report emphasized scientific and popular support for oceanographic research by the navy and a positive feeling that the setback delayed funding only until Congress met again.

In the same spirit, Thomas Vaughan of Scripps proposed that the NRC create a committee to study the submarine topography and structural history of the Caribbean-Gulf region. After the ICO, Vaughan had little trouble convincing the NRC to assemble a panel of expert scientists to offer advice on one of the regions named by the conference as a primary candidate for oceanographic exploration under navy auspices. The NRC's Division of Foreign Relations followed suit and commissioned its Committee on Pacific Investigations to continue even more intensely its cooperation with the navy in Pacific oceanography. The committee's executives, Herbert Gregory and the omnipresent Vaughan, had considerable experience working with the navy and believed that the goals set down by the Second Pan-Pacific Science Congress would provide a foundation for excellent and precisely focused programs of oceanographic research.

The political setbacks precipitated greater and more vocal support from the scientific community. After holding its own meeting during which it heard from Littlehales, Hayes, Ault, and Vaughan among others, the American Geophysical Union's oceanographic section joined the chorus and passed a resolution supporting the ICO's goals and appealing to the Congress and the president to supply the necessary funds. In addition, participants at a conference on the physical oceanography and marine meteorology of the northeastern Pacific held at Scripps in early November, 1925, applauded the navy's data-collection effort and the example it gave other federal agencies and private steamship lines.

Partisans within the navy began to do what they could even without the emergency funding. Hydro received permission to equip many vessels with thermographs to record water temperatures, and required officers to supervise the collection of water samples on routes regularly used by the fleet and its support services. Further development of the SDF commenced at the NRL, and the CNO sent the USS *Rainbow* to the East Coast for overhaul in preparation for conducting research at sea.[21]

The Coolidge administration's funding decision mobilized those within the navy and the scientific community who were determined both to attain the conference goals and sustain the navy's oceanographic commitment in financially lean times.

3

Disappointment and Persistence, 1926–30

■ RESOLVED: That a letter be written to the Secretary of the
Navy, by the Chairman, expressing appreciation of the services
of the Navy in Oceanography as well as of their future participation in oceanographic work.

—NAS COMMITTEE ON OCEANOGRAPHY, 12 October 1929

ONLY DIVINE INTERVENTION AND ADDITIONAL NAVAL APPRO-
priations could make the ICO program possible, and the probability of
both seemed equally remote. Rather than simply abandon apparently
worthy goals, both the navy and the civilian scientific community acted together to promote oceanography. In the process they displayed remarkable creativity as well as considerable adaptability.

Within its budgeted resources, how could the navy enrich its current program to address the ambitions expressed at the ICO? The Hydrographic Office
continued its work with the IHB and enhanced its survey program customarily responsible for generating charts, navigation aids, and information for international industrial and shipping interests.[1] Admiral Bassett and his successor, Rear Adm. Walter S. Crosley immediately ordered the crews of naval
research vessels to collect a wider variety of data and samples than their traditional surveying duties required. On the civilian side, ocean scientists struggled
to define and cope with oceanography's considerable scope while recognizing
the need both to train more qualified personnel and to create additional programs and institutions for education and research.

Disappointments and Obstacles

In spite of the determination evident at Hydro, the absence of presidential support and proper funding made the navy's limits painfully apparent. Available resources would not permit a comprehensive program. Most naval and private efforts thus remained within traditional mission and fiscal boundaries and oceanography's future course remained uncertain. Determined to heed the ICO's site recommendations as far as its resources would permit, the Hydrographic Office extended its enhanced mapping and oceanographic surveys into the Gulf of Mexico and Caribbean area. Regular line officers journeyed to Washington, D.C., for training at the Hydrographic Office and the Naval Observatory in data collection methods. When they returned to their ships, these officers participated in survey work and data collection on survey voyages or in the course of regular ship operations in strategically or scientifically important areas.[2]

During the two years after the ICO, Hydro had three surveying ships in the Gulf of Mexico–Caribbean area collecting oceanographic data in the course of regular surveying expeditions along Central America's east coast, South America's north coast, and in the vicinity of Cuba. For this work the navy used converted submarine tenders, colliers, yachts, or other auxiliary ships such as *Hannibal* (AG-1), *Nokomis* (PY-6), *Niagara* (PY-9), and *Bushnell* (AS-15).

A lawsuit briefly interrupted the effectiveness of this work by placing in question the navy's use of the SDF. When he first developed the oscillator the Hayes team used in this instrument, Reginald Fessenden was working for the Submarine Signal Company of Boston, and the assassination of Archduke Franz Ferdinand of Austria-Hungary, which unleashed the Great War, was still five months away. Twelve years and one world war later, Submarine Signal brought a series of suits against the federal government and a number of independent firms, including General Radio Company of Cambridge, Massachusetts, for using the Fessenden oscillator in several signaling and detection devices without formal permission.

The problem lay with the effect of the war on Submarine Signal's patent rights. After the defeat of the Central Powers, the United States and its allies seized a great number of German merchant ships and warships both in Europe and in American ports, some of which had a Fathometer on board manufactured by the Atlas Werke of Bremen, a Submarine Signal licensee. The federal government claimed both the ships and all on-board ordnance, systems, instruments and supplies, including the Atlas Fathometers found in many of the fifty-three merchantmen captured or stranded in the United States when the war began.

In its legal actions, Submarine Signal Company wanted compensation for the instruments produced by its licensee as well as clarification and reaffirmation of its patent rights to the Fessenden oscillator. It did not help matters that the navy waited nearly three years, from 1923 through the summer of 1926, to provide the company with a list of ships carrying its equipment, but the decisions handed down satisfied almost everyone concerned. The court guaranteed Submarine Signal's patent, but denied the company's right to financial compensation for the instruments seized from the Germans by the executive branch under congressional legislative authority. The navy had to pay royalties in the future for any continued use of the Fessenden apparatus in the SDF, but the court did nothing to stifle research and development—nor did it prevent Submarine Signal from profitably marketing its own Fathometer.[3]

While exhibiting the will and the technical means to follow the spirit of the ICO recommendations, the effort needed further careful coordination to avoid wasting assets like the SDF. Both the navy and civilian scientists addressed this problem by consulting prominent individuals and established scientific societies. On 9 October 1925, Frank R. Lillie, chairman of the Zoology Department at the University of Chicago and director of Woods Hole's Marine Biological Laboratory (MBL), informally approached Wicliffe Rose of the Rockefeller Foundation's General Education Board to discuss the conservation and use of the ocean as a natural resource. Their conversations revealed the need for cooperation between coastal biological stations like Scripps and other important universities and research establishments. Lillie, a marine embryologist and member of the NAS, also discussed with Rose the commercial support given oceanography by European countries heavily dependent upon harvesting the sea and the central coordination of their effort.[4] He wondered how to promote oceanographic research effectively without a strong central institution like the NAS, the Bureau of Fisheries, or the navy providing direction and administration. Lillie and Rose brought to the attention of their colleagues both the absence of an effective oceanographic center to help coordinate eastern seaboard research, a role performed informally by Scripps on the West Coast, and the necessity to take stock of both oceans given America's geographic position.[5]

Just like individual scientists and educators ready to take the initiative, NAS and NRC science advisers to the federal government played a vital role in plotting a course for American oceanography. They provided scientific information and encouragement as well as modest financial support. In its report for the period 1925–26, the NRC's Division of Geology and Geography lamented that the navy had little progress to show for the months following the ICO. Division chairman David White noted the absence of federal funding, reiterated the NRC's support for naval ambitions, and, because of the great value of the work, regretted the delay in inaugurating an oceanographic program.[6]

Coping and Cooperating

Without formal direction, the scarcity of resources combined with the determination to carry out the research recommended by the ICO to provide an alternative. Rather than the creation of a central administrative institution, the next decade witnessed the gradual development of a set of relationships, a *common practice* that enabled naval oceanography to grow and expand, coordinated by formal and informal ties and agreements. The ill-fated ICO pointed the way, and many naval and scientific leaders decided to follow as best they could with the resources at hand.

The largely ad hoc activities of 1925 and 1926 gave the growing relationship its character. Prominent individuals and scientific societies would facilitate cooperation and help build a common practice in the absence of a politically and financially able central authority. Established scientific societies stepped in offering an environment for the exchange of ideas, a meeting place to become closely acquainted, and an opportunity to forge the ties that made successful cooperative efforts possible. The activities of the NRC's American Geophysical Union (AGU) played this role effectively. At its annual meeting in April, 1926, the AGU provided a platform for reports on oceanographic work in and out of the navy. Thomas Wayland Vaughan offered his insights on the progress of the ocean sciences on the West Coast, while Henry Bryant Bigelow, Robert C. Murphy, and W. J. Peters did the same for the East. H. C. Dickinson reported on oceanographic research instruments developed by the Bureau of Standards, Lt. Comdr. Edward H. Smith expounded on the oceanographic side of the investigations performed by the U.S. Coast Guard Ice Patrol, and the NRL's ever-present Harvey Hayes offered a presentation on the latest developments in SDF technology. These meetings and others like them kept ambitions for oceanography fresh and hopeful, and research modestly productive.[7]

As chairman of the NRC's Committee on the Oceanography of the Pacific, Thomas Wayland Vaughan tirelessly took the concerns of his colleagues on the road in 1926 to the Third Pan-Pacific Science Congress in Tokyo. In a paper entitled "International Cooperation in Oceanographic Investigations in the Pacific," Vaughan insisted that in spite of differences between the goals of particular research voyages, all Pacific investigations as a matter of course should bring home the same basic data. These ventures were expensive undertakings, but the prompt analysis and sharing of data through publication would serve every nation exploring the largest portion of the world ocean. He also drew attention to the Tokyo congress's resolution urging national navies to offer the services of their submarines for gravity research and oceanographic exploration.[8]

Vaughan loved gardening and scientific activity. Contemporary photographs reveal him as a slender, professorial, rather elegant figure who could easily dominate a scene. He exhibited boundless energy at Scripps as director and scientist and as an amateur horticulturist in love with the beauty of the institution's setting. Throughout his career he remained a major player in many programs sponsored by professional geological and oceanographic societies. Although this energy was bridled in his later years by a treatable case of tuberculosis, he continued as a major influence in oceanography for many years. In Tokyo, he spoke forcefully before the congress emphasizing the utility of oceanography and the absolute necessity for each country's scientific community to solicit governmental support for research. He argued persuasively that "we live in an anthropocentric world, a world that evaluates human activity in terms of the benefit that it may confer on man. Scientific activity is no exception, and, in order to gain support for it, its cause must be presented to and plead before those who can render it the assistance it needs."

If limited financial support came from government, business, and private foundations, the scientific community had to make the most of it by avoiding both geographical redundancy and sharing the data collected. Regardless of the specific expedition goals, all research ventures should record depths, take temperature readings and water samples, perform chemical analyses of seawater, and make meteorological observations. For the biologists, every scientific foray should measure the number of plankton per standard unit volume of water, and general dredging would provide specimens for study all over the world. According to Vaughan, the ICES and the Marine Meteorological Observatory of Kobe, Japan, already sponsored such programs. He clearly wanted to interest national governments and their agencies in doing the same, especially navies, the merchant marine, and lighthouse keepers.[9]

Vaughan's colleagues echoed his statements and appeals. In its 24 April 1926 report to the NRC's Division of Geology and Geography, the Committee on the Submarine Topography and Structural History of the Caribbean-Gulf Region expressed its frustration at the lack of funding for the proposed ICO expedition and its concern that an important region south of the United States might go unexplored. A valuable analysis of sonic soundings recently taken in the area by the Hydrographic Office and the Coast and Geodetic Survey had yet to reach the publisher. Even worse, no federal agency or private institution had planned any gravity measurements in the Caribbean-Gulf region.

Research in that area as well as in the eastern Pacific remained a high priority for both the NAS and the NRC. The committee recommended that the NRC approach the navy and the Coast and Geodetic Survey to launch a joint effort to take gravity measurements according to the pendulum method developed by Felix Andries Vening-Meinesz of the Dutch Geodetic Committee. At the behest of William Bowie of the U.S. Coast and Geodetic Survey, Vening-

Meinesz agreed to take gravity readings during his submarine passage across the Caribbean Sea in the summer of 1926, en route to the East Indies via the Panama Canal. These measurements would not only provide information about the tectonic activity beneath the ocean floor in the Gulf and Caribbean, but they might also uncover oil deposits in a region of interest to American petroleum companies.

Since the United States could hardly depend on the kindness of a foreign scientist to supply more comprehensive gravity readings of the Caribbean-Gulf area, the committee recommended initiating an American project of the kind undertaken by Vening-Meinesz in the East Indies. They suggested that the navy provide a submarine to act as a stable submerged platform for measuring gravity variation with the Vening-Meinesz pendulum device. Surface turbulence rendered it impossible for the instrument to take accurate measurements from a regular ship. A submarine could descend to a depth at which surface disturbances would not interfere with the collection of accurate data. In closing his report, chairman W. P. Woodring also called the NRC's attention to the absence of seismological data for the region and the importance of such information to the navy: "The investigation of earthquakes should have direct appeal to the Navy, as naval vessels have several times been damaged by seismic waves in the region. The naval base at Quantanamo [*sic*] lies close to the edge of the northeast end of the Bartlett Trough, along which many earthquakes have originated."[10]

The activities of the NAS and NRC and many of the influential division and committee chairmen played an ever increasing role in forging the common practice that kept naval oceanography alive during the interwar period. In their reports, these scientists recorded research, lauded progress, urged federal support, facilitated international cooperation, communicated data and analysis, and employed naval personnel and vessels whenever possible.

Edited by the omnipresent Vaughan, the annual report of the NRC's Committee on Submarine Configuration and Oceanic Circulation[11] for 1926 illustrated the extent of naval involvement in oceanographic activity during this period. Of the nine committee members, two were scientists employed by the navy. George Littlehales served as one of the senior scientists at Hydro, and the professorial and pioneering Harvey Hayes hailed from the very active NRL. Littlehales supplied the committee with an extensive list of scientific and technical publications about the SDF, its ongoing development and use, and a complete list of soundings performed with this instrument from naval vessels — especially in the Caribbean-Gulf area and the northeastern Pacific Ocean. Hayes provided notes on the continuing refinement of the SDF to achieve greater penetration and precision. To this end, the Hayes team at NRL worked on augmenting with a human operator the purely mechanical device that recorded the data provided by the SDF. Preliminary research proved that the human ear was far more sensitive and accurate in determining the distinction between sounds of various sorts. Captain R. L. Faris of the Coast and Geodetic Survey contributed a commentary on the effectiveness of the SDF in coastal areas with depths greater than a hundred fathoms. The survey performed these soundings as well as taking water samples, plankton hauls, and temperature serial measurements along the Alaskan shore and the coast of the northwest United States. He also reported the first use of a radio-acoustic method of determining a ship's position. Then Lt. Comdr. Edward Smith of the U.S. Coast Guard provided an account of Ice Patrol activities, after which Henry Bryant Bigelow reported on a current oceanographic survey of the Gulf of Maine.[12]

Vaughan's committee report demonstrated the extent to which important fundamental research in cooperation with the navy had already begun in spite of disappointment after the ICO and limited funds. Both the NRC and NAS proved invaluable as a clearinghouse for available data accumulated by various institutions and agencies. Information about oceanographic investigations undertaken by Scripps between 1 July 1924 and 31 March 1926 was published in NAS and NRC reports for the benefit of interested scientists. These reports specified the regions covered and the types of data available. Vaughan included similar information on the eastern Pacific gathered by the Coast and Geodetic Survey as well as the temperature measurements, water samples, and plankton hauls done by naval vessels from March through November, 1925. Focusing on the effort in the Pacific both by warships and vessels converted for research, the committee report took note of more than a thousand SDF soundings taken on routes between San Diego and Guadeloupe as well as round-trip cruises from San Francisco to Hawaii, Australia, and New Zealand. Navy transports operating between Bremerton, Washington, and Balboa, Canal Zone, came away with 703 temperature records and forty-four water and plankton samples in 1925. By an agreement reached at the ICO the previous July, all of the navy water and plankton samples went to the Scripps Institution for analysis. The navy released the results as soon as Scripps scientists finished their work and financial resources made publication possible through the Hydrographic Office.[13]

Cooperation and sharing resources made sense and came close to accomplishing the ICO goals. In June, 1926, the NRC's Division of Geology and Geography passed a series of resolutions supporting research efforts in regions given the highest priority by the ICO and currently being surveyed by the navy. Chairman David White and his colleagues endorsed oceanographic research in the Caribbean-Gulf area and in two triangular sections of the Pacific, Panama–Hawaii–San Francisco and the Aleutian Islands–Hawaii–San Francisco. At the same time, Thomas Wayland Vaughan's Committee on Submarine Configuration and Oceanic Circulation entered into partnership with the De-

partments of State and Commerce, the navy, and local authorities to do oceanographic observations along the west coast of Central and South America.[14]

New Direction, New Partners

In the spring of 1927, this sense of partnership, joint responsibility, and the
need for bolder planning and direction led some of the most prominent men in
the ocean sciences to take action outside the realm of research. Building on a
foundation laid in his 1925 conversations with Wicliff Rose, Frank Lillie joined
forces with John C. Merriam, president of the Carnegie Institution of Washington, and Vaughan to ask the NAS to approach the problems and needs of
American oceanography in a more "active way."

Born in Toronto in 1870, Lillie did graduate work under Charles O. Whitman at Clark University and, after moving with his mentor to the newly established University of Chicago in 1892,completed his doctorate in 1894. At the
age of thirty-six, he succeeded Whitman as chair of Chicago's Department of
Zoology at Chicago and became director of the MBL at Woods Hole. For the
next half century, he spent every summer at Woods Hole. He was a man of initiative and already well accustomed to responsibility and action when he,
Vaughan, and Merriam formally suggested in April, 1927, that the NAS create
a Committee on Oceanography. Vaughan had already helped initiate the International Committee on the Oceanography of the Pacific, sponsored by the
Pacific Science Association.[15] All three of these scientists sought a more comprehensive approach, emphasizing not only the Pacific, but the world ocean.
They found others feeling the same way. Lillie recalled that he "presented the
problem briefly from the floor (at the annual NAS meeting) and I must confess
that I was surprised to find with what enthusiasm it was received."

The NAS responded to Lillie's suggestion by passing a resolution asking
the NAS president, former naval officer and Nobel laureate Albert A. Michelson, "to appoint a Committee on Oceanography from the sections of the Academy concerned to consider the share of the United States of America in a world
wide program of Oceanographical Research, and report to the Academy."
When he wrote to Wicliffe Rose on 11 May, Lillie reported the committee's
composition and asked Rose to consider funding all or part of the committee's
activities. Lillie accepted the chairmanship of a group of six individuals including Vaughan, Merriam, Bowie, Princeton biologist Edwin G. Conklin, and
Benjamin M. Duggar, professor of physiology and economic botany at the
University of Wisconsin.

A practical man with an ambitious agenda, Lillie always remained sensitive
to the need for adequate funding given the challenging mandate handed down
by the Academy. Without the support of the Rockefeller Foundation or some

other patron, the new committee's work could well end in the frustration experienced by its predecessor in 1919.[16]

Lillie decided early that the committee should step into the void left by the navy when the Coolidge administration failed to fund the program suggested by the ICO. The various agencies, research institutions, and universities involved in oceanography, separately or in cooperation, needed coordination to have the greatest effect. Successful research conducted thus far by the navy, the Carnegie Institution, Scripps, MBL, the Coast and Geodetic Survey, and the Bureau of Fisheries might have proved even more productive with effective central organization. In his first letter to members of his committee, Lillie argued that current conditions simply led to waste and retarded scientific advance: "As it is, American oceanographic activities have not been co-ordinated in any effective way with one another, and there is no central agency that could cooperate with international undertakings. In consequence, the interests of Oceanography have not advanced as they should in the United States of America in comparison with several other countries, even though we spend more money on various isolated projects within the field."

For the chairman, the province of the committee was clear. Those involved in oceanography in the United States had to forge links with colleagues around the world, thus permitting original and valuable work by all involved with reduced waste. If U.S. oceanographers sought integration in a worldwide program, American ocean science first needed central coordination through either a single institute or an organized group of government and private agencies. Lillie had already decided that he would not let this committee conclude its work without fundamental and constructive change.

Purpose and determination emanate from his letters to the committee members. Lillie expected them to play an active role in collecting information and becoming better acquainted with the current national and international assets and resources available to oceanography. He called their first gathering to coincide with the autumn meeting of the NAS at the University of Illinois in Urbana from 18 to 20 October 1927.[17]

Vaughan responded to his chairman's call for a meeting by sending his sincere regrets. Business at Scripps prevented his attending the Urbana meeting, but it did not preclude him from composing a characteristically detailed opinion of the major issues the group needed to consider. He showed particular interest in one of the major points discussed by Lillie and Rose in 1925. As director of the premier West Coast oceanographic institution, Vaughan argued very strongly for the creation of a similar research center on the East Coast. The new institution would require a laboratory building, a small staff of three or four trained scientists, and a research vessel. Vaughan, like Lillie, appreciated the value of current oceanographic research, but he also realized its often-random nature because of the informal ties that guided private and federal efforts. With major research institutions on both the Atlantic and Pacific coasts, Vaughan thought that it would "then become practicable to bring into more or less concerted efforts the activities of the marine stations and of a number of governmental organizations, such as the navy, Coast and Geodetic Survey, Bureau of Lighthouses, Bureau of Fisheries, and International Ice Patrol. I think it would also be practicable to develop a program, which would be complementary to the program of the oceanographic institutions in Europe."[18]

Organizing American oceanography in this manner would take advantage of both the measure of bipolar coordination afforded by Scripps and the new East Coast institution and the common practice that had sustained oceanography with some energy and effective results since the ICO.

Frank Lillie acted very swiftly. He immediately approached Wicliff Rose and proposed that the Rockefeller Foundation subsidize the new East Coast oceanographic institution. Speed proved essential, for Rose, a friend of the project and the committee, had announced his retirement in six to nine months' time. Rose insisted that the committee's application concentrate solely on the new institution, including the money needed and a proposed location. In his third circular letter to the committee on 21 October 1927, Lillie reported that in conference with Rose the latter insisted on specifics and commitment, leading to preliminary agreements on location and funding. The two men settled on a total of $3 million, one-third for a laboratory building and the balance to endow the new organization. Although he implied that the pressure of passing time made early agreements necessary, the pace of events served the chairman well. As an old MBL partisan, Lillie wanted to locate the new oceanographic institution in Woods Hole but expected some objection from his fellow committee members. At his meeting with Rose the two concluded that the small Cape Cod village and home to the Bureau of Fisheries, the MBL, and a sizable summer scientific community would serve the needs of a fledgling scientific center well. Although these decisions needed the imprimatur of Lillie's five colleagues and the NAS executive committee before formal consideration by the General Education Board, the chairman had the initiative and Woods Hole would clearly have the advantage over other sites.[19]

In the process of discussing the Atlantic coast station with Rose, Lillie did not want to give the impression that his committee had only one string to its bow. He wanted to protect the broad mandate granted by the NAS. Familiarity with the extent of oceanographic assets and problems in the United States would take time and money. Beyond that, Lillie wanted to make a general statement about these problems that could well form the basis for future national policy regarding the importance of oceanography for the nation. At their Urbana meeting, committee members authorized Lillie to ask NAS home secretary

David White for permission to apply to the General Education Board for $15,000 to permit the acquisition of a person who would represent the committee. The desired individual should be "a man of recognized scientific reputation and good judgement, who should receive from$6,000 to $8,000 for the year." This advocate and investigator would have a travel budget of $3,000, with an additional $2,000 for a secretary and an allowance for committee expenses and unexpected emergencies. He would conduct a preliminary survey of the field of oceanography in the United States to provide a context within which the committee might make recommendations regarding the East Coast institution, governmental cooperation in research, and the strengthening of marine laboratories.[20] In all of his discussions with Rose, Lillie made clear on behalf of the committee the important role of the government. The chairman knew that neither Scripps nor the new station would have the resources to accomplish the kind of program the committee and its investigator might propose. Close relationships with the navy and other agencies would bring support and critical resources.[21]

Lillie feared that the NAS executive committee might reject his request to make an application to the Rockefeller Foundation. The Academy took great care not to solicit funds from private sources, seeking to protect its impartiality as the nation's scientific consultant. He quickly assured White, through Wicliffe Rose, that the Rockefeller Foundation's General Education Board had already expressed interest in the committee's projects and wanted to help. On 13 December 1927, White informed Lillie of the executive committee's approval.[22]

From the beginning, the committee demonstrated dedication to a broad construction of the NAS mandate. Acting with the Academy's blessing and building on Vaughan's earlier observation that the committee lacked expertise in many areas vital to oceanography, Lillie and his colleagues suggested additional people to serve in an advisory capacity in spite of the potential expense. In this way, the committee members might come to know possible candidates for the directorship of the East Coast oceanographic institution as well as others who would provide the desired scientific depth to the committee's deliberations and recommendations. The first two men who came to mind were Henry Bryant Bigelow and the University of Toronto's A. G. Huntsman, director of the Marine Biological Station at Saint Andrews, New Brunswick. By the following March the committee had both expanded its list of advisers and resolved to commission an investigator to survey the current status of oceanography in the United States.

During a meeting at the Cosmos Club in Washington, D.C., on 22 April 1928, the committee formulated a list of candidates for these positions. The committee members had already determined the previous month that the $15,000 they had applied for earlier would simply not do the job. Lillie filed an application with the General Education Board for $75,000 to satisfy the cost of the investigator and any advisers whom they might choose. Given Rose's retirement plans, his assistant in these matters, Abraham Flexner, took charge of the new application. On the advice of David White, Lillie sent the paperwork directly to Rose and Flexner. The NAS executive board had already dispersed, so the committee had to rely on White to send NAS approval to the Rockefeller Foundation after the fact. In an 8 June letter from W. W. Brierley of the General Education Board, Lillie formally received the foundation's approval of the enlarged request.[23]

The Lillie Committee naturally reflected the social and professional preferences of ocean scientists and the university society in which nearly all of them resided. Not surprisingly, the advisers came from that same society. They resided in either academia or elite private research institutions. During the Cosmos Club meeting, committee members named nine candidates. Of these, only one came from a federal agency: N. H. Heck of the Coast and Geodetic Survey. The others came from research centers outside the government and reflected the deliberate diversity characteristic of the committee's short history: Henry Bryant Bigelow; John Fleming; George F. McEwen of Scripps; and W. F. Thompson of the College of Fisheries at the University of Washington; W. A. Clemens of the Marine Biological Station in Nanaimo, British Columbia; A. G. Huntsman, V. I. Pettersson of the ICES; and Harald U. Sverdrup, director of the Geophysical Institute of Bergen, Norway. Before the committee members left their comfortable accommodations at the Cosmos Club late on the evening of 22 April, the consensus favored consulting with Sverdrup, Fleming, and Petterssen.[24]

In July and August, 1928, with the committee's activities well under way and relatively well funded, Lillie invited his colleagues and some of the preferred consultants to a series of meetings in his favorite summer haunt, Woods Hole. He wanted to define more precisely oceanographic issues and problems and believed some time together in a congenial environment would facilitate the committee's work. Lillie convened the meetings at the MBL, and the village proved the perfect site for it was as bright, beautiful, and stimulating in the summer as it was gray, dreary, and often inhospitable in the winter.

In addition to the committee members, the chairman invited Harald Sverdrup from Norway, and sought opinions from Huntsman, Bigelow, H. V. Neal of the Mount Desert Island Biological Laboratory, E. L. Mark from the Bermuda Marine Biological Station, and L. B. Becking of the Jacques Loeb Laboratory in Pacific Grove, California. The committee also benefited from the scientific population in residence at the Bureau of Fisheries and the MBL in Woods Hole for their summer research. Lillie and his committee consulted with Elmer Higgins of the Bureau of Fisheries, Thomas Hunt Morgan from the California Institute of Technology (CalTech), and M. H. Jacobs of the MBL, along with scientists from leading American universities such as Yale, Princeton, Cornell, Johns Hopkins, and Columbia.

A number of important issues claimed the participants' attention. They discussed the site of the new East Coast oceanographic institution and debated Lillie's choice of Woods Hole. CalTech's Morgan—president of the NAS, Nobel Prize winner, and a regular at the MBL—suggested setting up shop in New York. He and others argued that Woods Hole winters did not appeal to many scientists, and the village's isolation from other laboratories could prove counterproductive in an institution designed to offer a center for East Coast oceanographic research. Lillie also sought out Harald Sverdrup and had various conversations with him between 20 and 25 August regarding this issue. Although no final decision on location emerged from these meetings, Woods Hole remained the preferred choice because of Lillie's authority as committee chair and earlier discussions with Wicliffe Rose regarding the village's suitability.

A particularly significant consequence of the Woods Hole summer conference emerged not from an intense debate or lecture but from a quiet Friday afternoon meal on 17 August. Lillie, Vaughan, and Edwin Conklin agreed around the luncheon table to nominate Henry Bryant Bigelow as an associate of the committee for one year dating from 1 October 1928. If he accepted, Bigelow would become the vital advocate and investigator responsible for the committee's final report on the present and future of U.S. oceanography. The committee approved the proposal during its afternoon session and, after some initial doubts, Bigelow accepted the appointment. His arrangement with the committee included an annual salary of $7,500 drawn from the Rockefeller funds. Vaughan left the conference by car that same day and, as the meetings and consultations ended, the remaining participants left Woods Hole or returned to their summer research. Over the next eighteen months Bigelow wrote the committee's "Report on the Scope, Problems, and Economic Importance of Oceanography, on the Present Situation in America, and on the Handicaps to Development, with Suggested Remedies."[25]

As the Woods Hole summer conference concluded, the navy continued its basic program of deep-sea soundings and hydrographic surveys. The Hydrographic Office accumulated sounding data using the SDF on board regular naval vessels between San Francisco and Hawaii, along the West Coast, in the Canal Zone region, around Cuba, and off the northern coast of South America. In addition, four hydrographic survey vessels, *Hannibal, Nokomis, Fulton,* and *Niagara* operated in the waters north of South America and in the West Indies throughout 1927 and 1928. Along with the data needed for charts, these ships collected SDF records, temperature and salinity measurements, and water samples that complemented topographical data taken from aerial survey photographs.[26]

In a particularly significant development, Hydro joined with the Carnegie Institution in Washington to sponsor an ambitious voyage to take gravity measurements in the Caribbean. Anxious to obtain accurate estimates of gravity variation in the Western Hemisphere, Carnegie and Hydro convinced Felix Vening-Meinesz that he should supplement his 1926 work in the Caribbean with a new, extended submarine research voyage. From the beginning, some naval officers readily recognized the credit the project would bring to the navy. According to the superintendent of the Naval Observatory, "it can readily be seen it is a most important step and one that cannot fail to bring credit to the submarine service as the essential agent of its accomplishment."

Vening-Meinesz was the designer of a pendulum gravimeter that could, in part, compensate for the motion of a ship. However, he took his most accurate measurements on board submarines submerged at sixty to ninety feet, where they were unaffected by the surface turbulence that plagued most vessels. To assist Vening-Meinesz and to learn the operation of his subtle gravity apparatus, Carnegie nominated Fred E. Wright, a petroleum geologist in its Geophysical Laboratory to participate in the gravity expedition. The Navy Department yielded to requests from Hydrographic Office and volunteered Elmer B. Collins, a Hydro senior scientist, and submarine S-21 for the project. Designed by the Electric Boat Company and built at the Bethlehem Steel Company in Quincy, Massachusetts, S-21 was a diesel-electric boat of 854 tons normal displacement commissioned in 1923.

On board S-21 in the West Indies and Gulf of Mexico in October and November, 1928, the Navy-Carnegie Gravity-at-Sea Expedition took SDF soundings and gravity readings at forty-nine stations arrayed closely together. This unique research revealed unanticipated stress in the ocean bottom north of Puerto Rico in the trench called the Nares Deep. Noting that the walls of the trench assumed a slope of as much as forty degrees at some points, the scientists discovered an unexpected weakness in the gravity measured at Nares. As they moved south toward Puerto Rico, gravity strength increased far beyond the expected levels. The conclusion reached by the Carnegie Institution, Hydro, and the NRC's Division of Geology and Geography had profound implications for the future. The NRC report stated: "These indications point to the conclusions that there is a horizontal pressure in a north and south direction in the ocean bottom in this region, causing a buckling to push Porto Rico [sic] up and the deep down." They found similar stresses in the central portion of the Gulf of Mexico, but not off the Mississippi River delta nor near the Bartlett Deep, south of Cuba. Observations such as these constituted, along with Hayes's work with the SDF, some of the very first steps in a slow, irregular, and often halting process of developing new ideas, instruments, and data into a geophysical research tradition.[27]

By the end of 1928, Lillie's committee began to provide the scientific direction for American oceanography that the navy had hoped to give after the ICO. Bigelow's project, a pledge from Vaughan to report on oceanographic facilities and resources around the world, and the progress made on the East Coast oceanographic institution permitted the committee to offer direction and professional support beyond the informal ties and agreements of the previous four years.

Recognizing the essential role the navy played in oceanographic exploration, Lillie and some of the committee members visited Secretary of the Navy Charles Francis Adams in Washington, D.C., during the autumn of 1928 to request a more precise definition of the navy's commitment to oceanography. Adams responded with the appointment of a committee to review the value of oceanography to the navy's mission and the degree of commitment current budget figures and force levels would make possible. The new Naval Committee on Oceanography commenced its deliberations in December, 1928, under the chairmanship of Rear Adm. Frank H. Schofield with Bigelow serving as liaison for the Lillie Committee. The Schofield Board produced a report on 22 March 1929 that offered guidelines for the practical extent of navy participation as well as a firm navy commitment to the research.[28]

The board broke the possible commitments the navy might make into two categories. In the ideal scenario, the navy would make the maximum investment. The most thorough oceanographic research required a special naval vessel designed and fitted out for this work. According to the board, this type of research included "a detailed examination of specific localities as to animal life, currents, temperatures, salinity of water at varying depths, temperature of water at varying depths, character of the bottom contours, of abnormal formations of the bottom, and allied subjects."

Hydrographer of the Navy Capt. Clarence S. Kempff, superintendent of the Naval Observatory Charles S. Freeman, and Schofield agreed with the board, but they set aside the recommendations, reminding Secretary Adams of budget realities and reduced personnel levels. Schofield made it clear that research of this sort was highly desirable if, at some future time, the navy could afford both the considerable expenditure and the commitment in ships and crews.

The second category the board considered had potential given both the fiscal situation and the precedent set by the navy's oceanographic activity over the preceding four years. While a specialized naval vessel represented the ideal, research conducted incidental to normal naval operations offered excellent possibilities. Scores of naval ships regularly plied the oceans of the world, and many of them could accommodate a scientist or a trained technician. Furthermore, many naval vessels came equipped with an SDF.[29] These resources offered the possibility of data collection on a scale never before contemplated. The board suggested regularly collecting information on depth, temperature, salinity, water color, currents, and ocean-bottom contours with a view toward accumulating enough information to permit a far more comprehensive knowledge of the oceans.

The navy to a large extent already played this role. The Schofield Board recommendations formalized and extended this type of commitment and practice. Specifically, the board recommended a series of steps that would provide ocean science with a constant flow of data from all over the world gathered by naval vessels and their crews. The board insisted on the publication of a world chart showing the unsurveyed portions of the ocean to exhibit the areas most in need of data collection. Another recommendation was that the commanding officer of each naval vessel capable of doing this kind of research be issued precise orders to undertake SDF soundings, air and water temperature measurements, water samples, and, when possible, particularly detailed observations of the very deepest parts of the ocean. The Hydrographic Office would coordinate the effort in a systematic way to gather the most vital and precise information. Schofield's report even went into detail on the need to ensure the careful calibration of the instruments, especially thermometers. Hydro would manage this as well, while serving as the central clearinghouse for the collected data, the publishing agent, and the liaison between the navy and the private institutions that frequently "worked up" or analyzed the information. The essence of the committee's recommendation made clear that with some flexibility and imagination naval assets could provide extensive coverage of the ocean. Naval vessels plying regular ship routes would serve oceanographic purposes and would make special route changes and unusual detours over the deeper parts of the ocean if time, personnel, and mission permitted.

Despite the austere fiscal climate, Admiral Schofield's and the board's final remarks echoed both the sentiments and the determination of the 1924 ICO. They concluded "that it is highly desirable for the Navy to undertake the class of work above indicated. While the direct and immediate contributions to oceanographic knowledge may not appear to be of great practical use, the Board is convinced that the possibilities involved . . . may result in valuable conclusions permitting weather, current, and other predictions of sufficient importance fully to justify the effort."[30]

As it prepared to deliver its final report to the NAS, the Committee on Oceanography monitored the deliberations of the Schofield Board and maintained close contact with the Navy Department. On 12 October 1929, the committee members visited Secretary Adams to report on the status of their projects and discussions, to thank him for his active commitment to oceanographic research, and to view some of the latest charts and maps produced by Hydro.

This visit gave Bigelow a brief respite from finishing the committee's comprehensive report for the NAS. The latter had received news of Bigelow's appointment and the committee's progress at its 1928 annual meeting, held between 19 and 21 November in Schenectady, New York. Two days short of one year later, on 17 November 1929, Lillie presented his final report, which the Academy promptly accepted and adopted. Bigelow's contribution provided one of the foundation stones of American oceanography by offering a succinct statement defining the science and its importance to the nation's defense and economy.

Bigelow called upon the United States to apply its personal, material, and financial resources more energetically to oceanography. If the country desired to reach a level of understanding already enjoyed by the states of northwestern Europe, then government and science had to educate the professionals and find or create the facilities, ships, and instruments to do the job. The report strongly supported creating the East Coast oceanographic station which, along with Scripps on the West Coast, would help coordinate the oceanographic activity of private institutions and federal agencies.

Lillie acted quickly to ensure the creation of the East Coast station so critical to the committee's plans for the future of oceanography. No sooner did the Academy approve Bigelow's report, than Lillie notified Max Mason, the new president of the General Education Board. Since Rockefeller Foundation funding was contingent upon Academy approval of the committee's plans, Lillie did not want to waste time insuring the availability of money to inaugurate the new research center's activities. He had already taken the precaution of informally retaining the architectural firm of C. A. Coolidge to compose basic studies and plans of the buildings and wharves the institution might need. Lillie's precisely laid plans left no doubt as to the location of the new institution. He first wrote to Coolidge in July, 1929, asking him to base his plans on the configuration of a piece of property on Bar Neck at Woods Hole. The chairman had planned and argued so effectively that the trustees of the Rockefeller Foundation had actually authorized its executive committee to fund the East Coast station on 14 November—three days before the Academy received Bigelow's report. Thus, when the report was accepted, the foundation immediately made $2.5 million available to create the Woods Hole Oceanographic Institution (WHOI). One million dollars of the total covered the erection of a completely outfitted laboratory building and construction of the R/V *Atlantis*. The second million took the form of an endowment that would become available upon completion of the building. The balance of the funds became the institution's operating budget, granted by the foundation in annual installments not to exceed $50,000 over a ten-year period. On 6 January 1930 the articles of incorporation were filed and the new board of trustees met on 15 January to officially elect Henry Bigelow as WHOI's first director. One of the navy's most productive future partners in oceanographic research thus became a reality.[31]

For Frank Lillie and his colleagues, submitting their report to the Academy and the beginning of work at WHOI did not mean the end of the committee's activity. Both Vaughan and Lillie involved themselves in ventures designed to promote more comprehensive oceanographic research in the Pacific. At their 27 April 1930 meeting at the Cosmos Club, the NAS committee members heard their chairman describe his impressions of the needs in that region and the interest expressed by Max Mason of the General Education Board in supporting West Coast research. Always conscious of the need to take advantage of opportunities, Thomas Wayland Vaughan made a series of proposals designed to draw funds and interest toward the West Coast. He moved that the committee ask the NRC to advise the federal government to remain active in the International Committee on the Oceanography of the Pacific, part of the Pacific Science Association. He further moved for the creation of a national oceanographic journal and the continuation of Pacific research begun by the Carnegie Institution's R/V *Carnegie*. The committee took no immediate action on the motions, but Lillie and the other members made no secret of their interest and support.[32]

Its considerable investment in the voyage of the *Carnegie* made the loss of the ship as great a tragedy for the navy as it was for the scientific community. During its seventh world cruise under the command of Capt. J. P. Ault, the nonmagnetic brigantine caught fire and burned on 29 November 1929 after an explosion occurred while loading gasoline at its berth in Apia, Western Samoa. Several crewmembers, including the captain, died in the explosion. This loss presented the oceanographic community with a major problem. The *Carnegie*'s ambitious research program came to an end with the ship and none of the major research institutions had a vessel to replace it. Even the navy could not find a suitable ship free to pursue the *Carnegie* program, which included collecting data for virtually every type of marine scientist, with a natural emphasis on the irregularities in the Earth's magnetic field.

Since it found bottom topography of particular interest, the navy had provided *Carnegie* with an SDF for the voyage. This device ran for 15,000 miles in the Atlantic Ocean and another 30,000 in the Pacific before the explosion arrested the process. Captain Ault shipped much of the collected data back to the United States at various times during the voyage, so most of the ship's survey and collection efforts survived. Discoveries pertaining to Pacific bottom irregularities included a rise reaching three thousand meters above the average ocean-bottom formations, which the *Carnegie* expedition named "Hayes Peak" after the NRL scientist and inventor of the SDF.

For a time the Hydrographic Office tried to convince the CNO to transfer the former presidential yacht USS *Mayflower* (PY-1) to the Shipping Board, which would legally permit the use of the vessel by the Carnegie Institution to continue the *Carnegie*'s research schedule. Budget limitations made this impossible in spite of a persuasive Hydro memorandum to the BuNav citing the importance of the work done by the lost vessel and its crew. Unable to offer a replacement, the navy did support later efforts by Scripps in cooperation with the Carnegie Institution to process the expedition data for publication by the IHB in Monaco. In research done on board regular naval vessels incidental to other obligations, the navy—in cooperation with WHOI, Scripps, and the Carnegie Institution—also took pains during the next decade to duplicate some of the routes the lost ship either covered or might have covered.[33]

At the same time the *Carnegie* disaster caused the navy and civilian science to scramble in search of alternatives to a promising research venture, the work of Rear Adm. Richard E. Byrd's Antarctic expedition brought favorable national attention to naval-sponsored science. Although the expedition concentrated on flights over the South Pole during January, 1929, and the knowledge and notoriety they might bring, oceanographic work went on at the same time. Along with careful surveys of the shelf ice, the navy explored atmospheric conditions. The expedition also employed the SS *City of New York* to take SDF soundings, especially at the entrance to the Ross Sea between New Zealand and Little America, all along the edge of the Ross ice shelf, and on a productive return trip near the Bay of Whales.[34]

4

Common Practice and Uncommon Business, 1930–40

■ I am mighty glad to hear that a [Navy] sub[marine] is detailed to the gravity work, and that [Richard] Field has the money to bring [Vening-]Meinesz over for it. Great things should result.

—HENRY BRYANT BIGELOW to WILLIAM BOWIE, 1 June 1931

I N THE PROCESS OF PLANNING SURVEYS AND COORDINATING data collection after the Great War, the navy Hydrographic Office carefully cultivated relationships with the directors of the two premier oceanographic institutions to obtain assistance, instruments, and advice. Thomas Wayland Vaughan of Scripps, Henry Bryant Bigelow of WHOI, and Rear Adm. Walter Gherardi, commander of Hydro from 1930 to 1935, recognized that the scarcity of funds, ships, and personnel made them mutually dependent. Only close co-operation would place within their reach the goals ocean scientists had pursued since 1924.[1]

In August, 1932, Gherardi asked Bigelow for assistance in obtaining some instruments Hydro needed in its surveys. Bigelow replied in early September that he had no spare instruments to loan, but perhaps he could help in another way. Recognizing that Hydro would have to spend about $2,000 to acquire the instruments it wanted, Bigelow offered WHOI's assistance in training naval officers to use the equipment. The director of the new institution reminded Gherardi that the "observations must be taken with great accuracy to meet modern standards." He wanted the admiral to assign one of his officers to

WHOI's R/V *Atlantis,* scheduled to depart on 20 September for the Gulf of Maine. If someone from the USS *Hannibal* (AG-1) went on the voyage, "he would see all the operations and have a chance to thoroughly familiarize himself with the whole procedure."

Admiral Gherardi wasted no time in accepting the offer. He wrote to Bigelow on 9 September that he found the suggestion very gracious and constructive and would certainly select a member of *Hannibal*'s crew to accompany *Atlantis* to the Gulf of Maine. By 14 September Hydro had ordered Comdr. Charles C. Slayton, commander of *Hannibal,* to join the WHOI expedition. Slayton's experiences with data collection methods and his participation in the activities of the cruise permitted Hydro to reap great benefit. As Gherardi commented to Bigelow in a letter penned on 30 September 1932, "There is nothing equal to the personal contact with those who have had continuing practical experience in the use of special apparatus and getting a first hand experience in its use."[2]

The Woods Hole Oceanographic Institution also worked very closely with Hydro in its effort to study the waters of the Caribbean. Bigelow and his staff helped Gherardi select the best and most scientifically significant temperature and salinity stations in the region. Columbus O'Donnell Iselin, WHOI's assistant director, provided the information requested by the navy. While Hydro naturally had its own idea about the schedule and route of its ships and the places where the navy needed data collected, Gherardi still wisely solicited Bigelow's assistance. He never missed a chance to augment and perhaps improve Hydro's selection process while strengthening established ties with Woods Hole. The advice and training would help Commander Slayton and *Hannibal* when they returned to the Caribbean in November, 1932. Friendship and cooperation would also enable Hydro to benefit from the research *Atlantis* and its scientific team planned to do in the Caribbean during their general oceanographic survey of the region from the Windward Islands to the Yucatan Channel in February, 1933. For Hydro, sowing the seeds of a long-term relationship made operational, scientific, and fiscal sense.[3]

Gherardi also cultivated a close professional connection with the Scripps Institution of Oceanography in La Jolla, California. He worked more closely with Vaughan than any other civilian scientific leader and both Hydro and Scripps profited immensely from the link. While all oceanographic institutions, universities, and seagoing businesses relied upon Hydro for essential navigation charts, only Scripps had the advantage of seeing the data before it went to the publisher. By agreement with Gherardi, Scripps received the oceanographic survey data directly from ships like *Hannibal* and *Nokomis.* Vaughan convinced the navy to permit Scripps scientists to perform the chemical and physical analyses necessary to render the information useful for chart composition. This gave Scripps personnel the first opportunity to use the information to augment their vision of the ocean and it gave Hydro a way of "working up" the data at minimal cost to the navy.

Data analysis formed only a small part of the Scripps-Hydro interwar relationship. Vaughan also provided Gherardi's people with suggestions for stations to occupy at sea. At these particular locations, both Hydro and the civilian scientists would find data on temperature, salinity, depth, bottom topography, and other types of knowledge that would provide information either indicative of a region or of particular interest for mapping or research purposes. Hydro did not always initiate the request for advice to plot station locations for a research voyage. Vaughan and his colleagues frequently offered their unsolicited advice, but the admiral repeatedly made it clear that he welcomed their requests and suggestions. He made it a practice to oblige Scripps and WHOI unless other commitments made it absolutely impossible.

The same policy applied to advice on scientific methodology and instrumentation. Often lacking sufficient funds to purchase adequate instrumentation for its research voyages, Hydro called upon private research institutions and other federal agencies for surplus items or equipment on loan. Scripps or WHOI might make the loan, help find surplus equipment, or, as in the case of Commander Slayton, offer to train personnel in lieu of actually furnishing hardware. Scripps offered all manner of instrumentation in response to Hydro requests during the interwar period, including apparatus for collecting specimens from the ocean bottom and bottles for storing the valuable water samples that would return to La Jolla for analysis. These activities helped provide the navy and commercial mariners with essential charts and critical information on the ocean environment. At the same time, civilian scientists found a valuable ally who helped them sustain their scientific research. The navy would discover soon enough that the alliance would prove as valuable in war as in peace.

Vaughan realized early the possibilities of the alliance with the navy and took measures to strengthen and exploit it. On 1 October 1932 he asked Admiral Gherardi if "it would be possible for one of our men to work on board one of your vessels while operating in the Gulf of Panama in order to carry out certain analyses." Along with the trip taken by Slayton on board R/V *Atlantis* at Bigelow's invitation, this request by Vaughan set a significant precedent for exchange of personnel for research and training purposes. Hydro immediately applied for permission to take a civilian scientist from Scripps on board *Hannibal* in the Gulf of Panama. In late November Gherardi received final approval from the Bureau of Navigators (BuNav) and informed Scripps on 9 December, writing to the institution's acting director, Eric G. Moberg, who was filling in while Vaughan traveled in Europe on business. Five days later, Moberg informed Hydro that he intended to send Richard H. Fleming to *Hannibal*. Fleming worked as a research assistant at Scripps while pursuing his doctorate in chemical oceanography at the University of California. Moberg guaranteed Gherardi that Fleming's work would not interfere with shipboard routine. Indeed, much of the young chemist's research would derive from the vessel's regular water sampling efforts.[4]

Despite some technical difficulties, the voyage proved very productive. Fleming left San Diego on 27 April 1933 on board the USS *Chaumont,* a navy transport of eighty-three hundred tons displacement built in 1920 by American International Shipbuilding Corporation of Hog Island, Pennsylvania. During the trip to Balboa in the Canal Zone, Fleming collected twenty samples of plankton and an equal number of ordinary seawater for salinity measurements. The crew also took water temperature readings at every *station,* the common scientific term for planned interruptions in the vessel's progress to permit experimentation.

After his arrival aboard *Hannibal,* Fleming took 423 oxygen determinations at eighty-five stations with samples obtained at a variety of depths from five to eighteen hundred fathoms. The variation of oxygen saturation in Gulf of Panama water proved remarkable. From little or no oxygen in some samples, others displayed as much as 200 percent. The highest concentration Scripps had discovered at any of its Pacific Ocean sites was 130 percent. In his report to Vaughan, Moberg quoted Fleming on his experience. Amazed by the oxygen content and the amount of plankton, the Scripps student described the Gulf of Panama as "soup," comparable to the Gulf of Georgia: "I have had a sample taken at practically all of the stations; there is a great variation in composition, some being practically all phytoplankton and others zooplankton. This certainly seems to be a most interesting region and I am sure an oceanographic boat could spend a year here quite profitably."

Gherardi found the voyage very fruitful and quickly thanked Scripps for providing the additional manpower and expertise. Fleming's work furthered his education, brought valuable data and knowledge to Scripps and Hydro, and did not interfere with the ship's survey mission for the navy. Everyone benefited from the young scientist's "piggyback" ride aboard *Hannibal.* Much to Moberg's satisfaction, Gherardi wrote: "Tentative arrangements are being made for a dynamic survey during the coming survey season of 1933–34, and I would like to be advised if the Scripps Institution of Oceanography would again assist us in titrating water samples and along other lines of endeavor germane to dynamic oceanography."

Piggyback science on navy ships, like that accomplished by Richard Fleming on *Hannibal,* continued, as would Scripps's analyses of basic seawater and plankton samples. With the Fleming voyage, Vaughan and Gherardi fortified significantly the Bigelow-*Atlantis* precedent for civilian-naval cooperation in field research.[5]

In the Pacific northwest, Thomas Thompson of the University of Washington quickly followed Vaughan's lead and asked the navy for both an SDF for the university's research ship *Catalyst* and space aboard a naval vessel participating in the impending survey of the Aleutian Islands in 1933. He met with success on both counts.

On 24 October, Thompson wrote John Fleming in support of his suggestions to the navy that *Ramapo* play temporary understudy for the geophysical research vessel *Carnegie,* lost in an explosion while replenishing fuel in Apia, Samoa, in 1929. Thompson also pledged University of Washington assistance in the sounding effort by committing *Catalyst* to work inside the continental shelf. The ship's echo sounder, a Submarine Signal Company Model 431, imposed this operational restriction because its range extended to only four hundred fathoms. For research in Puget Sound and much of the coastal passage to Alaska, the 431 provided excellent data. Because of its limitations, the staff aboard the *Catalyst* never bothered to use the device in the extraordinary depths beyond the continental shelf. Nonetheless, Thompson wanted a partnership with the navy. He offered the services of his colleagues and the university's vessel to the navy's Aleutian Islands survey of 1933.

Organized during the first quarter of 1933, Hydro commissioned the Aleutian Islands Survey Expedition, led by Comdr. H. A. Badt, to conduct a complete hydrographic survey of Adak Island and the surrounding waters. The USS *Argonne* (AP-4), *Swallow* (AM-4), *Kingfisher* (AM-25), and *Gannet* (AM-41) received instructions to supplement their hydrographic work with topographic surveys of not only Adak, but also Kanaga and Kagalaska Islands. The navy supplemented this effort with the same kind of aerial photographic work done by USS *Nokomis* (PY-6) and *Hannibal* in the Caribbean approximately four years earlier. Hydro approved Thompson's request to participate in the expedition and the University of Washington sent Clifford A. Barnes from Bremerton to the Aleutians and *Argonne* aboard the USS *Patoka* (AO-9). Before sailing, Thompson sent all of the equipment Barnes needed to the *Patoka* via the R/V *Catalyst*.[6]

Hydro and the commander of the expedition made a very strong appeal for time, ships, instruments, and funding based on strategic and tactical considerations. Commander Badt pointed out that countries like Japan frequently visited the area and knew far better than the U.S. Navy the choice passages and anchorages, as well as the challenging weather conditions.

Badt's detailed final report insured that Hydro would remain in the area performing passage, bay, and harbor surveys for a few years to come. He emphasized the lack of information on these islands as well as the absence of adequate naval assets and personnel to provide the navy with operating bases, station ships, or personnel with experience in the region. With eighteen weeks of work in the western Aleutians, the expedition personnel laid the foundation for naval expertise and future work. Furthermore, thanks to Gherardi, Thompson and his representative on the expedition, Clifford Barnes, could now count themselves among the few civilians acquainted with the Aleutians.

As a consequence of their participation, Hydro convinced BuEng to give the University of Washington oceanographers an obsolete SDF Model SE-1987 just removed from the USS *Detroit* (CL-8) for installation aboard the R/V *Catalyst*. This would allow Thompson's staff to take soundings well off the continental shelf. Unfortunately, it took BuEng months to find a more current model—and when it did, the university's research ship could not accommodate the system without facing prohibitive structural alterations. Although the navy could not equip *Catalyst* with an improved SDF, the university came away from the Aleutians experience with Hydro's confidence.

Nor was the University of Washington the only research center to see possibilities in a further survey of the Aleutians. At Scripps, Vaughan suggested that Pacific research also needed a series of soundings from near the center of the Aleutian chain, to the northern Hawaiian Islands. Although a shortage of time and resources did not permit him to fulfill Vaughan's request on the 1933 trip, Gherardi agreed with Vaughan's perspective and kept his letter on file, planning to make the string of stations at the earliest possible convenience.[7]

Hydro went back to the Aleutians during the spring and summer of 1934. Following Commander Badt's recommendations, the expedition, built around the USS *Oglala* (CM-4) and *Tanager* (AM-5), did triangulation work to plot coastlines accurately, took surveys of harbors and potential anchorages, recorded depth and bottom soundings, made magnetic field measurements, and conducted aerial photo surveys.

As the navy accumulated the data on the island chain, Gherardi found an opportunity to accommodate Vaughan's request for a string of vertical sections for temperature and salinity extending between Dutch Harbor on Unalaska Island south across the northern Pacific to Pearl Harbor in the Hawaiian Islands. At the conclusion of the Aleutians expedition, the USS *Bushnell,* a submarine tender and flagship of Submarine Division Twelve, planned to make the journey to Hawaii over the exact route that Vaughan had requested for his research. Admiral Gherardi succeeded in obtaining permission from both the CNO and Rear Adm. C. W. Cole, the new commanding officer of the submarine division, for *Bushnell* to perform the tests Vaughan had requested during the voyage southward. Furthermore, the Scripp's director obtained permission for one of his scientists to direct the work. The navy offered to transport Vaughan's candidate from San Diego to Dutch Harbor and then to Pearl Harbor aboard *Bushnell*.

Vaughan chose Roger Revelle for this job. Revelle, a University of California graduate student in oceanography and friend of Richard Fleming who went to sea aboard Hydro's *Hannibal* in 1933, graduated from Pomona College in 1929 and immediately initiated his graduate studies in geology at Berkeley. He moved down to Scripps in 1931 to work on the ocean-bottom samples sent back to the continental United States by *Carnegie,* and rapidly acquired significant expertise in ocean sediments. Earlier in 1934 he spent ten days with *Pioneer,* the U.S. Coast and Geodetic Survey's research vessel. On board the *Bushnell* from 17 to 24 August, Revelle made eighteen dynamic ocean surveys en route from Dutch Harbor—usually two each day, one hundred miles and twelve hours apart, at 5 A.M. and again at 5 P.M.

These dynamic ocean surveys differed from ordinary station taking because they provided additional water samples as well as temperature and salinity data from various depths to determine circulation patterns and water density. Usually, two "casts" were made in this type of survey, with water sampling bottles at different levels on the cable going over the side of the ship. The first cast had bottles at 25, 50, 75, 100, 150, 200, 300, 400, and 500 meters. Starting at 500 meters, the second cast went to 600, 800, 1,000, 1,500, 2,000, 2,500, 3,000, and 3,500 meters. Revelle's surveys crossed routes and stations in the north Pacific occupied at different times by both *Ramapo* and *Carnegie.* Vaughan designed the research surveys this way so Hydro and Scripps could use the data from all three vessels to confirm single or mutual observations and to correct the data for chemical and climate conditions.[8]

Revelle's work at sea from the Aleutians to Hawaii in 1934 confirmed, as did the research in the Caribbean, the cooperative nature of oceanographic exploration during the decade before Pearl Harbor. The navy and civilian science each made their distinctive independent contribution to a fruitful common practice. Driven by the imperatives defined in 1924 at the ICO, Hydro applied its resources to traditional surveys and an extended program of data gathering. Regular naval vessels obliged the Hydrographic Office and took thousands of soundings for depth and bottom topography during long voyages between the U.S. West Coast and the navy's distant bases in the central Pacific and Asia. As in the Gulf of Mexico and Caribbean, Hydro's survey vessels performed as much ocean science as the budget and naval strategic priorities would permit while compiling information for charts and maps to aid navigators and those responsible for estimating the strategic value of American possessions throughout the Pacific.

Their common practice was not merely a naval creation, but emerged also from the needs of the civilian scientific community. Leaders in oceanographic research like Vaughan and Thompson realized that only by sharing their insights, energy, ships, instruments, and talent could they achieve the maximum result from their modest resources. This inclination to recognize common ground and seek the navy's assistance created an informal system that propelled their cooperative research beyond the ordinary.

In each of the major Pacific ventures launched by Hydro, this common practice not only brought results, but also often led to potentially fruitful opportunities to advance ocean science even further. The 1934 Aleutians expedition began as the first survey to examine the strategic potential of a major string of islands in the often-inclement North Pacific. In the end, the expedition not only succeeded in its primary mission, it also prompted the navy's leadership to pursue further scientific investigation in the region. Hydro's experience and the advice received from Scripps and Washington, as well as the participation of their scientists, insured that time and resources were well spent.

Indeed, Hydro and its civilian colleagues rarely wasted any opportunities. The common practice made it natural for Admiral Gherardi to recognize in the *Bushnell's* 1934 return voyage an opportunity to grant the request for dynamic ocean soundings in the north Pacific made months earlier by Vaughan. No doubt the admiral took pleasure in granting his old friend wide latitude in personnel and technical matters on that leg of the *Bushnell's* journey. The navy and civilian scientists built their common interwar practice on this flexibility, their willingness to accommodate each other, and the regular data collection and analytical services performed by both naval and civilian agencies. Both Hydro and its colleagues outside the navy looked to serve their own interests by serving each other. In the economic environment of the Great Depression, they could together achieve things they could only dream of if forced to work independently of each other.

The significance and productivity of their common practice escaped no one, and few took it for granted. Participation was far too profitable for all concerned. On 2 July 1934, Vaughan wrote to Albert Barrows, executive secretary of the NRC, reporting on a meeting of the Committee on Oceanography of the American Association for the Advancement of Science (AAAS) attended by his assistant director. While commenting on the various issues and activities of members and their institutions, Eric Moberg, who attended the meeting for Scripps because Vaughan was ill, particularly noted the presence of a navy captain from the Hydrographic Office. Captain R. S. Culp expressed Admiral Gherardi's thanks for the cooperation of private research institutions, particularly Scripps, and made a point of pledging "the continued willingness of the Navy to cooperate."

The word *cooperation* did not fully reveal the extent of the dynamic, remarkably productive, and largely informal system they had created since the ICO in 1924. Just a few days before the AAAS gathering, Richard Fleming had returned the Scripps analysis of temperature and salinity data taken by nine different Hydro vessels working along the U.S. West Coast. Revelle departed for the Aleutians hours before Culp's comments at the meeting. No sooner had the captain's remarks concluded, than Vaughan once again put in a request for additional dynamic soundings and ocean surveys. Both sides of this relationship repeatedly demonstrated comfort with regular interaction and high mutual expectation.

This time Vaughan asked Hydro for naval aid to perform independent or piggyback research in an area bounded by the Great Circle route between Hawaii and the Canal Zone on one side, with the third point of the scalene triangle at Los Angeles–San Pedro. He also commented on the desirability of using a submarine in the Pacific for gravity measurements. Gherardi's initial response revealed that Hydro could not find ships to take these measurements because projects already under way in the Pacific and Caribbean had nearly exhausted the navy's resources. However, discouraging realities had rarely deterred either Hydro or the civilian scientific community in the past. By this time, their natural reaction compelled them to try and find a way to do the necessary research. With expensive and demanding gravity work ongoing with special instruments and submarines in the Caribbean, Gherardi encouraged Vaughan in his other requests by suggesting that he have the NRC approach the secretary of the navy. The admiral rarely closed the door on any valid project, and this advocacy had worked in the past.[9] With his five years as hydrographer of the navy coming to a close in 1935, Gherardi had worked with Scripps many times and had come to know Vaughan well. The admiral knew that the Scripps director would relentlessly pursue any suggestion that the navy might provide transport, funds, or other forms of support for oceanography. Thus, Gherardi fully expected the navy secretary to receive a Vaughan-inspired letter in support of his new request for soundings and surveys between Hawaii, southern California, and the Canal Zone. On 17 April 1935, Secretary of the Navy Claude Swanson received just such a letter, not from the chairman of the NRC, but from William Campbell, president of the NAS. Campbell supported Vaughan's request, asked for more naval assistance with Pacific research, and in general urged the navy to continue its vital work in oceanography. Gherardi certainly knew his man. This kind of personal knowledge made fruitful interwar oceanography possible.

Uncommon Business

The varied and productive ocean surveys performed by Hydro during the 1930s addressed the general needs of the navy and the scientific community. For the former, surveys in strategically important areas took priority and governed the extent of Hydro's cooperation with civilian science. The navy thus placed its interwar survey emphasis on the triangular region between the U.S. East

Coast, the Caribbean Sea, and the Panama Canal, as well as a similar geometry in the Pacific between southern California, Hawaii, and Alaska.

Beyond the productive routine surveys, unique projects, innovative methodology, and special institutional agreements, the common practice paved the way for progress in compelling oceanographic particulars. In most cases, Hydro's ability to go to sea and its determination to continue oceanographic investigation placed the navy on the cutting edge of pioneering research as sponsor, participant, and facilitator.

The Navy-Princeton Expedition of 1932

By the early 1930s, gravity research was certainly on the cutting edge of the infant geophysics discipline. On board the S-21 in 1928, a team from the Carnegie Institution and the navy led by Felix Vening-Meinesz, Elmer Collins, and Fred Wright discovered magnetic anomalies in the Gulf of Mexico and Caribbean. This sparked a review of the current perspective on local magnetic and gravitational phenomena as well as the accepted views on isostasy.

Unfortunately, the 1929 R/V *Carnegie* disaster cast a shadow over research on the Earth's gravity and structure. When that graceful sailing vessel burned in Apia Harbor, so too did a valuable pendulum device from the Netherlands created to measure anomalies in the Earth's magnetic field. The ship's destruction left incomplete a Pacific research program designed by John Fleming and the Carnegie Institution's Department of Terrestrial Magnetism that had called for gravity measurements to complement those taken off the southern coast of the United States by the S-21.

The research current set in motion by the S-21 and *Carnegie* led William Bowie, chief of the Geodesy Division of the U.S. Coast and Geodetic Survey, to attempt a terrestrial supplement to the Carnegie Institution's program. Without a ship immediately available to take his curiosity to sea, Bowie sent observers from his office to the West Indies. He asked them to take gravity and magnetic measurements on the islands of Cuba, Haiti, and Puerto Rico designed to broaden the value of the S-21 data and provide it with proper context.

When the loss of their ship retarded the Carnegie Institution's ambitions, Bowie and the Geodetic Survey joined forces with Prof. Richard Field of Princeton and Arthur Keith, chairman of the NRC's Division of Geology and Geography. They hoped to persuade the navy to contribute the services of another submarine to continue the gravity program begun by the S-21. In a letter to Henry Bigelow dated 16 May 1931, Bowie informed the new WHOI director that Field and Keith had met with Admiral Gherardi and requested another submarine for gravity work at sea. The admiral supported the suggestion, and a formal proposal drafted by Field and his colleagues to the secretary of the navy soon went forward with Gherardi's endorsement. Reluctant to leave anything to chance, Bowie wrote his note to Bigelow hoping to take advantage of the latter's friendship with Secretary of the Navy Charles Francis Adams. Bowie felt that the submarine gravity project would need all of the support he could find among those prominent both in politics and in gravity research.

Although Bigelow thought he lacked the stature to approach Adams on Bowie's behalf, Professor Field's reputation and the opinion of Admiral Gherardi obtained the approval sought by the civilian scientists. For Field, Bowie, and their colleagues, this geophysical expedition held great promise. The ocean kept 70 percent of the Earth's crust from their view and prevented the kind of detailed study they pursued on land. Beyond the revolutionary images of the bottom offered by Hayes's SDF, the data obtained by Vening-Meinesz with his pendulum apparatus on board Dutch submarines in 1923 and 1926 revealed intriguing and remarkable magnetic field anomalies in the southwest Pacific and in the Caribbean. His work aboard the S-21 with the same instrument in 1928 only further enticed the fledgling geophysical community. Additional research conducted from the relatively quiet and stable platform of a submarine would almost certainly provide insights into the movement of the Earth's crust and role of the mantle beneath.[10]

Rather than merely providing a submarine, the navy went several steps further. Gherardi quickly realized the significance of the work, and wanted Hydro involved. His office found a suitable location at the NRL for the expedition's initial base of operations. After the Vening-Meinesz gravity apparatus arrived in the United States from the Netherlands, the navy transported the instrument to the NRL for calibration by its inventor and two other members of the project's scientific team: Harry Hess of Princeton and Thomas Townsend Brown of the NRL. A river steamer carried the instrument and accompanying personnel to Norfolk, Virginia, where they all watched as navy yard personnel loaded the device aboard the USS *Tarbell* (DD-142) for a four-day journey to Guantanamo Bay Naval Base in Cuba under the command of Lt. Comdr. H. K. Fenn. As *Tarbell* steamed south, the submarine S-48, assigned to the expedition by Secretary Adams, left Coco Solo in the Canal Zone for a four-day voyage to rendezvous with the destroyer in Cuba on 4 February.

By early 1932, S-48 had already qualified as an eleven-year veteran of the submarine force. The Lake Torpedo Boat Company of Bridgeport, Connecticut, built the vessel and the navy commissioned it in 1922. The S-48 had a normal displacement of 903 tons, direct-drive diesel propulsion, and a top surface speed of 14.5 knots—which the ship rarely reached, even on its best days. After the S-48's arrival, the submarine tender USS *Chewink* (AM-39) joined the submarine's scientists and crew in their preparations. They installed equipment, calibrated instruments, tested SDFs, loaded supplies, and made ready for departure.

The *Chewink* and S-48 put out of Guantanamo on 7 February 1932, setting their course south and east on the first leg of the journey. They steered for the Bartlett Deep south of Cuba, where they made the expedition's first undersea gravity station. They repeated this routine every sixty miles or four to five hours, depending on their speed and progress. The way the scientists numbered the stations demonstrated their feeling that this research voyage continued where Meinesz, Collins, and Wright had left off in 1928. The S-21's expedition staff took forty-nine gravity stations, so the 1932 effort began its station count with the number fifty. After completing the first station, the ships turned to the southwest and made three stations along the south shore of Jamaica. Moving then to the north and west, they took six more readings just above the Bartlett Deep, measuring the deepest part of that trench via SDF at twenty-four thousand feet. The balance of stations on the first leg took the two ships around Cuba and north toward Key West, Florida, for a short rest and a chance to develop the film used by the Vening-Meinesz gravity device. The five pendulums that formed the heart of the instrument inscribed their courses on film by means of light reflected from mirrors on each pendulum bob. While two pendulums monitored the attitude of the device itself, the other three oscillated in the same vertical plane. Scientists then mathematically reduced the data from these last three to create two hypothetical pendulums, free from the forward motion of the vessel. This device and method thus determined the intensity of the Earth's gravity at a particular point without any interference. The leg of the trip from western Cuba to the Florida Keys also offered an opportunity to experiment with the SDF and the effect of surface reflection and bottom composition on the returning echo. Stations sixty-nine through seventy-six brought the vessels back to Guantanamo and ended the first "loop."

The second loop began on 25 February and required the expedition to invest all of its time in the vicinity of the Bahamas, taking measurements near the Turks and Caicos Islands, Eleuthera, Cat Island, and San Salvador among others. Scientists on board the vessels accumulated twenty-three hundred miles of SDF profiles displaying the region's bottom topography and twenty-seven gravity stations before the project concluded in Miami on 19 March.[11] Some were sorry to see the voyage end, especially the inquisitive young lieutenant who kept the S-48's erratic diesels operating. Roughly twenty years would pass before Lt. Hyman G. Rickover would encounter a scientific problem with comparable challenge and appeal.

This project revealed as much about civilian science and its priorities as about the importance of naval involvement. Field, Bowie, and Vening-Meinesz, among others, saw an opportunity in the new science of geophysics to make fundamental statements about global processes that affected the Earth's configuration. They had the desire, the talent, provocative magnetic anomalies begging for analysis, and new instrumentation, making the research potentially profitable. Following on the heels of Vening-Meinesz's work in the 1920s, the Navy-Princeton project onboard the S-48 continued identifying the major problems, setting priorities, perfecting methodology, and gaining experience. Geophysicists had clearly begun both to define themselves and their work and to establish research traditions and expectations.

The navy played no small part in this process. A navy scientist developed the SDF, all of the ships selected for important gravity and sound work belonged to the navy, and the most geophysically experienced officers and crews, next to those on the *Atlantis* or *Carnegie,* manned naval submarines, survey ships, and destroyers. With contributing scientists from naval laboratories on board each major gravity expedition organized in the United States, the navy was creating its own commitments and traditions as fast and as firmly as the scientists.[12]

The Navy–American Geophysical Union Expedition of 1936

By the mid-1930s, the desire to accumulate data about the Earth's magnetic field had assumed a life of its own. The possibility of discovering the laws governing the behavior of the mantle as well as those that might explain both regional variations in gravity and their significance proved compelling. The curiosity, the theoretical foundation, and the instrumentation existed to pursue this type of research. Without breaking stride, the scientific community continued geophysical research after the S-48 docked in Miami.

As Field, Bowie, and Gherardi had organized the 1932 undertaking in the Caribbean, Henry Bryant Bigelow and the fledgling Woods Hole Oceanographic Institution chose to sponsor Australian explorer Sir George Hubert Wilkins, who planned to study the Arctic in a submarine. He intended to take Fathometer readings, do research on the ice, and perhaps make a submerged voyage to the pole. The navy removed the submarine O-12 from the active list on 29 July 1930 and transferred responsibility for the vessel to the Wilkins team via the U.S. Shipping Board. The submarine also underwent modifications at the Philadelphia Navy Yard in preparation for the Wilkins expedition. While the scientists rechristened the O-12 *Nautilus* after Jules Verne's formidable fictitious submarine, the yard altered one of the ship's forward compartments to provide a means of equalizing the internal and external pressure.[13] This modification permitted the crew to open a hatch directly to the sea, making it possible to lower sampling bottles, reversing thermometers, bottom samplers, and plankton nets while submerged. In addition, WHOI employed money from the institution's Rockefeller Foundation endowment to purchase a Vening-Meinesz gravity-measuring instrument for the voyage. Save for the one lost on *Carnegie,* no other American scientific organization possessed this device, making gravity research dependent upon the participation of the instrument's inventor, who would bring the apparatus with him.

Wilkins's attempt at submerged polar research achieved only minimal success. Before heading for the top of the world, he made a June, 1931, transatlantic crossing to Bergen, Norway, where Harald Sverdrup of the Christian Michelsens Institutt and other expedition members came aboard. Unfortunately, when *Nautilus* arrived at the edge of the ice pack in mid-August, Wilkins and his company discovered they had lost one of their control surfaces. A forward dive-plane had come loose and carried away during the trip north, making diving under the ice impossible.

Although they could now only speculate about the possible difficulties a submarine might face under the ice, the *Nautilus* scientists could still look closely at the general conditions on the edge of the polar region. Wilkins and his expedition journeyed along the rim of the ice pack, occupying nine oceanographic stations. The scientific team took numerous deepwater samples at depths down to three thousand meters. They also gathered nine bottom samples, eight of them in the form of cores twelve to eighteen inches in length. In addition, the staff recorded three hundred SDF soundings, made seven gravity measurements, frequently collected plankton, and stopped to take one magnetic station on the ice surface.

In addition to the data collected and crew experience gained, the cruise demonstrated the continuing potential of submarines as research platforms. Both the Wilkins project and the series of gravity-research voyages culminating in the 1932 Navy-Princeton effort made the relative quiet of the ocean below the surface a desirable work environment. Undisturbed by surface turbulence, oceanographic instruments provided more reliable and significant data.

Wilkins certainly made the most of an unfortunate situation and, writing to Bigelow on 7 October, Sverdrup claimed that they were "not dissatisfied with the results." Engaging in considerable speculation, Sverdrup and Wilkins erroneously concluded from their limited experience that submarines could navigate under the ice with only minimal difficulty, especially during the more favorable climactic conditions of the summer months. It would take another quarter century to appreciate the challenges posed by under-ice operations.[14]

At the conclusion of the Wilkins venture, the crew unloaded all of the scientific instrumentation at Bergen, including WHOI's valuable Vening-Meinesz gravity-measuring apparatus. As the expedition's scientists departed for their home institutions with the accumulated data, Bigelow reminded them of their promise to permit WHOI to publish their findings in the initial publication in the institution's first scientific series. As for the *Nautilus,* it reverted to the designation O-12 and returned to naval service.[15]

For some of the most important scientists on the geophysical side of oceanography, WHOI's publications seemed less important than getting the gravity-pendulum apparatus from Bergen to the United States. After playing a critical role in the Navy-Princeton expedition, Richard Field immediately initiated planning for further Caribbean research with a gravity-pendulum device. He had little hope of borrowing the WHOI instrument if it continued to lay idle in Bergen. He thus quickly initiated an effort to bring it home.

This was easier said than done, however. The WHOI could not bear the import customs duties if Sverdrup shipped the device from Bergen to Woods Hole via Boston or New York. Furthermore, the original agreement with Wilkins placed the gravity device at his disposal whenever he required it. Field proposed asking Hydro to bring it into the country under the assumption, supported by the State Department, that Gherardi could avoid the customs duties if WHOI made the apparatus a gift to the navy. While Bigelow initially agreed with this approach, he had to reconsider because of WHOI's obligation to Wilkins. At the time, rumors suggested that the Australian explorer had secured partial funding for a return to the Arctic, making an instrument transfer to the navy unlikely.

While this conundrum continued, the gravity device lay deteriorating in storage. Bigelow and Field tried for years to find a solution, but WHOI could not pay the import duties and for months Wilkins's plans remained uncertain. Meanwhile, the Norwegians accidentally gave one of the precision chronometers designed to complement WHOI's gravity device to the master of a hydrographic survey ship in need of a timepiece. To make matters worse, the U.S. Navy refused to pay the customs fees. For a while, Bigelow contemplated loaning the instrument to Prof. J. Stanley Gardiner of Cambridge University. At one point in this frustrating process, Bigelow wrote to Sverdrup "that there is no power, human or divine, which can get that tiresome gravity machine into the United States." For many months Sverdrup himself became the instrument's caretaker for lack of a suitable alternative.

Tired of endless negotiating and plotting, Bigelow and his board of trustees concluded that another Wilkins expedition seemed unlikely and placed the instrument in the custody of the U.S. Coast and Geodetic Survey. Indeed, Hydro and the Coast and Geodetic Survey could not afford to bring the machine into the country until mid-1936. However, from Field's point of view it was better late than never, for his often-delayed geophysical expedition to the West Indies was about to begin.[16]

Commencing with his leadership in organizing the 1932 Navy-Princeton expedition, Richard Field had assumed a role of authority in interwar geophysical oceanography. From his position on the Princeton faculty and as chairman of the AGU's Committee on the Geophysical and Geological Study of the Ocean Basins, he became the principal sponsor and promoter of promising methodologies and the instruments designed to support them. His belief in the importance of submarine geophysics as a new and important avenue of research both strengthened his resolve to support fellow scientists and heightened the intensity of his search for a way to bring the gravity-pendulum apparatus into the country. In a letter to Bigelow written on 23 June 1933, Field argued: "If we have the equipment we can keep up the proper interest and personnel. If, however, we do not have the equipment, I am afraid that the interest and personnel will be lost, for the nature of the operations is such that spasmodic attempts are both exceedingly costly and difficult to organize."[17]

Promising instruments, new techniques, and hard-won results captured both his attention and enthusiastic patronage. Thus, simultaneously with his efforts to bring the gravity-pendulum device into the country, he applauded the research and innovative methodologies of N. H. Heck of the Coast and Geodetic Survey and W. Maurice Ewing of Lehigh University in submarine structural geology. Ewing planned to apply inland seismic techniques to deep-sea research, and Field saw great potential in his methodology and research. More than a half-century later, one of Ewing's graduate students, Allyn Collins Vine, recalled Field's leadership and initiative and Ewing's promise and ingenuity: "I think Field put it so beautifully that geology had gone about as far as it could go on land work, and we needed to get the perspective of the much larger ocean. This is what persuaded Field to look around for some promising young geophysicists to start doing it. So I think it was the other way around. It was Field looking up Ewing and not Ewing looking up Field."[18]

Working with Heck and drawing from standard geological research techniques, Ewing developed a method for employing dynamite as a wide-spectrum sound source to determine both the topography and composition of the ocean bottom by reflection and refraction. The latter technique employed the pace at which sound traveled through the bottom sediments before reflecting back to the surface to determine their nature and constitution. In 1934 and 1935, Ewing received support from the Geological Society to study this methodology as well as the sediments. The Lehigh University physics instructor, a freshly minted physicist from Rice University in Texas, undertook a seismic traverse of the Atlantic coastal plain. He literally brought his terrestrial technique ever closer to the ocean until he felt prepared to take his traveling laboratory and his young assistants on board ship for experiments at sea.

Ewing's relationship with Field helped convince those not comfortable with dynamite that a refraction technique employing explosives could work on board ship. As Ewing continued his research across the coastal plain in the summer of 1935, he proposed to WHOI an experimental voyage on board *Atlantis* designed to measure the thickness of the unconsolidated sediments along the continental shelf near the opening of the Chesapeake Bay. With vivid images of the institution's elegant ship blown into shrapnel by a cargo of Ewing's dynamite, Henry Bryant Bigelow immediately consulted WHOI's insurance company. Field quickly assured Bigelow that WHOI's support of Ewing would help refine an important methodology in submarine geophysics. With that, the WHOI director negotiated with his insurance company and managed to avoid the threatened $200 premium increase that initially followed his announcement that Ewing wanted to carry dynamite.

The Lehigh Physics Department gave Ewing leave for October, 1935, to carry out his plans on *Atlantis*. He and Heck had more than sufficient experience, having spent parts of June and July at sea on a Coast and Geodetic Survey vessel based at Norfolk testing the methods they proposed to use aboard *Atlantis*. By the end of the voyage, Field was elated and Bigelow converted. The latter wrote to Field while the Princeton professor was preparing for the gravity cruise to the Caribbean and Gulf: "The reports that have come to me of Ewing's cruise show the method to have been so practical at sea that I begin to think seriously of the future program along this general line. . . . It seems essential to establish whether different people (of course adequately trained) would arrive at the same interpretation. Iselin has spoken to me briefly of the proposal to raise money to send a ship out especially for submarine geology for a couple of years, mentioning someone's suggestion that Woods Hole Institution be asked to operate it."[19]

In December, 1935, Field notified Bigelow that he planned to invite Maurice Ewing to join the Navy-AGU gravity expedition scheduled for the end of the following year and he hoped that Woods Hole would continue its support of the latter's geophysical research once the anticipated gravity work concluded. Given Bigelow's reaction to *Atlantis*'s October trip to the Chesapeake, Field's letter did not betray any doubt that WHOI would welcome a continued association with Ewing and Lehigh.

In early 1936, Field and his colleagues were ready for the West Indies and gravity research. He announced the justification, purpose, and itinerary of the Navy-AGU expedition in his report to the annual meeting of the NRC's American Geophysical Union as chairman of the Geophysical and Geological Study of Ocean Basins Committee. The Princeton geologist noted that the expedition would explore the western and southern portions of the West Indies to determine the shape and direction of the foredeep and the extent of the avenue of high negative gravity anomalies discovered by the 1928 and 1932 expeditions.[20] Hydro assumed direction of the expedition under Capt. Lamar Leahy, with Ewing, Princeton's Harry H. Hess, and Lt. A. J. Hoskinson of the U.S. Coast and Geodetic Survey as shipboard scientific staff.

Field argued persuasively that the recent development of new geophysical and oceanographic techniques made possible an expedition with enormous potential. The SDF made sounding and positioning more accurate than ever. Recent experience with submarine-gravity surveys demonstrated the value of the methodology developed by Vening-Meinesz and the accuracy of the instrumentation he created. Cooperation with the navy's submariners considerably increased the power and value of the new technology. Field felt compelled to observe that "although these surveys require a submarine, the United States Navy has and is continuing to cooperate in a splendid fashion." In addition, the technology for sampling the ocean bottom had improved considerably. Henry Stetson of WHOI proved that dredges could effectively collect lithologic samples from submarine canyon walls and several new core samplers became available. From Field's point of view, all of these developments, as well as the effective translation of terrestrial seismic refraction techniques for shipboard use at sea, made it the perfect time to continue the West Indies gravity work.[21]

By mid-1936, the essential assets for the expedition began to assemble. The American Philosophical Society had already pledged the $2,000 necessary to provide stipends for Ewing and Hess. In a letter to Captain Leahy, Maurice Ewing reported that WHOI's gravity apparatus would arrive in the United States from Norway during the third week of June. He immediately peppered the captain with questions about power sources on board the submarine and the best place to install the Vening-Meinesz device. On 29 July the navy finally informed Ewing that the chief of BuNav had assigned the USS *Barracuda* (V-1) to the expedition.

In a venture of this sort, the details could spell the difference between success and failure. To enhance the precision of the gravity-measuring device, Hess and Ewing procured a few timing devices vital for taking gravity measurements. Several stopwatches came from the U.S. Naval Observatory, and an essential crystal chronometer developed by Bell Telephone Laboratories at Ewing's request took the place of the chronometers used by Wilkins when he employed the WHOI gravity system on his Arctic cruise.

It did not bother Ewing that Bell Laboratories would not have the chronometer ready for use until 1 September 1936. Actually, the timing was excellent because Ewing devoted the rest of the summer to restoring the gravity device. After years sitting in storage in Norway under Sverdrup's watchful eye, the precision device showed serious signs of deterioration. Without frequent use, this was unavoidable. While patiently erecting, restoring, and calibrating the apparatus in the Coast and Geodetic Survey's gravity room in July, Ewing wrote to Vening-Meinesz for advice. He had difficulties aligning the pendulums, the soldered joints attaching the bobs to the pendulum arm had decayed, and metallic corrosion had taken its toll on a variety of components.[22] With guidance from the Netherlands and the assistance of Bowie and Field, Ewing had the device ready by the scheduled departure time.

Beginning at Coco Solo in the Canal Zone on 30 November 1936, the Navy-AGU expedition wound its way across the northern coast of South America and then through the eastern islands of the Caribbean. In December, the *Barracuda* made oceanographic and geophysical stations at Trinidad, Barbados, Martinique, and Saint John, Antigua. The expedition concluded in the first half of January with more stations at Basseterre, Saint Christopher and Saint Thomas in the Virgin Islands. After measuring the acceleration of gravity at fifty-one sea stations and in nine harbors, *Barracuda,* its crew, and the scientific staff left Saint Thomas on 9 January 1937 and five days later arrived in Philadelphia, their final destination. In addition to a wide array of sounding and sampling data, the expedition determined the true extent of the strip of negative gravity anomalies discovered north of Haiti and Puerto Rico by the 1928 and 1932 expeditions. In a preliminary report to the AGU, Ewing commented that the anomaly strip "extends around the convex side of the Lesser Antillean Arc, having a total length of more than 1200 miles."[23]

In less than a decade, geophysics, a critical contributor to oceanographic knowledge, had established a significant research tradition complete with interested practitioners, effective instrumentation, and particular regions of the globe yielding remarkable new insights into the natural world. Thanks to Field's leadership and the navy's enthusiastic participation, a more complete geophysical picture of the world's island arcs, particularly in the Caribbean, began to emerge. Each time a significant expedition went to sea, the navy contributed political support, management, innovative instruments like the SDF, and vessels well suited as survey ships and instrument platforms, as well as skilled personnel.

Rather than wait long enough to catch his breath, a week after the expedition ended in Philadelphia, Richard Field immediately resumed his promotion of the Ewing-Heck effort to employ seismic refraction in the deep ocean. His correspondence with Henry Bigelow, Columbus Iselin, and Maurice Ewing guaranteed that *Atlantis* would carry the latter to sea in the summer of 1937 to continue the effort to perfect refraction methods for deep-sea research. The agreement between these men, who all served on Field's AGU committee, represented WHOI's introduction into some of the most fruitful techniques and research projects in the history of early oceanography and geophysics. With Ewing the relentless field scientist taking the lead along with Field, Iselin, and Bigelow, they mapped out a plan to refine research techniques, develop instruments, foster mutual cooperation between research institutions, and flesh out the state of geophysical knowledge to test the most important geophysical hypotheses of the time. Field stated it best when commenting on the results of the

Navy-AGU expedition: "It has already been reported to me that . . . the strip of negative anomalies is found to parallel to the island arc. While we had expected that this would be the case, we now know it is a *fact;* and thus we can say that both the East Indian and West Indian arcs are remarkably similar both as to pattern and *depth* of the major deformative movements of the crust (sial). All of this will have a direct bearing on the 'permanency of the ocean basics controversies,' and especially the Wegener Hypothesis (comparison of Atlantic and Pacific conditions)."[24]

Using every available opportunity for fieldwork, Ewing continued his research on the coastal plain from 1937 through 1940, armed with explosives and assisted by two remarkable Lehigh students. Fifty-three years later, pioneering oceanographer and submarine geophysicist J. Lamar Worzel recalled working with Ewing in 1937, his sophomore year at Lehigh, in conjunction with graduate student Allyn Collins Vine, who had arrived earlier from Hiram College in Ohio. Like Ewing, whom he called "Doc," Vine brought great physical and creative energy and unbounded enthusiasm to the project. These qualities would help him develop into one of the world's foremost authorities on underwater sound transmission, an instrument designer of amazing ingenuity, and the father of the *Alvin* deep submersible. Recalling events in 1937, Worzel remembered his first encounter with "Doc" as being so appealing that he swiftly found himself caught up in Vine's enthusiasm. On his way down to the New Jersey shore from Bethlehem, Pennsylvania, he sat just a few inches away from a load of dynamite in "Floozy Belle," Ewing's archaic truck. According to Worzel, Ewing and Vine:

> were doing explosive work in New Jersey, testing out the ocean bottom seismographs that they had built, but they were testing it out in New Jersey with the help of the geologist of New Jersey. They were using it there as a means of making sure everything worked. We made two lines in the course of that work. One was called Barnegat Bay line, and the other was called the Cape May line, and those were the two end points of the line across the coastal plain. At that time, nobody knew what happened under things like the coastal plain, how much sediment there was. Were there three feet or 50,000 feet? Nobody knew. That was interesting information. Ewing was trying to extend this out into the deep ocean at the time. They had a former student who had agreed to come work with them and help drill shot holes. This was manually done at that time and he didn't show up for one weekend so they were shorthanded. So Allyn Vine said to me, "Hey, you want to have some fun?" And I said, "Sure!" He said, "Well, come with us. We're going down to New Jersey and we're going to drill shot holes and we're going to set off explosives, and

it's going to be a lot of fun." Well, the long and short of it was that I did, and that was the end. I was hooked. I worked with him on all the New Jersey work and then the following summer I went to sea with him for the first time and so on. I never varied after that; I was always interested in geophysics.[25]

Ewing needed Field's support because building a full-scale, ship-deployed seismic apparatus for dynamite refraction shots proved very difficult.[26] The Lehigh team members had to develop all of their own instruments and to overcome many misconceptions about the behavior of explosives at great depths. Ewing, Vine, and Worzel continued the New Jersey research well into 1939. This work provided a chance to test many of the components Ewing took to sea. The amplifiers, circuitry, geophones, and detonators worked well on the Barnegat Bay line in the winter of 1937–38, and the team completed the Cape May line the following winter.

Taking the system to sea proved far more difficult. Between the two seasons it took to complete the New Jersey lines, Vine, Norman Webster—another Ewing student who later worked briefly for DuPont—and Worzel boarded *Atlantis* with their teacher to test the apparatus at sea. With *Atlantis* during Cruise 81 from 8 to 21 September 1938, Doc's team had even less success on this, Worzel's first voyage, than Ewing, Vine, and Webster had during the summer of 1937. They experienced difficulty trying to tow their rather complex system from a ship. Their Ocean Bottom Seismograph (OBS) consisted of a bank of galvanometers, a camera, batteries, amplifiers, and an automatic bomb-shooting mechanism built into the cavity of a section of gun barrel obtained from the Bethlehem Steel Company. Three TNT bombs set five hundred feet apart on the same chain preceded four geophones attached at fifty-foot intervals. Then came the gun barrel with all of the control, recording, and timing mechanisms. Ewing set the detonation times to allow the entire array to sink to the bottom before the timer set off the bombs in a preset sequence. Once the barrel went over the side, and with it the last fifty feet of chain, the crew connected the final link to both a lead weight and a spool of wire rope long enough to permit a journey to the bottom without breaking or falling short. They hoped the large weight would absorb any tugging on the array from the ship at the surface, but that very problem emerged as their most relentless foe. Instead of the gear resting on the bottom and performing as intended, the ship's slight movements on the surface dragged the instruments and explosives around. As Worzel recalled, this and other less complex problems persistently plagued them. They "found that the connecting leads to the geophones and the bombs would often get tangles and knots tied in them when we were lowering them over the side even though we took great precautions to try to make sure they went down straight. . . . This of course meant there was a fair chance of someday blowing the equipment off the line or damaging the equipment beyond repair from our own explosions, which we thought was unsatisfactory."

These difficulties led Ewing and his assistants to abandon the linked-array concept. They designed new bomb containers and seismograph-geophone packages as free-floating units. After building four of these devices, the team used anchors ballasted with salt to provide weight, sufficient bottom time, and the capability of returning to the surface unaided. When seawater eventually permeated the anchor, mixing with and dissolving the salt, the buoyancy provided by a gasoline-filled float overcame the resistance of the salt solution and the device floated back to the surface.

In the course of refining their methodology they discovered that even the experts knew very little about explosives under great pressure. Initially advised that the TNT would not explode reliably or at all at great depths, Doc and his students housed it in pressure vessels in their early experiments, only to find through experience that they needed no protection at all for the explosive. Only the detonator needed a guard to avoid having the wires sheared off by the compressing action of water pressure on the TNT. Ewing approached friends at DuPont and together they developed a reliable shielded fuse.

In a like manner, most of Ewing's shipboard methodology and supporting instruments grew out of experience at sea and earlier coastal sediment work. Accidents often supplied answers to stubborn questions. In the summer of 1939, the Lehigh group discovered that an oscillograph that operated well in the lab would perform poorly at sea because their housings would compress in the deep ocean, moving the instrument and distorting the tracing. During Ewing's Guggenheim Fellowship year in the winter of 1939–40, they realized that some of the aluminum containers actually began collapsing in a barely perceptible way, crowding the devices inside.

The lesson that new methods and instruments demanded attention, precision, and delicacy quickly became apparent aboard *Atlantis* during another piggyback cruise in the summer of 1940. Doc and his students deployed the seismic equipment more than thirty times and barely managed a handful of acceptable records. The cameras that recorded the oscilloscope results failed to advance photographic paper through the device to capture the data. Each time he deployed the equipment, Ewing insisted upon new Burgess camera batteries to insure a full charge. On one lowering, Worzel accidentally left a depleted Eveready test battery in the camera to run the small electric motor that moved the light-sensitive paper. In spite of the weak charge in its battery, the camera worked very well. Thus by chance, Doc's group discovered that while Burgess batteries lost much of their capability at low ocean-bottom temperatures, the Eveready batteries showed greater reliability. All of these lessons helped refine a valuable methodology necessary for an important line of research, but they also consumed great amounts of time and money. By the time Ewing's techniques began to show results in the autumn of 1940, he had already received a leave of absence from Lehigh University and moved with Vine and Worzel to Woods Hole to participate in the earliest navy-sponsored oceanographic research designed to combat the U-boat menace in the North Atlantic.[27] He and Columbus Iselin, who replaced the retiring Bigelow as WHOI's director in 1940, anxiously realized that antisubmarine warfare would, unfortunately, divert them from preferred research for some time to come. From then on, time and personnel might prove scarce and precious, but they would not suffer from a shortage of work or money for the foreseeable future.

The advent of war in Europe did not arrest oceanographic research at sea. While many oceanographers, especially along the East Coast, turned their attention toward the battle raging between the Royal Navy and the German U-boats in the North Atlantic in 1940, others took advantage of some final peacetime opportunities. Motivated by President Franklin D. Roosevelt's desire to place the United States in the company of nations investigating Antarctica, Rear Adm. Richard E. Byrd led a second expedition to the region in November, 1939, placing the navy at the head of a multiagency effort to reestablish an American presence at the bottom of the world. The U.S. Antarctic Service took primary responsibility for the navy-led project in cooperation with the NAS/NRC, and the Departments of State, Interior, and Treasury. Employing the USS *Bear* (AG-29) and the Coast Guard cutter *North Star* (WPG-59), the expedition remained in the southernmost portion of the globe until the spring of 1941. Byrd set up his eastern base of operation on Stonington Island and Little America III, his western base, on the Ross Ice Shelf.

Although the project yielded mountains of data in a variety of different scientific disciplines, the results gave the appearance of administrative chaos when scientists did not wait for a single comprehensive publication from the joint effort. At the project's conclusion there appeared an avalanche of articles and reports in a variety of scientific journals. Byrd himself discussed the possibility of publishing with the American Philosophical Society, to the dismay of the NAS, which sought a more coordinated effort.

In the end, responsibility for the publication of data and a narrative describing the expedition fell into the NAS's hands almost by default. The debate over coordination of results and publications ended when the Japanese attack on Pearl Harbor in December, 1941, suddenly gave Byrd, the expedition scientists, and the U.S. Navy a very different focus.[28]

5

Research, Relationships, and Policy, 1930–40

■ There were those who protested that the action of setting up N.D.R.C. was an end run, a grab by which a small group of scientists and engineers, acting outside established channels, got hold of the authority and the money for the program of developing new weapons. That, in fact, is exactly what it was.

—VANNEVAR BUSH, president, Carnegie Institution of Washington; Director, OSRD (1941–45)

THE EXPERIENCE OF NAVAL AND CIVILIAN SCIENTISTS IN THE years following the Great War revealed that the navy lacked not only political influence and financial resources, but also a comprehensive policy for applying scientific research to naval problems. The Schofield Board, for example, met simply because Frank Lillie suggested to Secretary Wilbur that the navy needed a policy on oceanography to guide the Hydrographic Office. Hydro frequently performed oceanographic investigations in conjunction with private research institutions both before and after the 1929 Schofield recommendations. Thus, nothing of substance changed. The committee simply defined upon request the current extent of the navy's commitment to a particular line of applied research in the ocean sciences. In a similar way, the Interagency Conference on Oceanography emerged, not from a naval science policy, but from the immediate political and budgetary considerations that drove Harvey Hayes to write his landmark memorandum.

The interwar navy needed a better peacetime image. Neither the Navy Department's involvement with the National Academy of Sciences nor its common practice with civilian science emerged from a defined plan or policy.

Rather, research prospered as either an important political tool or an expression of interest in a particular promising technology with immediate application to the navy's mission. While both of these addressed important naval needs, neither promised a continuing commitment to scientific research as a unique and important activity of value to the navy.

Science had not yet acquired an independent identity within the Navy Department. In Progressive Era America, the general public and industry measured scientific developments by their utility. Applicability to daily problems and the promise of increased leisure time thanks to technological innovation inextricably bound science and engineering. Although scientific theory provided the essential foundation for many remarkable and laborsaving devices, it was engineering that turned theory intopractical application and brought these changes to the public.

Thomas Edison embodied this period and its philosophy. In his proposal for what became the Naval Research Laboratory made during his tenure as chairman of the Naval Consulting Board, Edison argued that science and engineering research did not belong in the new facility. In October, 1915, he suggested that the navy would draw from generally available scientific knowledge to develop war machinery and instruments for testing and production. If science had a role, it was subordinate to engineering and the genius that sparked useful innovation.

With the emphasis placed on utility, engineering, and application, the navy's lack of a science policy does not seem strange. In spite of the research mandate given the NRL by the majority of the NCB, scientific inquiry belonged to the universities or, perhaps, to private industry, but not to the navy. Naval technical leaders saw engineering solutions as their primary job. In this way their ideas would go to sea and serve the fleet.

After 1917, the Great War introduced the U.S. Navy to the possibility that a commitment to applied and, in some cases, basic long-term scientific research could provide some useful dividends. The U-boat war demonstrated the increasing sophistication of naval warfare and the possibility that only scientific research could unearth the equally sophisticated answers to naval defense. When Harvey Hayes composed his memorandum to John Halligan, he wanted to sustain his research into sonar and submarine detection and provide a high-profile peacetime role for a struggling navy. At the same time, he realized that the navy had to encourage oceanography. This multidisciplinary science held the key to many of the navy's most pressing operational problems. In a seemingly endless and difficult process of debate and compromise between 1930 and 1940, ocean science began to acquire for itself an identity separate from naval engineering and worthy of a policy providing direction, programs for research, and independent administration.

At the beginning of the decade, those bureaus concerned with technical matters and their view on the importance and nature of naval scientific research formed the core of the problem. During the interwar period, most naval scientific activity came under the authority of the Bureaus of Construction and Repair and Engineering, and this lasted until the two were combined into the Bureau of Ships (BuShips) in 1940. The Bureau of Engineering, for example, funded nearly 80 percent of the activity at the NRL from 1923 until the bureau assumed direct control over the facility in 1931. Bureau direction lasted until the secretary of the navy and his science adviser assumed responsibility for the laboratory in 1939.

The bureaus controlled scientific research within the navy and narrowly defined this type of work. Until 1939, they reduced much of their scientific effort to testing extant systems and exploring a few promising projects like sonar, radar, and radio. Immediate utility thus ruled the day, and scientific research formed just another component of a general engineering effort to support the fleet. In these circumstances, the individual bureaus, and in some cases activities within the bureaus, defined daily, and by default official, naval policy on scientific research.

Activities like the Hydrographic Office within the BuNav demonstrated a clear understanding of the difference between engineering and scientific research as well as the need for careful planning. Hydro surveys employed trained scientists and used professionally recognized data collection and evaluation techniques. To the frequent regret of the NRL staff and personnel at other naval and private research centers, BuEng policy placed this type of activity both in bondage to naval engineering and toward the bottom of the priority list.

Research, Tradition, and Policy

What kind of research did the navy support? The needs of the fleet and affordability determined the type of research that traditionally received the Navy Department's imprimatur. Precisely defined applied research that directly affected the navy's mission usually received attention and funding. The Hydrographic Office thus had little difficulty justifying its plans and needs. For the navy at large, improved data gathering techniques, new survey methods, new instruments, better survey ship conversions, and improved technologies for constructing and printing charts needed little justification or political salesmanship. Any officer who had to navigate in foul weather or in hazardous waters appreciated the results of this investment.

Basic or fundamental research did not fare as well. To invest in promising long-term research without the expectation of attaining early a precisely defined goal went against strongly held traditional views on the utility and purely applied purpose of naval scientific endeavor. Exceptions to this rule occurred only when particularly able scientists and engineers within the navy engaged in what sociologist Donald McKenzie has called heterogeneous engineering. Displaying considerable entrepreneurship, scientists like Field, Ewing, Vaughan, and Bigelow presented an effective case for their research to the bureaus, the CNO, and the General Board and sold their projects to those officers and politicians who held the purse strings. When Harvey Hayes first suggested that the navy make a commitment to oceanographic research, he had peddled his ideas to naval authorities as a political strategy to achieve a specific set of goals for the NRL, the Hydrographic Office, and a rapidly diminishing fleet in an era of arms reduction. Theodore Roosevelt, Jr., aggressively pursued Hayes's proposals less as a commitment to oceanography than a wise political step by a struggling, treaty-era navy. Rarely the center of attention, basic research followed in the train of political initiatives or promising applied-research projects persuasively championed by able practitioners.

This research emphasis did not change between 1930 and the beginning of World War II. Less certain to produce immediate dividends, a commitment to basic research required greater resources, trust, and the navy's acceptance of scientists—naval or civilian—as full partners in wartime. The Great War clearly demonstrated the increasingly significant role of science in warfare. Acquiring the necessary trust and resources would take time.

Why was trust an issue? In many ways the most scientific of the services, the navy rarely looked beyond its own domain for solutions critical to its mission. With the bureaus, the NRL, the David Taylor Model Basin, the Naval Academy, a postgraduate research and education program, and a host of research centers around the country, the fleet had certainly embraced science. However, the Navy Department clearly defined the terms. America's first Nobel Prize, won in 1907 by Naval Academy graduate and physicist Albert A. Michelson, certainly indicated the quality of basic naval education in the sciences. But the Nobel laureate was forced to resign eight years after receiving his commission in order to pursue full-time the basic research that enhanced humanity's knowledge of light. Thus, while training within the navy offered great possibilities, traditional naval attitudes restricted officers, scientists, and contractors to research with immediate and obvious naval applications. These circumstances severely limited research horizons and reflected the navy's natural inclination to insist upon complete control of activities it viewed as exclusively naval.[1] Most scientists found this customary view professionally suffocating, and many of their colleagues within the navy had to struggle to survive and prosper in that environment.

Between 1930 and 1935, the policy-formulation process, if not the emphasis on pure utility, began to change, and these modifications affected every type of

naval scientific endeavor. Without political support and adequate financial resources, the navy found it increasingly difficult to encourage and control scientific activity affecting warfare at sea. Oceanography was no exception. The navy
could not take the lead in shaping policy after the failure of the ICO program.
Well endowed by the Rockefeller Foundation, Frank Lillie and his Committee
on Oceanography placed the NAS in a position to assume the role of informally
coordinating oceanography in the United States. After 1930, any significant
efforts to formulate a policy on research with naval applications would have to
come from a joint effort by the navy and the NAS.

Lillie's committee never broke stride after the creation of the Woods Hole
Oceanographic Institution in 1930. Drawing on the continued support of the
Rockefeller Foundation's General Education Board, the MBL director and his
colleagues appointed an array of subcommittees to study particular oceanographic problems. In 1931, with the approval of the NAS, the committee published Henry Bigelow's report on the state of oceanography with Houghton
Mifflin of Boston. It also commissioned two reports. The first would study international cooperation in solving oceanographic problems. In a 19 December
letter to the secretary of the General Education Board, W. W. Brierley, NAS Assistant Secretary Paul Brockett mentioned that Lillie hoped to draft Thomas
Wayland Vaughan to prepare this volume. The other report would explore the
state of oceanographic research at American universities.

In spite of the positive tone of Brockett's letter and his gratitude for the
foundation's financial support, he also betrayed the concern of some within the
NAS that the committee had already taken too long to complete its work.
While Lillie understood that the committee's activities had consumed a great
deal of time, he would not hesitate to use more months and dollars to complete
the ambitious program he and his colleagues had in mind. The financial accounting for the 1931 calendar year showed that publishing the Bigelow report
and sending copies to targeted individuals cost the committee $1,964 and re-
imbursing Vaughan's travel costs amounted to $160.64. Lillie told Brockett
and his fellow committee members that before completing their work he expected to consume another $20,000 of the $60,000 made available by the
Rockefeller Foundation.[2]

Lillie's ambitions clearly demonstrated his alarm at the vacuum created by
the ICO's demise and the knowledge that the navy and American universities
needed help in planning and coordinating oceanographic research. He intended to provide American oceanography with the program and direction
that it lacked. The navy could come along for the ride.

Science, Roles, and Missions

Like Frank Lillie, some naval officers and scientists acutely felt the absence of a policy that would better exploit scientific knowledge. In a speech before the American Association for the Advancement of Science presented in 1928, Capt. Clyde S. McDowell, an engineering duty officer (EDO) and inspector of naval materiel in San Francisco, argued for a determined effort to develop a close relationship between officers and scientists. This would encourage the application of scientific developments to the navy's advantage.

A Naval Academy graduate commissioned in 1906, McDowell's extensive experience made his appearance before a distinguished body of scientists a significant event. He was one of the few naval officers who felt at home in that company, having served as part of the navy's Special Board on Submarine Devices during the Great War. After a brief term at the Engineering Experiment Station in Annapolis, McDowell worked at the headquarters of the General Electric Company in Schenectady, New York, successively as naval inspector of machinery, engineering materials, and ordnance. As a naval representative and authority on electrical machinery for a sectional committee of the American Institute of Electrical Engineers he also knew well the purpose and procedures of the NRC and the NAS.

True to naval tradition, McDowell told his audience that naval officers should direct cooperative research. From his point of view, however, this did not preclude the development of a relationship with civilian scientists characterized by trust, as long as both parties made a commitment to personal and professional understanding. McDowell reminded his listeners of the importance of science in the 1914–18 conflict, the potential for cooperation, and the immediate opportunity to refine the relationship between the navy and civilian science in peacetime.

He suggested revisiting the wartime practice of offering naval reserve commissions to scientific experts as a method of developing a productive postwar partnership. In this way, an organization like the NRC could recommend candidates in particular specialties that might work with naval research personnel to address problems of importance to the fleet. These specialists would first visit ships, make short cruises, and familiarize themselves with naval tasks. Afterward they might serve active-duty assignments on board ship or at the NRL. McDowell also suggested introducing civilian technicians to navy research in this manner, encouraging communication between naval activities and professional scientific and engineering societies, and creating naval research fellowships at the best technical schools. The captain intended this program as a way of promoting and refining a relationship between the navy and civilian science in peacetime in spite of the depleted state of the navy's interwar resources.[3]

A General Board investigation of NRL research activities in 1932 indicated that the navy's research program had degenerated rather than improved in the months after McDowell's speech. As the most important naval research establishment, the NRL provided an excellent indication of general conditions in the navy's research community. In its report, the board noted that during the course of its investigation questions arose regarding the laboratory's intended mission as opposed to its current use, which could represent "a departure from the original purpose for which it was established." When they proposed the creation of the NRL in 1916, the majority of the NCB's members suggested that the navy emulate the successful applied-research laboratories of private industry. General Electric, directed by Willis R. Whitney, and Elmer Sperry's modest electrochemical laboratory in Washington, D.C., both founded in 1900, offered excellent models.[4] The board noted in its communication to Secretary Adams that "instead of being engaged solely in research and attendant experimental work, [NRL's] activities have been extended to include service test work and even production. This expansion, to the detriment of research, has been brought about by undertaking production in an attempt to supplement maintenance and reduce overhead. This condition now exists to such an extent that test and production work for the Bureau of Engineering constitutes a large portion of the Laboratory's activities."

These conclusions certainly suggested that BuEng had diverted the laboratory from its intended purpose, depriving the navy of a sorely needed research capability and thwarting the NCB's original intention. If not for this diversion, the navy might have developed a more effective naval research organization with the NRL as its foundation.

This General Board report advised the secretary to allow the NRL to return to its original purpose. Research, experimentation, and a limited commitment to production of instruments and devices in their early stages seemed the best set of priorities. Thus, just one year after BuEng assumed control of the NRL, the board strongly recommended centralizing naval research under the CNO to avoid the dilution of the navy's diverse scientific resources by BuEng.

In challenging the bureaus' hold on scientific research, the General Board never suggested that the management of the NRL should pass to a civilian naval employee with a scientific background. Like McDowell, the board advised the appointment of a line officer to direct the NRL and demonstrate official control over research and development. Although the NRL's various divisions and many individual activities at centers like the EES in Annapolis had a great degree of autonomy, the presence of a naval officer at the helm reminded one and all of the laboratory's purpose and the navy's scientific tradition.[5]

When former commander of the U.S. Fleet and senior board member Rear Adm. Jehu V. Chase sent the views of the General Board to the secretary, he no doubt realized the challenge it posed to BuEng's authority. In the trenches at the NRL, the navy's scientists and technicians also wanted to challenge the bureau's authority. Hard times made persuasion difficult. In addition to disturbing bureau policies, by 1930 the Great Depression had imposed heavy burdens on the entire country and forced the laboratory, along with the rest of the navy, to endure some drastic budget cuts. Already diverted from much of their research activity, the NRL staff faced substantial program, salary, and staff reductions scheduled for 1932 and dismal prospects for the future.

Commander Edmund D. Almy gave voice to the anger felt by those involved in naval science at the NRL. In a 28 June 1933 letter to former NCB member and Roosevelt confidant Thomas Robins, Almy lamented the diversion of NRL and other naval scientific resources. The laboratories found themselves involved in activities better suited to engineers, industry, and private scientific or technical contractors. As deputy director of the NRL from March, 1930, through early 1932, when he began a brief term as director that lasted through May, Almy witnessed these disturbing developments firsthand.

Convinced that this situation deserved the attention of experts able to suggest remedies, Robins sought the opinion of the NRL's Lynde P. Wheeler, assistant superintendent of the laboratory's Radio Division. Robins needed Wheeler's informed opinion for confirmation and context. The Consulting Board veteran was about to assume the position of executive secretary to the Science Advisory Board, a New Deal agency charged with providing scientific consultation services to federal agencies. He had to understand the situation at the NRL. During their discussion, Wheeler largely confirmed Almy's observations and Robins's fears, and shortly thereafter the NRL scientist himself became a victim of the personnel cutbacks at the laboratory.[6]

As Robins pursued Almy's concerns, Isaiah Bowman, the newly elected chairman of the NRC, looked into conditions at the NRL in an effort to appraise himself of the state of naval research. After speaking with Wheeler, an acquaintance from his Yale University days, on 17 July, Bowman concluded the NRL had "a problem of great intricacy and delicacy." Wheeler convinced his colleague that the "laboratory is now under the control of the Bureau of Engineering and its work has been reduced to that of testing, with long-range research practically eliminated." Bowman quickly recognized that a closer working relationship between the NRC and ranking naval officers offered the only way to effectively address the problem. He wanted to propose a committee within the NRC to confront this problem and promote a better and more actively constructive relationship with the navy. In a memorandum he prepared on his 17 July interview with Wheeler, Bowman noted that the CNO, Adm. William H. Standley, seemed anxious to discuss the problem and willing to work with the NRC. Meanwhile, Bowman looked forward to a dinner at the

Wheeler's scheduled for 23 July. He wanted to talk further with the former NRL scientist and hear the views of another dinner guest, Harvey Hayes.

Two days after his interview with Bowman at the NRL, Wheeler composed a confidential memorandum on the flaws in the existing administration and co-ordination of science in the navy. He concluded that the navy did not strive for the same efficient use of its scientific resources that it regularly demanded from its ships and crews. "Under modern conditions of rapid developments in science and engineering," wrote Wheeler, "the present naval organization does not ensure adequate efficiency in respect to personnel and cannot ensure it in respect to material."

Wheeler pinpointed those elements of naval tradition, policy, and admin-istration causing problems at the NRL and throughout the naval scientific community. For one thing, undergraduate training stressed engineering over a more comprehensive education. Emphasis on the latter would include a deeper appreciation of the fundamental sciences that repeatedly provided en-gineers with their best practical analytical tools. In addition, the specialized graduate training of naval officers did not compare well with that enjoyed by their colleagues in private industry. Although the policy of rotating officers in and out of positions on a regular basis did provide broad experience of the navy, it did not permit the most efficient utilization of the scientific and engi-neering expertise acquired by some officers. When the navy found its experts engaged elsewhere, some private company usually benefited as the recipient of a lucrative development contract. Wheeler insisted that this wasted resources. It also retarded the development of the best instruments and ship designs at the NRL and other facilities. As far as he was concerned, "the policy in regard to research is such as to make it certain in the matter of new devices and meth-ods (which may affect strategy and tactics profoundly) that our Navy's devel-opments will lag behind not only similar developments in the industrial world, but also those of foreign navies which maintain research organizations. It is axiomatic that failing adequate research facilities of its own directed to-ward seeking naval applications of modern scientific and engineering ad-vances, our Navy is bound to become a follower and cannot be a leader in its technical equipment."

Policies and administration of this sort only wasted money, time, and re-sources while producing inferior results. Wheeler suggested that similar prob-lems at the army's McCook Field, Monmouth, and Edgewood laboratories in-dependently confirmed that diverting resources and staff from their mission did not work for any of the military services.

Wheeler did not dissect and criticize naval research policy and administra-tion without an alternative proposal. He had too much invested in the process not to take the lead in offering alternatives. Both for scientific advice and to ensure that the NRL could return to its original purpose, Wheeler suggested re-habilitating the NCB as a full-time naval science adviser. If this proved imprac-tical, the NRC could assume the same role. He also proposed that the bureaus relinquish control of naval scientific activity to a director of naval research—a civilian scientific professional responsible to the secretary of the navy. The NRC and the White House could take responsibility for the nomination process. This structure would liberate research from the bureaus and permit direct fund-ing by Congress. Wheeler described his recommendations as a collection of general characteristics necessary for a productive research policy. The General Board could employ these suggestions and others to formulate a coherent plan for the future. At the very least, Wheeler believed that any new policy required two basic features to succeed: (1) continuity in administration in order to en-sure continuity in research policy; and (2) undivided control of research funds by the responsible research administration.[7]

Lynde Wheeler's ideas fell on a few sympathetic ears within the NRL, the Navy Department, and the civilian scientific community. On the same day Wheeler prepared his memorandum, Frank B. Jewett of the American Tele-phone and Telegraph Company put pen to paper and decried the liabilities of the navy's personnel rotation system in a letter to Captain McDowell, then in-spector of machinery at the Westinghouse Electric and Manufacturing Com-pany in Essington, Pennsylvania. To Jewett, who succeeded Frank Lillie as NAS president in 1939, these circumstances mitigated against consistently having the navy's best people in the most important design and research posts.

Aware of the communication between Almy, Rear Adm. Harold Bowen, and Wheeler, and encouraged by Jewett's support, in Captain McDowell in early 1933 embarked upon his own crusade to restructure naval research activ-ity along lines similar to those suggested by Wheeler. In an unofficial letter to Max Mason of the Rockefeller Foundation, McDowell explored the possibility of creating an organization to coordinate naval research modeled on the New London Experiment Station established during the Great War. Proposed in June, 1917, as a way to address the German U-boat threat collectively and sci-entifically, the station emerged from a joint effort by the navy; the NRC's first chairman, George E. Hale; and the NRC's Committee on Submarine Detec-tion, chaired by Robert A. Millikan of CalTech. On leave from the University of Wisconsin and working at New London during the war, Max Mason had provided the creative genius behind several generations of the navy's so-called M devices (multiple-tube, passive submarine sensors).[8]

McDowell also approached William D. Coolidge, director of General Elec-tric's research laboratory. Responding to McDowell's letter on 14 July, Coo-lidge applauded the possibilities of a national defense research board and looked forward to working with an activity that would manage the relationship between the private sector and the federal government's various bureaus and agencies. From industry's point of view, an organization of this sort could help eliminate redundancy in federal agency research and might induce the various government activities to pool their resources for important projects they could not independently afford.[9]

Coolidge's response and Mason's enthusiastic support for the project led McDowell to approach Isaiah Bowman of the NRC. Their discussions began in New York City during June, 1933, and spilled over into a second conversation in Washington on 19 July—the same day Wheeler finished his memorandum on the reform of the navy's research organization. From that meeting emerged the idea of creating a NRC-sponsored Naval Research Committee to help formulate research policy and provide advice. A naval officer would serve as director, and the NRC would nominate its members. Excited by the prospect, Bowman reported that "Captain McDowell drafted a memorandum on the spot and agreed to put it into official form and submit a copy to Admiral Standley and also to me."[10]

Minutes after leaving McDowell on 20 July, Bowman met with the CNO. He felt a little embarrassed by his enthusiastic reception. Admiral Richard H. Leigh, chairman of the General Board, met Bowman in the reception room and immediately whisked him into the CNO's office past others waiting to see Standley. McDowell brought the admiral up to date on his conversations with Bowman earlier in the day, and Standley received the NRC chairman with great interest. Bowman reviewed the NRC's functions and the possible use of his organization to create a body that might place the nation's scientific resources at the navy's disposal. He did not hesitate to mention others within the NRC, like Robert Millikan, who supported his efforts. Under the new Roosevelt administration, the navy might expect an increase in available resources for ship design and construction, and Bowman suggested that the NRC could offer the best advice on how to exploit available science and technology. The navy would hardly want the "quick out-turn of a navy" that would immediately stand in the shadow of more advanced weapons and warships operated by foreign fleets. Standley agreed to the appointment of a naval officer as liaison between the Navy Department and a division of the NRC responsible for helping the navy wisely invest its resources and manage its plan for naval research.[11]

Encouraged by his meeting with the CNO, Bowman placed the navy's lack of a coherent research policy and its need for advice in the context of an ambitious plan he had contemplated for some time. The day following his meeting with Standley, 21 July 1933, the NRC chairman submitted a proposal for a Science Advisory Board (SAB) to Daniel C. Roper, the secretary of commerce and chairman of the Industrial Recovery Board created by the New Deal's National Recovery Act.[12] As part of the NRC, the SAB would function as a vehicle of the administration and assist federal agencies with formulating and executing research programs.

Franklin Roosevelt's personal friend and president of the Chemical Foundation, Francis P. Garvan, recommended early in 1933 that the government create a board of scientific advisers with authority to coordinate the use of the country's scientific resources. Bowman hoped to build on this interest and the positive response he received from Admiral Stanley by inducing Roper to sponsor the proposal to the president. When the secretary gave no indication of understanding the need for further coordination of scientific resources, Bowman feared for the success of the project.

Later in the day, Bowman received a fortuitous telephone call from Secretary of Agriculture Henry A. Wallace. After Wallace requested NRC assistance in his effort to modernize the Weather Bureau and select its new director, Bowman took the opportunity to mention the SAB proposal and asked for the secretary's support. As manager of one of the government's largest research programs, Wallace embraced the idea, mentioned that in the past he had toyed with a similar proposal, and energetically supported Bowman's recommendation before the president the following day. The NRC chairman left his office with a sense that his proposals might indeed reach the president, but he had no idea just how fast events would move.

Bowman traveled to Lynde Wheeler's home on the evening of 23 July to enjoy a good meal and to refine further his naval proposals in conversation with his host and guest Harvey Hayes.[13] After-dinner conversation quickly turned to the situation at the NRL and the absence of a coherent science policy for the navy. They agreed on approaching the president, perhaps through Thomas Robins, and then arguing their case before the House Committee on Naval Affairs. All agreed that a scientifically qualified civilian should take charge of naval research because the "control of young and ignorant project officers . . . is having a paralyzing effect." No one in the Wheeler home that evening could or wanted to shut the navy out of its own laboratory. However, tight control and engineering priorities had deprived research projects of competent scientific leadership and diverted the NRL and other scientific activities from their intended mission. They also noted that Captain McDowell was working on a general improvement scheme that would enhance the relationship between the navy and civilian science. A productive future for the navy lay in cooperation with the private sector, not in isolation from the wider scientific community.[14]

Secretary Wallace moved faster than Bowman had reason to hope or expect. The secretary communicated with the president immediately after his telephone conversation with Bowman, and the president responded seventy-two hours later by asking Wallace to draft an executive order. Since the NRC was nominally an arm of the National Academy of Sciences, Bowman spoke with NAS president William W. Campbell on 26 July, receiving his approval to prepare the order for Wallace to submit to Roosevelt. Wallace ensured both Bowman and Campbell that the president would not act on the measure until the following month. This comforted Campbell, who feared that Bowman's SAB might encroach upon the advisory role played by the Academy. He wanted more time to discuss the roles and missions of the two scientific advisory groups.

To the dismay of many in the NAS and NRC, Roosevelt signed the order creating the SAB on 31 July. The board came into existence without the careful deliberation on respective roles that scientists like Campbell, Millikan, and John Merriam of the Carnegie Institution, desperately wanted.

Roosevelt appointed fifteen of the most distinguished scientists and engineers in America to the SAB. His first nine choices included Bowman; Campbell; Merriam; Millikan; Frank Jewett; celebrated engineer Gano Dunn; Charles F. Kettering, president of the General Motors Research Corporation; and C. K. Leith from the University of Wisconsin. The president chose Massachusetts Institute of Technology (MIT) president Karl T. Compton to serve as chairman. After hearing criticism that the selections did not include any biologists, chemists, or medical doctors, the president made a second round of appointments, naming six more members to redress the imbalance. Frank Lillie thus joined the board along with two medical doctors and three others specializing in chemistry, biology, and botany.[15]

The SAB was the scientific face of the New Deal. Roosevelt's program for recovery promoted an activism that asked science to play a greater part in the nation's welfare and defense. The new agency rarely waited for a government activity to request assistance, nor did it recoil at the possibility of becoming identified with a particular political view or program. Indeed, before the SAB's term expired in 1935, Chairman Compton proposed a controversial $16 million scientific recovery program to promote research in important areas of science, the national economy, and defense.

Many of Compton's colleagues genuinely feared the sort of active, political commitment embodied in the SAB. Although their reaction may seem archaic and patrician sixty years later, their image of the purpose and function of science definitely influenced interwar policy formation. Prominent scientists trembled at the prospect that their research might become something less than they believed it to be: the familiar, nonpartisan quest for understanding that motivated them to choose their life's work. Thus, many leaders within the scientific community, like Campbell and Jewett, took up the New Deal challenge and participated in the SAB as much to influence the way science and scientists might change over the coming years as they served the country with their scientific expertise.

In spite of doubts, this environment offered the navy an opportunity to work with the SAB and NAS to compose more precisely a general policy toward the role of science in naval affairs. Unfortunately, broader and more effective cooperation and coordination with the private sector proved difficult, as the civilian scientific community soon found itself in the midst of a rancorous debate between the Academy, the NRC, and the new Science Advisory Board. As these three agencies vied with one another for the job of premier science adviser to the government, the civilian scientific community went through a particularly painful process of defining roles in the general effort to ensure the nation's well-being during domestic and international crises.[16]

For the NAS, created as an advisory body to the federal government during the Civil War, working with the SAB presented a problem of jurisdiction and philosophy. Which agency would rank first as the government's source of scientific advice? William Campbell detected a possible conflict on this issue in his conversation with Bowman before 31 July. Campbell also feared that Compton's SAB had the insider's advantage. The board was a Roosevelt creation and a New Deal initiative, while charter restrictions limited the NAS to an impartial advisory role. This distinction quickly laid bare the concerns held by some prominent scientists that, using the SAB, the federal government would interfere in scientific practice to such an extent that freedom of inquiry would become politicized and compromised. They viewed science as an unpolitical activity, admirable and nonpartisan, and they dearly desired to preserve these qualities.

Politically conservative scientists like Campbell wanted to define carefully the nature of legitimate research and the extent of its commitment to governmental problems. For them, science showed no partiality and had no political purpose; it was a search for understanding. They had no difficulty offering advice when the government requested assistance. Indeed, they valued that function. A commitment to work within the government, like that made by SAB members, might also require political allegiance and could well redefine their role as scientists.

In many ways, the goals of the civilian scientific community resembled those of the navy. Both wanted to exert complete control over the extent and purpose of their commitments. Without a mutually satisfactory compromise, the relationship between them that took root in 1917–18, so critical for an effective partnership in war, might wither before the country and the navy profited from its benefits. Naval and civilian advocates of this partnership, like Captain McDowell and Isaiah Bowman, refused to consider such a possibility. An effective relationship would only emerge from flexibility on both sides.

Despite the almost immediate clash between the SAB and the NAS, the navy began submitting requests for assistance. Scientific activity within the navy needed better coordination and a more effective research policy.[17] Before the new year began, the navy, NRC, and SAB created working committees to provide a means of effectively communicating and cooperating. Millikan, Jewett, and Leith made up the SAB group, while the NRC permanent committee drew members from four of the council's divisions. The latter combined in late 1933 with its SAB counterpart to constitute a joint liaison committee to work with a group of officers in the Navy Department appointed by Admiral Standley and led by Capt. Edward J. Marquart.[18] The CNO called this group the Liaison Committee on Naval Research.

This arrangement demonstrated a subtle tug-of-war between two groups eager to communicate but reluctant to surrender the small measure of autonomy required to build a truly cooperative relationship. Captain McDowell had already offered his ideas to Standley. In his dinner conversation with Wheeler and Hayes in July, Isaiah Bowman learned of the captain's efforts to formulate a proposal that would draw civilian science and the navy into active and productive scientific cooperation. On 25 July, two days after the dinner at Lynde Wheeler's home, McDowell produced a plan, sending copies first to the CNO and then to Bowman.

The captain proposed a Naval Research Committee to work within the NRC, chaired by a naval officer and a staff of scientific experts from "each of the basic sciences." The Navy Department would nominate the chair for the council's approval and the latter would also appoint the members of the committee for specific terms. For McDowell, this seemed the best blend of naval and civilian authority in a committee to provide consultation services and advise the navy on science policy. In nominating and appointing the chair, the Navy Department would have to alter its standard personnel practices and agree to an extended tour to avoid losing the expertise of the officer chosen when it came time for rotation. McDowell thus addressed the destructive side of the navy's personnel system and confronted concerns expressed by Jewett and Almy.[19] He also suggested that the navy grant the members commissions as lieutenant commanders to integrate them better into the service they had to advise.

Consciously or not, the captain's scheme perfectly illustrated the dilemma of the civilian scientific community and the navy in a period of transition. Thoroughly mixing the naval and civilian influences on the committee might pave the way for effective cooperation. However, selecting the chair, and not simply a member, from the naval service and proposing commissions for all participants suggested that the navy's traditional desire for complete control died hard.

To McDowell's dismay, Standley found even these modest recommendations a bit revolutionary. The liaison committee he created to work with the joint NRC/SAB group in the final weeks of 1933 took little notice of the

McDowell recommendations and sought a more traditional and segregated arrangement. Consisting of bureau representatives led by the head of the materials division of the CNO's office, Standley's Liaison Committee on Naval Research existed more to foster carefully the connections between the NRC and the navy, rather than to quickly integrate the two.

Nonetheless, the very existence of the navy committee and the NRC/SAB group demonstrated a willingness to build a relationship and with it a constructive naval policy toward research and more effective scientific programs for the fleet. In a letter to SAB member Frank Jewett, Captain McDowell singled out, not the fate of his proposal, but the determination, motives, and intent of the participants as the most important factor. He admitted "that there are certain difficulties which will have to be surmounted in getting some working scheme underway, and that this will include such questions as patents and financing, as well as the obtaining of a close knit liaison. I believe, though, that this can all be overcome provided the effort seems as important to the Advisory Board [the SAB] and the National Research Council as it does to me."[20]

The two committees met for the first time on 2 January 1934. Despite repeated suggestions by Millikan—who served as chairman of the joint NRC/SAB committee—that the meeting should take place at the NAS building on Constitution Avenue, the CNO insisted they gather at the Navy Department. There the scientists met Standley, liaison committee chairman Captain Marquart, the bureau chiefs, and officials from the CNO's office.

After this initial contact, the navy lost no time exploiting its newfound resource. Before the beginning of May, Captain Marquart delivered eighteen consultation requests on a wide variety of issues from mechanical respirators to the brittleness of battleship steel and armor. The NRC/SAB committee acted quickly on all of the requests, recommending preferred approaches to the naval problems and experienced scientists from major universities to serve as consultants. Bowman felt that the committee had an obligation to respond quickly and said as much in a 24 February 1934 letter to Millikan. Speaking for the NRC, Bowman commented: "We have the feeling that we are, in a sense, on trial and that it is important for us to reply to some of the questions as speedily as possible and to indicate that in the case of others some time will be needed."

The only major problem Bowman could foresee involved time and money. He had no trouble thinking of scientists who would advise the navy upon request, but their advice and guidance could take time. In most cases, individual scientists did not have the resources to depart from their funded research without remuneration. To address this problem, Bowman resurrected Captain McDowell's suggestion that the navy commission any willing and qualified scientists. This might induce their home institutions to grant them leave, and the navy would pay their salary and expenses. A number of the NRC/SAB committee members, including Chairman Millikan, had served in the armed forces in such a capacity during the Great War.[21]

Questions about implementation began to arise regularly in late 1934. Everyone involved expected them because neither the navy nor civilian science had seriously tried to build formally on the foundations laid by the NCB nearly twenty years earlier. Under Standley's conservative leadership the navy pursued a more effective way of employing science to benefit the fleet. On the civilian side, Bowman and Compton saw the creation of the SAB and the navy's requests as an excellent opportunity to forge a workable and productive relationship.

More important, in a letter sent to Millikan at CalTech on 13 November, NRC executive secretary Albert Barrows insisted that the navy had no desire simply to pick up where the NCB had left off after the Great War. Naval ambitions for scientific research now went much further. Barrows felt that the navy wanted "a different set-up for developing these problems, and they want to create an organization to work during peace times which could also be relied upon to furnish technical advice and guidance for the critical investigations arising in an emergency. I believe that it is important that we meet this proposal promptly and aggressively, if we are to make good on our offers of assistance to the Navy." Barrows assured Millikan that these ideas had actually emerged in a conversation with Lieutenant Commander Dietrich earlier in the year.[22]

Barrows wrote to Millikan again on 26 December to report that the NRC had sent Isaiah Bowman's suggestion regarding commissions for scientists to the CNO as a first step in responding to the navy's request for a viable plan to mobilize scientific resources in both war and peace. He also wanted to inform Millikan of two other preliminary ideas. Some NRC members thought that placing naval officers as interns at scientific and industrial facilities around the country would encourage mutual understanding and a collegiality that would expedite scientific mobilization. Another suggestion sought to place naval personnel within the major professional technical and research societies to encourage the same type of familiarity and cooperation.

In addition to private consultation and cooperation, the joint NRC/SAB committee took the initiative on research projects of medium to long duration. In 1934, Ivey F. Lewis, head of the NRC's Division of Biology and Agriculture, began planning a study of methods to retard the fouling of ship's bottoms by marine organisms. To help in the effort, the division suggested recruiting oceanographers, marine biologists, and fisheries scientists, including Henry Bryant Bigelow and Thomas Wayland Vaughan. Although it proved a difficult project to fund during the depression, the NRC and the navy launched the project three years later at Scripps on the West Coast, and then in 1940 at the Woods Hole Oceanographic Institution on the East Coast.[23]

While the civilian scientific community debated the nature of the navy's proposed scientific mobilization, the service took steps to create a more permanent and effective system to help the secretary shape policy and direct naval research. In 1934, RearAdm. Stanford C. Hooper assumed command of the Navy Department's new Naval Research Committee as well as the technical division of the CNO's office. Hooper, a pioneer in the naval application of radio communication, became the fleet's first wireless officer in 1912 and compiled an enviable record in communications technology before assuming these responsibilities. While Hooper's command coordinated naval research activity, Marquart's Liaison Committee on Naval Research continued its fruitful interaction with the NAS.

On 9 November, Lt. Comdr. R. W. Gruelick, who replaced Dietrich in 1934 as Marquart's assistant, offered for consideration an informal memorandum detailing yet another alternative framework for the mobilization of naval research in peacetime that would easily cope with the onset of an emergency. The scheme called for a new Navy Department agency staffed by about a dozen officers matched with civilian technical or scientific partners. They would represent various general technical fields of naval concern such as navigation, metallurgy, or submarines. This agency would coordinate the research effort, while the NRC would be a consultant and resource bureau. The council would provide general scientific advice, the names of qualified investigators, and explore other ways to address particular problems as the need arose. Some of the officers would naturally come from the bureaus, and additional experts in aerial warfare would come from the National Advisory Committee on Aeronautics.[24] Those in charge would have complete authority to set priorities for the use of naval resources and facilities and guide any project to completion without interference from the bureaus. In addition to tackling important problems, these men would also represent their command or agency in the coordination effort. Gruelick thought that the problems already proposed to the NRC by the Liaison Committee on Naval Research could serve as a beginning for this more permanent arrangement.[25]

From Bowman, to Wheeler, to McDowell, to Barrows, Gruelick's proposal was the last in a long line of recommendations that made their mark on the draft plan for research mobilization finally formulated and circulated by the NRC/SAB committee in 1935. In its plan, the NRC/SAB committee insisted that the navy commit itself to both basic and applied research, for both could have a significant effect on naval operations. The system adopted to mobilize scientific and technical resources for the navy should also encourage suggestions for important avenues of research from naval officers, civilian scientists, and interested citizens. Naval authorities could not afford to waste any resources or ignore avenues of inquiry from which the fleet might derive profit.

The committee thus brought to the navy's attention the diversity of human and physical resources available. These included thousands of scientists, well-equipped laboratories and libraries, and scores of technical societies nation-wide.

The proposed plan included five components to exploit these resources. A permanent Science Advisory Committee was the first and most important. It would take up residence within the NRC or SAB and consist of five or more eminent scientists to advise the Navy Department, to help develop specific recommendations concerning research, and to provide important personal contacts with people qualified to formulate effective solutions to naval problems. These elders within the scientific community would also coordinate relations between the navy and the NRC and attempt to enhance communication and cooperation with the War Department.

The influence of proposals by McDowell, Bowman, Gruelick, and others became obvious in the other components of the NRC/SAB plan. The committee suggested creating a consultant group of fifteen "younger" scientists, all appointed by the secretary of the navy on the advice of the permanent Science Advisory Committee. These consultants would work within the Navy Department as commissioned reservists. During periods of active duty, they would become more familiar with the navy as well as the requirements and problems of fleet operations. Working within the system would soon bring valuable first-hand knowledge of naval technical and scientific needs. In seeking solutions and an easy cooperative relationship, the Navy Department might choose to attach this consulting group to its Liaison Committee on Naval Research or one of Admiral Hooper's activities. The NRC/SAB committee also wanted to make it mandatory for these younger officers to visit naval and industrial laboratories for an extended period to become acquainted with and involved in projects seeking scientific solutions to naval problems.

As the Navy Department altered its approach to research and set about better coordinating scientific projects, the committee also warned that innovations frequently emerged in the course of productive research and thus the issue of patents could not be ignored. At the beginning of the decade this sore point had often soured relations between naval and civilian staff members of the NRL. A successful reorganization of naval research had to include a mechanism to address the issue of patent awards.[26]

The plan did not bring as much criticism as it did applause. F. W. Willard of the NRC's Division of Chemistry suggested that implementing the plan might politicize the research process and deprive the council of its liaison function with the nation's professional societies.[27] But his was a voice in the wilderness. Lieutenant Commander Gruelick reported to NRC executive secretary Albert Barrows that many important figures within the navy reacted favorably to what became known as the "Navy Plan." Gruelick discussed the scheme with his superior, Captain Marquart, with Capt. Bruce L. Canaga, director of the Central Division of the CNO's office, and Rear Admirals Walton R. Sexton and John W. Greenslade of the General Board. The only substantive questions raised involved reserve rank and duration of duty in peacetime as opposed to war.[28] On 18 March 1935, the NRC/SAB committee officially submitted the plan to the Navy Department as a program civilian science could support and recommend.[29]

As this plan went into effect, the civilian scene changed considerably. The SAB's authorized three-year term expired and the president extended its life until December to allow the staff to bring projects already under way to a successful conclusion. The ambitious and expensive national research plan that SAB chairman Karl Compton had submitted for the president's consideration came to naught and, as the new year dawned, only the NAS and NRC remained to work with the navy.

Painfully aware that the activities of the SAB and its work with the NRC had created a rift in relations with the NAS, the latter's executive board elected a new president in late 1935 who would also serve as chairman of the NRC. As the new NAS president, Frank Lillie directed both consulting agencies for a time, giving them one voice and one policy. The initiative and independence of Isaiah Bowman and the conservatism of William Campbell were exchanged for the single firm hand and extensive administrative experience of the University of Chicago–based director of the Marine Biological Laboratory.

Lillie continued the activities of the old NRC/SAB joint committee, as well as the NAS Committee on Oceanography. The former now became the Army and Navy Department Subcommittee of the NAS Government Relations Committee. During the years 1936–39, it nominated scientific candidates for reserve commissions and contacted universities to determine if they had research programs of interest to the navy. The Committee on Oceanography, Lillie's old NAS group founded in 1927, continued as before, only now under the able chairmanship of WHOI's Henry Bryant Bigelow.[30]

The Committee on Oceanography carried on until 1937, when Lillie declared its objectives met and requested dissolution. The committee had founded the WHOI and published its estimate of the present and future possibilities for American oceanography as penned by Bigelow in 1931. It inspired ambitious private and federal programs beyond the work already done with the government by Scripps, and consulted with the NRC and foreign experts such as Sir John Murray, Harald Sverdrup, and Fridtjof Nansen. In 1937, the committee finally received and published Thomas Wayland Vaughan's *International Aspects of Oceanography*, a work much delayed by the author's illness. In a final letter to Raymond B. Fosdick, president of the Rockefeller Foundation's

General Education Board, Lillie thanked the board for the funds and support that had made the committee's ambitious program successful.[31]

While Lillie's term at the NAS/NRC provided the strong direction, continuity, and calming effect needed by both organizations, world events in 1938–39 again introduced uncertainty.[32] Frank Jewett of Bell Laboratories succeeded Lillie as NAS president in 1939 and quickly realized that Adolf Hitler and German National Socialism posed a threat that could easily spill over European borders. A veteran at the NAS and NRC, Jewett knew all too well the difficulty of properly advising the federal government and the yet primitive state of the plan for scientific mobilization to assist the navy.

The appointment of Rear Admiral Bowen to direct the NRL with collateral duties as the secretary of the navy's science adviser did not improve matters. On 8 December, the admiral took charge of another new research policy activity, the navy's Council for Research, composed of representatives from each bureau with technical and scientific needs. With the Liaison Committee, Hooper's research activities, and Bowen, the navy scene seemed rather complicated and the ultimate line of authority in scientific matters too difficult to follow.[33]

Scientific mobilization, in war or peace, would be extremely difficult and complex. The navy had come to terms with the need for a science policy and a commitment to naval research. Moreover, the civilian scientific community had emerged from its internal rancorous debates ready to serve and advise. Unfortunately, selecting those willing to serve with reserve commissions and creating working relationships with a number of universities according to the NRC/SAB plan seemed an intimidating task made more complex by an awkward naval research administration. The critical world situation begged for speed, and those working on the problems of naval research could make no promises.[34]

Frank Jewett found this situation particularly disturbing. The NAS/NRC organization seemed to him far too cumbersome to serve the country well in peace or war. In a letter to Arthur L. Day, director of the Carnegie Institute's Geophysical Laboratory, Jewett declared that something had to change. After thanking Day for assuming the acting chairmanship of the NAS's War and Navy Department Subcommittee, Jewett commented: "In connection with this whole matter of advice to the Government, it seems to me that with the requests coming thick and fast, and destined probably to come thicker and faster in the immediate future, you, [Ross G.] Harrison [chairman of the NRC] and I, with such others as have had experience, will need to get together shortly and see if we cannot formulate a more efficient and less cumbersome setup for the Academy and Council than the one we now have, and one which will be less confusing to the Departments of Government. It would seem to me that we ought to be able to do this in time to present the matter to the Academy or its Council at the time of the Fall meeting, and if possible secure their approval."[35]

Although Jewett fully intended to discuss reorganization matters with Harrison and Day, he had already contributed to a far more radical solution to the potentially awesome problems of naval policy and national scientific mobilization. Since 1937, he had served on the NRC's Committee on Scientific Aids to Learning.[36] Despite the country's tilt toward isolationism, Jewett and his committee associates thought that the United States should prepare to confront any threat Hitler might pose. This subject frequently dominated conversation before and after meetings when Jewett gathered with his colleagues on the committee, including Vannevar Bush, chairman of the National Advisory Committee on Aeronautics; James B. Conant, president of Harvard University; Richard C. Tolman, dean of the CalTech's graduate school; and Karl Compton, MIT president and former SAB chairman.

Concerned that the working relationship between the army, navy, and the NAS/NRC militated against mobilizing science in a timely fashion, these five men went to Franklin Roosevelt through his close personal adviser, Harry Hopkins, and proposed creation of an independent mobilization agency called the National Defense Research Committee (NDRC). After the president scrawled "OK—FDR" on the Hopkins-endorsed four-paragraph proposal, Vannevar Bush, acting as the spokesman for the group, took steps to set up the administrative structure that would permit the government to tap quickly the nation's vast scientific and engineering resources. Bush, Jewett and their associates sought to bypass both the existing NAS/NRC committees and military and naval activities for the sake of speed and simplicity. Reflecting on the radical move thirty years later, Bush commented: "There were those who protested that the action of setting up N.D.R.C. was an end run, a grab by which a small group of scientists and engineers, acting outside established channels, got hold of the authority and the money for the program of developing new weapons. That, in fact, is exactly what it was. Moreover, it was the only way in which a broad program could be launched rapidly and on an adequate scale. To operate through established channels would have involved delays—and the hazard that independence might have been lost, that independence which was the central feature of the organization's success."[37]

Beginning in May, 1940, the NDRC formulated mobilization policy and served as the organizer and administrator of oceanography projects done by scientists in and out of the navy. Bush, Jewett, Tolman, and Conant—acting with the president's consent—took the initiative and bypassed the slowly developing relationship between the navy and NAS/NRC characterized by efforts on both sides to give up as little control as possible over projects, methods, and professional independence.[38] In the end, the NDRC rendered the established links between science and the navy obsolete and made further painstaking debate irrelevant.

Interwar Observations

O CEANOGRAPHY CAME OF AGE IN BOTH EUROPE AND THE United States by serving a critical purpose. Just as fisheries management and economic necessity drove oceanographic research in northern Europe, the German U-boat threat during the Great War mobilized critical resources necessary to initiate and support large-scale American oceanographic studies. As underwater sound emerged as the most promising method of submarine detection, the NCB and the NRC soon discovered that effective anti-submarine warfare regularly required basic and applied research in some aspect of oceanography.

After the war, the navy turned to oceanography as a way to survive the peace and contend with contracting budgets. The creation of the International Hydrographic Bureau and the worldwide activities of the Hydrographic Office kept the navy involved in oceanography in the early postwar period. While certainly significant, these efforts did not emerge from the background of naval affairs to take center stage until the Hayes memorandum of 19 February 1923. The former Swarthmore physicist offered a practical, concrete program to broaden the navy's popular and political appeal by demonstrating that it could educate and help feed Americans in peace as well as destroy an enemy in war. In offering this perspective to Capt. John Halligan in 1923, Harvey Hayes sought to induce the navy to work for a deeper appreciation of its operating environment. While it seemed natural for a navy to explore the ocean, these proposals also served Hayes and his underwater sound program in a pragmatic way. If oceanography captured the imagination and resources of the navy, Hayes's own underwater sound program at the EES would have an adequately funded new facility at the NRL in Anacostia to continue ocean research with concrete ASW applications.

In 1924, the Navy Department embraced the suggestions made by Hayes and the NRC's Andrew C. Lawson by committing itself to oceanographic research at the Interagency Conference on Oceanography. With the ICO, the navy took its first step toward formulating a coherent research program to support its mission.

Despite frustrating responses from President Coolidge and the Congress to Capt. Frederic Bassett's request for support to carry out the ICO proposals, the fiscally trying interwar years marked the beginning of the navy's commitment to the ocean sciences. Supporters of the navy's fledgling oceanographic effort averted potential disaster in 1924 by departing from normal institutional relationships to obtain maximum results from combined naval and civilian assets. Consultation, informal agreements, and personal contacts took the place of formal policy and consistent, dedicated funding. This *common practice* evolved during the interwar period to sustain and further crucial research, especially in the Pacific and Caribbean–Gulf of Mexico regions.

The growth of this informal matrix of relationships to support the ocean sciences grew with careful and persistent cultivation. Even with promising wartime results and the recognition of potential postwar political benefits, most naval officers, like most national leaders, gave no indication that they perceived any continuing operational applications for oceanography after hostilities ceased in 1918. For the average officer, peace returned oceanographic investigation and analysis to the map and chart makers. While depth soundings and more accurate navigational aids naturally had great naval and commercial value, even this activity fell under the shadow of depleted naval budgets, further obscuring the potential oceanography had exhibited during the war.

Only the civilian scientific community, those officers convinced of oceanography's usefulness, and Hayes's small, underfunded group at the NRL detected potential and tried to extend the advantages oceanographic knowledge could offer to warships at sea. Within this company it did not take long to realize that ad hoc arrangements and informal agreements could not provide completely for the future of oceanography in the United States. Pursuing ICO goals, coordinating available funds and assets, and integrating the diverse ambitions of interested groups required some central direction. With Congress and the president tying its hands, Hydro could not take the lead. By 1925, prominent civilian scientists like Thomas Wayland Vaughan, Frank Lillie, and their colleagues, working through the National Academy of Sciences, appreciated the dilemma and took action. Their efforts gave life to the NAS Committee on Oceanography, providing a small measure of the central coordination for American oceanography as envisioned by the ill-fated ICO.

Although faced with severe fiscal limitations, the navy persisted in its involvement with oceanographic research. The Lillie Committee kept the navy informed while naval officers, crews, and instrumentation participated in a variety of ocean-science activities ranging from the *Carnegie* soundings to the gravity work on S-21 and S-48 to the Byrd Antarctic Expedition to the thousands of soundings and samples taken by naval vessels on routine missions around the world. The links forged at the ICO in 1924, the activities of the NAS in conjunction with the navy, and the informal relationships established between Hydro and the civilian research community all contributed to a productive dialogue and helped sustain interwar activity. Cooperative agreements and regular consultation between Hydro, civilian institutions, and occasionally reluctant operating forces permitted ambitious and effective ocean science in the strategically significant Gulf of Mexico, Caribbean, and Canal Zone regions despite considerable financial and cultural obstacles. In this case, for both civilians and naval officers committed to oceanography, necessity was the mother of the common practice.

In addition to the challenges of promoting and mobilizing oceanography and continuing the common practice, the navy also played a critical role in creating a new research tradition on the geophysical side of oceanography. The Navy Department supplied destroyers, survey ships, and submarines as well as experienced personnel and administrative services for Wilkins's and Byrd's polar expeditions. The gravity and survey expeditions of 1928, 1932, and 1936–37 benefited from naval interest in the revealing and potentially revolutionary geophysics of the Caribbean area. The Navy-Carnegie, Navy-Princeton, and Navy–American Geophysical Union expeditions helped perfect new methodologies, advanced the state of instrumentation, and provided experience and training for leaders in a significant and fertile field. The common practice placed essential human and material resources at the disposal of talented, curious, and ambitious scientists—both naval and civilian.

Oceanographic surveys and sounding programs fueled by the maturing civilian-naval common practice continued until the advent of World War II. The navy clearly provided most of the means for the work accomplished all over the globe, but the civilian community furnished much of the determination and expertise. Thomas Wayland Vaughan and Thomas Thompson had well-defined ideas about the most significant oceanographic research the naval-civilian partnership needed to accomplish. Rarely did they resist communicating their ideas to Hydro, the secretary of the navy, the NAS/NRC, the U.S. Coast and Geodetic Survey, or any other agency that might sponsor their purpose. From the navy's viewpoint, this vision and ambition proved profitable and complemented the plans formulated by the Hydrographic Office. Rear Admiral Walter Gherardi and Capt. Lamar Leahy listened carefully to civilian advice, combined their own plans with those of their civilian colleagues, and together they accomplished an extraordinary amount of science with severely limited resources. Furthermore, their policies and methods ensured that progress would continue after an officer's tour of duty at Hydro. The ease of transition from the Gherardi-Vaughan era in Pacific research to that of Leahy and Sverdrup demonstrated the strength of the ties and the productivity of the relationship.[1]

Unfortunately, as late as 1930, the Navy Department still did not have an official science policy to coordinate basic or applied scientific activity. Unlike scientists and officers involved in oceanography's common practice, most naval authorities did not see the need. Manifesting the last traces of Progressive Era philosophy after the Great War, naval scientific endeavor broadly defined was viewed as part of a general effort by the service's engineering talent to support the fleet's mission. Convinced that select companies and private universities had already done the necessary basic research, the Navy Department had only to fashion civilian initiatives into more capable instruments of war. With very few exceptions, naval research and development belonged to the engineer and the applied scientist.

Navy-sponsored science had to declare its independence from engineering. Each had to achieve independent recognition. The Great War at sea reaffirmed the importance of having engineers skilled in system development aboard ship. At the same time, antisubmarine warfare activities conducted between 1914 and 1918 graphically demonstrated the need for an intimacy with the ocean environment only a scientist could provide. In the long term, the navy could neither encourage nor make the desired advances in oceanography or fully exploit them without a precisely defined official science policy. The Hayes memorandum, the ICO, the Schofield Board, and the Lillie Committee—all contributing to the evolution of the common practice—marked a first step. They slowly but clearly demonstrated both the exciting possibilities and the severe limitations of the unofficial and informal, as well as the difference between science and engineering and the value of both to the navy's mission.

Once the Navy Department addressed the need for an official policy governing naval scientific activity, it nearly drowned in the number and variety of schemes submitted by officers and scientists. The efforts of Isaiah Bowman, Lt. Comdr. R. W. Gruelick, Capt. Clyde McDowell, and the NRC/SAB committee demonstrated the difficulties in changing the naval perspective on scientific activity, formulating a coherent policy, and formalizing a productive relationship between the navy and civilian scientists.

The desire for control constantly complicated matters. Both the navy and civilian science had concerns about developing official policies and relationships that naturally never emerged in the informal environment of the common practice. Officially, each wanted to retain complete authority over its activities to an extent that would often interfere with the emergence of a productive dialogue. The latter demanded a certain degree of flexibility and a willingness to trust and compromise. These two distinct communities of practitioners needed to bridge differences in professional language, standards, traditions, and mission. The navy customarily wanted to minimize civilian influence in its activity, and many conservative leaders within civilian science sought to distance themselves and their research from the federal government's political meddling and partisanship. This fear of partisanship also fueled the conflict over the Science Advisory Board in the early 1930s. Even with the NRC/SAB committee cooperating fully with the navy, the unresolved open debate over philosophy, politics, and scientific roles persisted. For a time, these conditions retarded the formulation of a well-defined naval science policy and an effective dialogue between the navy and the scientific community that might take their relationship and activity beyond the common practice.

When the outbreak of the European war in 1939 focused attention on the absence of both a coherent scientific policy and an effective official mechanism linking science and the armed forces, President Roosevelt, Harry Hopkins, Vannevar Bush, James Conant, Richard Tolman, and Frank Jewett created the National Defense Research Committee. In so doing, they completely suspended the normal course of events. The NDRC bypassed working committees and naval activities burdened with political agendas and cultural baggage. Frankly confronting the possibility of American involvement in another world war, Vannevar Bush and his colleagues muted the seemingly endless debate and dragged both the armed forces and civilian science into one of the most successful and productive collaborations in history.

Its swift birth and temporary nature notwithstanding, the NDRC's roots ran deeply in the prewar scientific and defense communities. Informal interwar cooperative ventures like the common practice in oceanography provided an incentive based upon common need that contributed significantly to the cohesion and character of the wartime organization. Bush created a formidable system to manage scientific mobilization in 1940. The strength and effectiveness of the NDRC in the ocean sciences actually rested upon a strong, unbureaucratic mix of professional and personal ties between civilian scientists and their counterparts in and out of uniform working in the navy labs, bureaus, and at sea. In this sense, the interwar experience set a precedent in form and practice for the NDRC, providing a foundation that would serve naval oceanography in a remarkable way from Pearl Harbor to Tokyo Bay.

R/V *Atlantis* in rough north Atlantic seas just before World War II. The photo was taken looking aft from a position roughly amidships. Courtesy Woods Hole Oceanographic Institution.

Thomas Wayland Vaughan, director of the Scripps Institution of Oceanography, at his desk, 1 August 1932. Courtesy Scripps Institution of Oceanography Archive/UCSD.

The staff of the Scripps Institution took time to pose with director Thomas Wayland Vaughan in the early 1930s. Behind the director to the right stands Roger Revelle, the tallest figure in the back row. To the left of Vaughan in the striped tie and dark jacket is Richard Fleming. Both emerged from World War II as national leaders in oceanography. Courtesy Scripps Institution of Oceanography Archive/UCSD.

Harvey C. Hayes, physicist, Navy sonar specialist, and creator of the SDF. He is pictured here as wartime chief of the underwater sound division of the Naval Research Laboratory, c. 1942. Courtesy U.S. Navy.

Rear Admiral Walter Gherardi, head of the Navy Hydrographic Office 1930–1935 and coauthor, with Scripps's T. W. Vaughan, of the common practice that sustained interwar seagoing ocean science. Courtesy U.S. Navy.

J. Lamar Worzel, geophysicist and student of Maurice Ewing at both Lehigh and Columbia Universities, holds one of the bombs used in the early Ewing refraction research at sea, c. 1938. Courtesy Ewing (Maurice) Papers, The Center for American History, University of Texas at Austin.

Allyn Vine, in the dark jacket, at sea on a research cruise with his mentor Maurice Ewing, c 1938. Courtesy Ewing (Maurice) Papers, The Center for American History, University of Texas at Austin.

Professor Maurice Ewing of Lehigh University preparing one of his early underwater cameras for testing on board WHOI's R/V *Atlantis,* c. 1938–1939. Courtesy Ewing (Maurice) Papers, The Center for American History, University of Texas at Austin.

This is a 1940 view of the dock area behind the Bigelow Building (right), the main laboratory of the Woods Hole Oceanographic Institution. The image was probably

snapped from the end of the Martha's Vineyard–Nantucket Steamship Authority dock just a short distance away. Courtesy Woods Hole Oceanographic Institution.

Allyn Vine loads an automatic camera assembly into a pressure vessel designed to protect it in the ocean depths, c. 1940. Vine worked and studied with Maurice Ewing at Lehigh University and came with him to do research at WHOI before and during World War II. Courtesy Woods Hole Oceanographic Institution.

Frank Jewett of ATT Laboratories as president of the National Academy of Sciences, 1940. Courtesy National Academy of Sciences.

The Norwegian Harald Ulrich Sverdrup served as wartime director of Scripps and elevated that laboratory to an oceanographic institution of world rank.

Columbus O'Donnell Iselin, director of the Woods Hole Oceanographic Institution, at his desk in 1940. Courtesy Woods Hole Oceanographic Institution.

Finding a Niche, 1940–41

All we want here is to be given a chance to get to work.

—COLUMBUS O'DONNELL ISELIN, DIRECTOR, WHOI, 22 August 1940.

I F THE CREATION OF THE NDRC PROVIDED A MEANS FOR MOBI-lizing science in view of the increasing probability of American involvement in the war, defining the best use of each scientific specialty still presented an intimidating problem. How should the committee best spend the $4.8 million authorized by Roosevelt for its use in 1940? Vannevar Bush and his colleagues first had to determine the most pressing tasks and the types of knowledge and expertise best suited to address them.

During those last months of peace, civilian oceanography worked with the navy and the NDRC to identify available talent, to allocate resources, and to define its role in the country's preparation for war. This effort included an ex-traordinary amount of personal persuasion and creative entrepreneurship on the part of leading oceanographers, for neither they nor the NDRC controlled their own destiny in these matters. If war came, the navy would occupy the front line. It became the task of oceanographers in and out of the naval service to demonstrate their value to skeptical officers whose war-fighting priorities lay elsewhere. For the latter, tried and proven weapons technologies and engi-neering techniques had far greater value than science practiced in laboratories far away from the heat of battle. In their prewar work with the navy, experienced translators like Thomas Wayland Vaughan had already discovered the inadequacy of applying engineering and scientific methodologies to the physical world alone. As sociologist Donald MacKenzie noted in his history of Cold War missile-guidance systems, "People had to be engineered too—persuaded to suspend their doubts, induced to provide resources, trained and motivated to play their parts in a production process unprecedented in its demands."

Vaughan became a pioneer of this heterogeneous engineering process in the ocean sciences. Both his example and the coming of war called forth a new generation of translators determined to promote a productive dialogue between naval culture and the scientific community.[1]

Products of an educational process that imprinted individuals with a strong sense of naval culture, officers molded at the U.S. Naval Academy largely viewed themselves as engineers applying modern technology to a style of warfare their community had virtually defined. As the war began, the services provided by translators proved one of the most effective ways to bridge the gap between officer and oceanographer, between engineer and scientist, indeed, between technology and science. Either scientists with a talent for engineering solutions or engineers comfortable with scientific concepts and theoretical literature, translators became a critical channel of communication. In suggesting solutions from engineering practice or scientific research they promoted the flow of ideas across a distinct but ill-defined boundary between two communities of practitioners. After 1940, advanced instrumentation developed or refined by translators complemented verbal persuasion. Instruments became an effective method of demonstrating the complementarity of the two communities and fostering comfortable interaction. Translators helped scientists effectively apply their knowledge of the natural world to the problems of naval war in a way the officer-engineer could readily appreciate. In this way, oceanography's translators equipped scientific and engineering entrepreneurs with valuable instruments used not only to collect data and solve immediate wartime problems, but also to provide tangible evidence of oceanography's strategic and tactical utility. In so doing, they hoped to win political support and initiate a fruitful dialogue between naval and scientific communities based upon communication, credibility, and trust.[2]

Mobilization provided different experiences for the navy's two major civilian oceanographic institutions. The Scripps Institution, still under the direction of Vaughan's successor, Harald Sverdrup, prepared for war under an agreement between the NDRC and the University of California that created the university's Division of War Research (UCDWR) on 26 April 1941. The UCDWR took up residence with the U.S. Navy Radio and Sound Laboratory (USNRSL), established on 1 June 1940 at Point Loma near San Diego. Thus, much of the wartime activity undertaken by the university in cooperation with NDRC or BuShips obliged the Scripps faculty to divide its attention between projects at home in La Jolla and a few miles down the coast at UCDWR. Sverdrup, trapped in the United States as the Germans overran his native Norway, remained at Scripps for the duration of the war courtesy of the University of California, which granted him an unrestricted extension on his contract as director.

An Early Translator at Work: Columbus O'Donnell Iselin

Created a decade earlier and not formally affiliated with any university, in 1940 the Woods Hole Oceanographic Institution took on the appearance of a dedicated entrepreneur. While the NDRC struggled to organize and acquire sufficient operating resources, Columbus Iselin energetically wrote memoranda and walked the corridors of government agencies in Washington, D.C.

This Harvard graduate, expert seaman, and student of WHOI's first director, Henry Bryant Bigelow, succeeded his mentor in his thirty-seventh year. Raised in a patrician environment, Iselin came from a modestly wealthy New York family that derived its fortune from diverse interests, including a small boat-building company. The new WHOI director displayed both a naturally affable manner and a professional competence that led people to enjoy his company, trust his judgment, and seek his advice. Fifty years after he first arrived at WHOI, renowned physicist and engineer Allyn Vine enjoyed recalling the director's habit of "walking the ship," drawing attention both to Iselin's frequent practice of walking about the institution to stay in contact with his people and the director's considerable skill as a seaman. His relationship with the people at WHOI led to busy, informative, and insightful personal communication and correspondence. He wanted his scientists and technicians to keep him informed, but their conversations and letters went far beyond scientific progress reports. They provided personal reflections and local color as well as observations on working with the navy and the difficulty and satisfaction of doing oceanographic research.

Iselin shared their concerns and knew well the difficulties of conducting scientific research at sea. Shortly after WHOI opened its doors in 1930, he made his mark at the institution as an expert seaman. Bigelow had the research vessel *Atlantis* built in Denmark with funds earmarked for the purpose by the Rockefeller Foundation. Iselin sailed the vessel back across the North Atlantic armed with only a small crew and his own navigation and sailing skills.

In 1940, Iselin worked to convince the navy that oceanography in general, and his institution in particular, could play a critical role if the United States again went to war. Unlike the Scripps Institution, attached to its University of

California parent, WHOI was alone. If the young institution expected to survive, its director had to attract patronage. Having worked with the navy in the late 1930s on underwater sound and the discovery of the afternoon effect, Iselin knew the cultural barriers and understood the possibilities and limitations of the interwar common practice. He also appreciated the resources that the navy would have at its disposal if the country went to war.[3]

For Iselin and others, events in Europe provided a considerable incentive. The Nazi war machine had already brought France to its knees, and the Battle of Britain raged in the skies over the United Kingdom. Admiral Karl Dönitz, commander of Germany's U-boat force, ordered his submarines from Kiel and Wilhelmshaven to French Atlantic ports like Brest, Saint-Nazaire, Lorient, Bordeaux, and others from which he could easily reach the East Coast of the United States. Each day, actual combat seemed closer. As a naval reserve officer, Iselin believed—as did the NDRC—that the time to address the threat had come. Besides, working with the navy might bring additional recognition and considerable resources to Woods Hole.

Iselin began his campaign by discussing with the infant NDRC and the Naval Research Laboratory the types of problems oceanographers were best equipped to attack. As early as 5 August 1940, just thirty-four days after Vannevar Bush held his first NDRC meeting, Iselin addressed a "Memorandum to the NDRC concerning the research facilities available at the Woods Hole Oceanographic Institution." The memorandum's content and tone suggested that the navy at large and the civilian oceanographic community had little knowledge of one another: "[N]ot only is the Navy unaware of the modern developments in this field of science, but the handful of oceanographers only very imperfectly understand the naval problems for which their specialized knowledge might prove valuable."

Unlike the mutually beneficial scientific activities pursued by Scripps director Vaughan and Hydro's Rear Adm. Walter Gherardi during the preceding fifteen years, Iselin now spoke in terms of an applied science that would enhance the offensive capability of warships and their crews. The relationship developed by necessity during the interwar years provided valuable training for scientists and technicians, improvements in navigation and mapmaking, and increased useful scientific knowledge. By 1940, Iselin felt that the volatile international situation might provide opportunities for remarkable advances in science and engineering in support of the armed forces.

Thanks to Iselin's determination to help shape the role oceanography would play in the coming conflict, WHOI had already begun work with both the navy and the Coast Guard. Collaborating with Coast Guard oceanographers Floyd Soule and Clifford Barnes, Woods Hole scientists now regularly collected large amounts of seawater from the Gulf Stream. With the Germans overrunning Europe, the source of "Copenhagen Standard Seawater," which acted as the control substance in many ocean-science experiments, had suddenly disappeared from the free world and the scientific supply markets. At WHOI, staff members filtered the Gulf Stream water, stored large quantities of it, and then sealed into glass tubes precisely measured amounts of the fluid with the proper balance of required standard elements. In addition, BuShips supported a Woods Hole project led by Bostwick Ketchum to study both the performance of antifouling paints on ships and the organisms that collected on ships' bottoms. As any shipmaster knew only too well, a thoroughly fouled bottom shaved too many precious knots off a vessel's speed. For a destroyer chasing a submarine while protecting a convoy, and even for the ships in the convoy, slower and less efficient movement would certainly cost lives and money.[4]

Iselin's 5 August memorandum suggested additional ways in which oceanography and WHOI could provide essential services to the navy. He reminded Bush and the NDRC that Woods Hole already had a productive relationship with the navy. The work on fouling simply added to a recent history of cooperative work with U.S. naval forces, including various underwater sound and submarine detection experiments between 1936 and 1940 with USS *Semmes* near Guantanamo Bay, Cuba, and off New London, Connecticut. Iselin also reminded the NDRC that WHOI had a number of naval reserve officers among its regular and visiting staff, emphasizing the institution's commitment to the navy's problems.

As consultants, the institution's able community of permanent and visiting university-based scientists could provide advice critical to successful naval operations. Naval officers might also come to WHOI for education in the strategic and tactical opportunities provided by the ocean environment and for training with instrumentation that could provide vital information in submarine warfare and ASW. Informed WHOI scientists would stand ready to compose reports on operational problems and ocean characteristics that might take the navy years to distill from the extant and often cryptic professional literature. In the same vein, Iselin suggested his staff could prepare special studies for the navy on demand. Woods Hole had expertise in ocean currents, the thermal structure of the upper three hundred feet of the ocean—where U-boats could effectively hide—and meteorology, especially fog and storms. In addition to an excellent staff and a significant array of scientific services, WHOI could also take its people to sea aboard the R/V *Atlantis,* the best civilian seagoing laboratory then available.

Iselin also made a point of noting Germany's recent progress in the ocean sciences. Thanks to the expedition of the R/V *Meteor* from 1925 through 1927, America's most likely Atlantic adversary had considerable scientific expertise, valuable practical experience, and some of the best data from parts of the At-

lantic already recognized as war zones. The scientific community lauded this
interwar voyage as the best-organized and most productive oceanographic project since the nineteenth-century British *Challenger* Expedition. The rump navy
left to the Weimar Republic by the Treaty of Versailles provided the fully manned
Meteor, a converted Imperial Navy gunboat, to the University of Berlin for research in the central and southern Atlantic. Iselin suggested that the U.S. Navy
should avoid underestimating the operational applications of oceanography and
the hazard of paying too little attention to naval warfare's natural environment.
He intimated strongly that Germany would not make the same mistakes.[5]

While the NDRC was probably the most important agency for scientific
mobilization, Iselin did not spend all of his time in the nation's capital waiting
for interviews with Vannevar Bush. Next to the fouling work WHOI did for
BuShips, Iselin felt that the navy should build on the foundation recently laid
by cooperative research with WHOI on submarine detection and underwater
sound. During the interwar period the navy had achieved great success with its
own program at New London, which, along with its director Harvey Hayes,
migrated to the navy's Engineering Experiment Station in Annapolis, and then,
in 1923, to the new Naval Research Laboratory in Anacostia. On Monday, 26
August 1940, Iselin visited NRL assistant director and navy liaison officer to the
NDRC, Comdr. Robert P. Briscoe, to discuss the progress made in physical
oceanography and ways in which WHOI could work with the NRL on underwater sound research. The letter Iselin wrote arranging the meeting demonstrated his belief that, outside of Hydro, even the people at the NRL might not
completely appreciate the nature of oceanography and the advantages it could
provide. The visit also served Iselin in another way. Given the NRL's history of
underwater sound research, the officers and scientists there might recognize
more quickly the strategic and tactical value of his proposals. In the second
paragraph of his initial letter to Briscoe, the WHOI director indicated that he
wanted "to explain briefly the progress in physical oceanography during the last
few years so you will have a clearer understanding of the sort of problems we
can most effectively attack. . . . In thinking the matter over . . . it seems likely
that a thorough investigation of the transmission of sound in sea water will
prove our most useful contribution to your regular research program."[6] This
suggestion certainly struck a resonant chord at the NRL.

Iselin needed Briscoe and the people at the laboratory to help him influence
naval decision making and better define a truly effective place for oceanography
in the mission of the armed services. Their support would become more critical as the role of oceanography in general and WHOI in particular began to
take definite shape. Both civilian and naval scientists would have a major say in
determining how their skills would affect the course of possible American involvement in the war.

If Briscoe, Harvey Hayes, and the NRL staff became valued allies in fulfilling these aims, Iselin's relationship with NAS president Frank Jewett proved even more critical. A Woods Hole trustee, founding member of the NDRC, and one of the nation's premier engineers, Jewett strongly supported Iselin's efforts. When he could escape the demands of the NAS and his professional home at Bell Laboratories, Jewett traveled to Martha's Vineyard, where he and Iselin frequently met. The latter lived on the Vineyard year-round and had a residence on Lake Tashmoo adjacent to Vineyard Haven, where Jewett had his own retreat. Every day, Iselin spent thirty minutes crossing 4.5 miles of Vineyard Sound to the director's office at WHOI in his small boat *Risk*—a name that frequently prompted his scientific colleagues to comment that he had a realistic perspective on life at sea.

In the summer of 1940, Jewett and Iselin corresponded frequently, exchanging ideas on the best place for oceanography in NDRC's plans. In a 13 August letter to Iselin, Jewett noted that in his discussions with Bush and Karl Compton, the three agreed that, if war came to the United States, the WHOI staff should remain at Woods Hole and on task for the duration. They even suggested that Iselin's institution might well play the role of the central oceanographic laboratory in a network of facilities devoted to U-boat detection and underwater sound problems.

During August, Iselin kept Jewett informed of his discussions in Washington and the growing opinion that, after the fouling project for BuShips, the best possible application of WHOI's skills lay in underwater sound. Iselin met separately with Commander Briscoe at the NRL and Adm. L. O. Colbert at the U.S. Coast and Geodetic Survey, and each confirmed that underwater sound was a major priority. On one of his trips to Washington in late August, Iselin, with Lehigh University's innovative geophysicist and translator extraordinary Maurice Ewing in tow, presented WHOI's underwater sound proposal to a number of naval officials. These included Rear Adm. George S. Bryan of Hydro and, according to Iselin's occasionally imprecise language, "the various sound men in the Navy Department." By the end of the month the WHOI director was sure that he had the necessary support to add a navy sound project to the fouling research already under way at his institution. After the Washington visit, he informed Jewett that all WHOI needed was NDRC approval and funding. In an upbeat tone, Iselin wrote, "I believe we are ready to go to work, if you can find us some money."

Iselin pushed hard at the end of August, hoping he could obtain NDRC approval by the end of the month for WHOI's plan to initiate a major investigation of underwater sound for the navy. At the conclusion of the NDRC's 29 August gathering, the young WHOI director's efforts were rewarded. The good news came in a short handwritten note: "Committee this morning approved two year program as presented. Work can proceed full speed at once. I expect to be at Vineyard Haven on Saturday and Sunday and will get in touch with you regarding details of contract. F. B. Jewett."

In telling Commander Briscoe of the NDRC's decision, Iselin mentioned that the committee insisted that a disinterested party evaluate WHOI's capabilities for the sound work. This requirement did not pose an obstacle to an NDRC contract with the navy, but rather ensured that Bush's office would not encounter accusations of potentially illegal nepotism. With Iselin's primary advocate and WHOI trustee Frank Jewett on the NDRC, Bush wanted an independent opinion. He thus recruited CalTech's Max Mason to visit Iselin and the institution as the underwater sound work began. Mason first consulted with naval personnel in San Diego regarding navy's needs and expectations before setting off for Woods Hole. Iselin, confident in his people and program, told Jewett that he had begun recruiting additional personnel and assembling instrumentation and that he expected to launch the program by 16 September.[7]

The Colpitts Committee and Organizing for ASW

As Iselin pushed WHOI programs forward, the navy took action to define better the nature of the submarine-detection dilemma and the extent of the fleet's readiness to tackle the problem. At the request of Secretary of the Navy Frank Knox and the secretary's technical aide and director of the NRL, Rear Adm. Harold Bowen, the NAS created a committee to explore submarine detection and the current state of American scientific and technical knowledge in ASW. Admiral Bowen submitted the secretary's formal request on 16 October 1940, the NAS established the committee in November, and the group held its first meeting at the Academy in Washington on 9–10 December. Edwin H. Colpitts, former vice president of Bell Laboratories, chaired the committee consisting of William D. Coolidge, director of research at General Electric; Harry G. Knox, secretary of the committee and president of the Dymonhard Corporation of America; Vern O. Knudsen, dean of the University of California, Los Angeles (UCLA) graduate school; and Louis B. Slichter, an MIT geophysics professor. They conducted a careful two-month investigation of the problem and the navy's readiness to cope with it.

From the beginning, the NDRC, the NAS, and Colpitts himself displayed sensitivity toward the navy's awkward position. They did not want to suggest that the fleet could not yet carry out its ASW mission. Just before the committee's work began, Frank Jewett advised Max Mason: "So far as getting full cooperation from the Navy people is concerned, I think that the thing to be stressed in the preliminary sessions is that the aim of the Committee is to be

helpful and not critical or protagonist for something which the Navy is not doing and which the Committee thinks should be done."

At the end of its investigation in January, 1941, the group issued its "Report
of the Subcommittee on the Submarine Problem," which became widely
known as the Colpitts Report. The committee's evaluation of America's ASW
capability made it clear that the navy desperately needed to advance its knowledge of the ocean environment. Admiral Bowen immediately rejected the report as something he would find difficult justifying to the Navy Department.
However, prominent scientists like Jewett and Mason, the latter having assumed the presidency of the Rockefeller Foundation, gave the report their endorsement and complete support, more than offsetting Bowen's objections.[8]

The Colpitts Committee suggested that current tools and methods simply
would not do the job. The navy's ASW procedures employed state-of-the-
art supersonic echo-ranging equipment that usually worked well up to several
thousand yards and under very favorable conditions. But in all too many instances the nature of the water and weather rendered the ASW vessels blind.
Further oceanographic research and instrument development seemed absolutely necessary. In addition to better training for sound equipment operators,
the Colpitts Report recommended long-range research in underwater sound
transmission and reception, as well as the development of magnetic, micro-
wave, and radio-acoustic devices to supplement standard sound equipment.
The navy and civilian science also had to determine the complete spectrum of
underwater sounds emitted from various types of ships under as many conditions as possible, both to improve silencing in friendly vessels and to permit the
detection of noisy enemies. Most important, the Colpitts Committee urged the
navy to study the "sound propagating properties of oceanic waters over the entire frequency range likely to be involved in the use of detecting devices." This
task included studies already under way to support the navy and the Merchant
Marine, as well as research into variations in temperature, salinity, convection,
density, sound scattering, attenuation, and reverberation.

The Colpitts Committee appealed to the navy and civilian scientists for research that would provide effective oceanographic data and instruments for the
ASW and prosubmarine effort if the United States entered the war in which the
U-boat already played an extremely important part. Sounding a call to arms for
scientists, Colpitts and his colleagues emphasized "that a much more comprehensive and fundamental research program is needed. . . . The gravity of the
emergency is such that the present research facilities and personnel are wholly
inadequate. We need the best talent of the country. In these days of aroused pa-
triotism, that talent is available."[9]

The findings of the Colpitts Committee resonated perfectly with Columbus
Iselin. The recommendations on sound and detection research confirmed the

line of reasoning and research the WHOI director had advocated since the previous summer.[10] When the Colpitts Report received the support of so many scientists vital to the navy's research effort, the General Board held a meeting with Vannevar Bush and Frank Jewett to discuss the ways in which the navy might pursue some of the goals suggested by Colpitts and his colleagues. As a result, Rear Adm. Samuel Robinson, the BuShips chief, formally requested NDRC assistance. Bush initially asked Jewett's Division C, the NDRC's Communication and Transportation Division, to take up the challenge. After the creation of the Office of Scientific Research and Development (OSRD) on 28 June 1941, the subsurface warfare problem became the focus of Division C's successor, NDRC Division 6, headed by John T. Tate of the University of Minnesota. The OSRD assumed the role of parent agency for the NDRC, guiding many of the committee's scientific developments through the prototype phase and into production. Thus, by mid-1941, Iselin and other scientists determined to investigate sound in the sea had complete institutional support both from BuShips and the NDRC divisions responsible for all tasks related to undersea warfare.

As the Colpitts Committee investigated and its conclusions took effect within the navy and NDRC between December, 1940, and April, 1941, the work on underwater sound and submarine detection continued at WHOI. On 1 February, Iselin and Ewing produced the final draft of their pioneering work, "Sound Transmission in Seawater." This study provided the context for much of their subsequent work and provided the navy's underwater sound schools with the latest view both on the physics of sound in seawater and the oceanography of the ocean's surface layer. Other components of the WHOI program involved conducting tests with the bathythermograph (BT), ensuring that the navy deployed the device on as many ships as possible, and training officers to act as BT observers.

Although he found individuals within the navy's operational forces interested in applying oceanography to combat problems, Iselin felt acutely the isolation of a scientist not yet accepted as an integral part of a war-fighting team. After sending *Atlantis,* under the command of Capt. F. S. McMurray, to Puerto Rico on Cruise III to conduct research with the BT under the direction of WHOI scientist Alfred Woodcock, Iselin himself set out for Key West via Washington, D.C.[11] In a letter to Ewing written from his favorite Washington haunt, the Hay-Adams House, Iselin expressed the realization that reluctant senior naval officers like Bowen and line officers at sea would have to change their ways and attitudes. The WHOI director reported to Ewing that he

> saw Dr. Jewett yesterday morning and had a long talk with him. Apparently about ten days ago Bowen asked him to write a report on what should be done about underwater sound. This was partly the

result of your agitation but mainly the result of the report of the Colperts [*sic*] Committee. They had reported that the sound schools were hopeless and that nobody in the navy knew anything about underwater sound. Jewett wrote an 11 page letter to the secretary of the navy pointing out the danger to this country of submarines and advocating that a properly organized attack be made on the whole problem, for which our work is fundamental. He suggested that a senior officer be put in charge with power to act and make changes in equipment and technique. As Jewett expressed it "this is a very hot potato." The navy will have to do something. They just can't forget the whole business.

While working with the BT in the Caribbean and composing the final draft of their sound-transmission report, Iselin and Ewing discovered that the Colpitts Committee had good reason for its doubts about the navy's sound schools. Iselin observed that introducing fleet officers to scientific tools differed significantly from working with those attached to more scientifically informed commands like Hydro or the NRL. The latter understood the potential of employing the characteristics of the ocean environment as a shield, or perhaps even a lethal weapon. For the former, only steel, explosives, and seamanship provided protection or the means to attack. As yet, many saw no cause to combine the two views.

After a few days at Key West's venerable Casa Marina Hotel, Iselin joined Lt. Comdr. William Pryor, commander of the USS *Semmes* and only recently liberated from the sound desk at BuShips, to train ASW sonar operators for the Key West sound school. On 8 March the WHOI director penned a revealing letter to Maurice Ewing lamenting the awkward situation at the school. Commander Pryor and the students cooperated and listened, but Iselin knew that his message did not get across. For them, discovering the submerged vessel seemed less a complicated problem in sound, seawater, and ordnance than a virtuoso ship-handling performance by the captain. The short ranges employed in test runs off Key West assured a quick and familiar echo for the inexperienced operators, and many officers still emphasized discovering a submarine's initial position by observing the track of its torpedoes. For the entire time it worked with Iselin off Key West, *Semmes* never started its approach from more than three thousand yards. This scenario hardly emulated combat conditions. Dönitz's U-boats would rarely give the U.S. Navy the opportunity of finding them so close or in shallow water, and following a torpedo track was a dangerous gambit, even if the commander's ship-handling skills were above average. In addition, the scenario employed by the school prevented an adequate demonstration of the BT's capabilities because the exercise area did not provide water deep enough to demonstrate the ocean's stratification and the possibilities for both the hunter and the submarine. Iselin drew some interesting and understandable conclusions from his experience in a letter for Ewing: "All of this would leave one most discouraged with the Navy were it not for the fact that very obviously they are all able and know their business. The answer seems to be that none of the officers are interested in improving existing techniques. They are only keen on perfecting the execution of the present limited employment of sonic equipment. They have not been ordered to do research. They have been told to train and drill the men in the established procedure. The "Semmes" is only experimental in that she tries out new auxiliary gadgets. She does not try out new methods of locating a sub."

Iselin placed his hope in the officers. He and Ewing, who also worked between Puerto Rico and Key West with *Semmes* between February and July, 1941, would have to convince and educate if they expected to initiate a change of mind among those giving orders and writing doctrine. They had to convert even someone as informed and interested in oceanography as Lieutenant Commander Pryor, who typically tended to leave the science behind once he went on board ship to train for war. Iselin commented in a communication to Ewing that Pryor always ordered his crew to lower the BT, but he never made any use of the data the instrument provided. The sound school's new commander, Capt. Charles D. Swain, seemed more determined to integrate the BT into the curriculum. Iselin hoped that he might persuade senior officers like Swain to reshape the school's program and conduct their training in more realistic hunting conditions, which would, in turn, better display the BT's advantages.

Iselin's 1941 experience at Key West taught him an important lesson about cultural resistance within the navy to any change that might alter the traditional ways of fighting a naval war. For more than a year, Iselin had argued, debated, informed, and proposed with varying degrees of success. Finally, his Key West experience led him to the conclusion that, as a civilian, he had exhausted his initiatives. For better or worse, naval officers would have to take the next step. He said it best in a letter to Ewing: "It seems to me that we have learnt one important thing during the past months. We are a research outfit and the Navy is not. We had better stick to research and not try to be too practical. In other words, from now on I believe we should put all our effort into investigating sound transmission in sea water and wait for the Navy to ask us for help in their practical problems. We have given them the report and a few instruments. That will be enough until they beg us for more."

In the end, those facing the U-boat threat would have to make the leap of faith and welcome oceanographers as part of their team. From what they had read about the European war and observed in working with the fleet, both Iselin and Ewing knew they would not have long to wait.[12]

Meanwhile, NDRC Division C and BuShips proposed a formal national organization to address the subsurface-warfare problem. As early as February, 1941, Frank Jewett suggested that any formal organization must include an oceanographic laboratory to provide the pure scientific knowledge and expertise necessary for any ASW program to succeed. As the United States witnessed increasing German U-boat success in the North Atlantic, completely unaware that only the code-breaking efforts of the British provided a slim margin of safety, the need to confront this menace daily grew more important. For Jewett, Bush, Iselin, Ewing, Vine, and others, the U-boat became America's problem long before Pearl Harbor. On 17 February, Jewett outlined the formal objectives for an ASW oceanography laboratory. Scientists with a talent for engineering like Allyn Vine and J. Lamar Worzel should develop instruments capable of sound analysis from the lowest to the highest frequencies on the spectrum. They had to examine scores of ships and submarines to determine their sound signatures and recommend methods for noise reduction. Studies of attenuation and reverberation as affected by temperature, salinity, and density would help fill out the training of sonar operators once the laboratory turned the knowledge into instructional materials. For Jewett, the laboratory component of the underwater sound organization not only provided much of the raw knowledge but also helped apply these insights to warfare.[13]

Jewett's vision of a laboratory would dramatically affect the future of WHOI and play an important role in the navy's response to the Colpitts Report. On 10 April 1941, the chief of BuShips, Admiral Robinson, asked the NDRC to recommend a comprehensive approach to the submarine-detection problem designed specifically to address the shortcomings emphasized by Colpitts and his colleagues. Eight days later the NDRC produced a recommendation that the admiral immediately began to implement. Frank Jewett, chairman of Division C, composed the requested NDRC response to the U-boat menace, assisted by John Tate. Since Tate had left for the United Kingdom a few days earlier to take stock of the Royal Navy's ASW effort, Jewett drafted Edwin Colpitts himself to initiate the program. Colpitts would eventually play the role of Tate's assistant director in Division 6.[14]

The Five-Legged Creature

The 18 April plan described a five-branch organization that NDRC and the navy would have to invent quickly with very few existing resources. Jewett, Tate, and their colleagues, feeling a sense of urgency, designed and implemented this amazing five-component program in roughly six weeks with a budget of only $3 million.

A central control group would administer the entire operation from offices at 172 Fulton Street in New York City. The group consisted of Tate, Colpitts, and Elmer Hutchisson as technical aide. They received assistance from a number of scientists who acted as regular consultants. They included Tate's choice of Gaylord Harnwell and H. Nyquist, as well as William Coolidge, Ernest O. Lawrence, Max Mason, and George B. Pelgram, appointed to Division 6 by Vannevar Bush.

Two other legs of this American ASW creature touched both the East and West Coasts. On 16 May 1941 the navy broke ground at Fort Trumbell near New London, Connecticut, for the U.S. Navy Underwater Sound Laboratory (USNUSL). This facility would serve as the organization's Atlantic Coast laboratory. Lieutenant J. B. Knight assumed command of the finished facility and Columbia University directed the scientific and technical staff under the immediate supervision of T. E. Shea, vice president of Electrical Research Products, Incorporated. Columbia signed a contract with the NDRC to establish a Division of War Research to direct the program at USNUSL among other defense projects. Both the destroyer *Semmes* and the USS *Sardonyx,* formerly the yacht *Queen Anne,* docked at the USNUSL pier. The $175,000 needed to build the laboratory, dredge the piers, and purchase the *Queen Anne* came from Admiral Robinson at BuShips.

The Pacific Coast laboratory formed a new addition to the USNRSL established at Point Loma the previous year. The University of California signed an agreement with the NDRC similar to that adopted by Columbia when it established the UCDWR at the Point Loma laboratory to work on ASW and prosubmarine research for the duration of the war. Captain Wilbur J. Ruble, director of the USNRSL, provided a point of contact with the navy. Otherwise, Vern O. Knudsen of UCLA directed the facility's research for the university and was succeeded on 1 April 1942 by Gaylord Harnwell. The UCDWR also solicited support in oceanography from Harald Sverdrup and his staff at Scripps. The oceanographers employed the Scripps research vessel *E. W. Scripps,* and the navy purchased the 135-foot motor ship *Stranger* and renamed it USS *Jasper* before assigning it to the UCDWR.

Yet another component of the new ASW organization supported both of these laboratory activities with consultation, instrumentation, and materials. Classified as "other cooperating laboratories," this limb initiated or extended contracts with a variety of laboratories and companies, like Western Electric, Radio Corporation of America (RCA), General Electric, and Harvard's new underwater sound laboratory under the direction of Frederick V. Hunt. The NDRC approved the arrangement on 30 June and "other laboratories" received the imprimatur of the newly created OSRD on 18 July 1941.[15]

The all-important oceanographic laboratory described by Frank Jewett the previous February formed the final leg of the ASW organization created by the navy and NDRC in 1941. Now Iselin's work over the past year paid off. Woods Hole filled this void with a commission retroactive to October, 1940. Oceanography had found its niche in preparations for war, but in this case Iselin had virtually defined the role it would play and cast his own institution as the headliner. Doubtless, he owed a great debt to his patron, Frank Jewett. However, the WHOI director's early and tireless activity in Washington and Woods Hole went a long way toward determining the most useful tasks for oceanography even before the NDRC had time to organize itself. Then, the talents of people like Vine and Ewing demonstrated that the small institution on Cape Cod could assume a potentially vital responsibility if war came to the United States. "This network of laboratories and facilities had for its objective, the most complete investigation possible of all the factors and phenomena involved in the accurate detection of submerged or partially submerged submarines and in anti-submarine devices," wrote Jewett. "While the detection of submarines operating as surface craft permits the use of physical phenomena other than those adaptable to under water detection (e.g., micro-waves), this memorandum is concerned primarily with the latter problem."[16]

Circumstances changed completely on 7 December 1941. The United States instantly became a partner in the war against the Axis and confronted scores of critical strategic and tactical problems in two oceans before the smoke cleared from Pearl Harbor's Battleship Row. The navy faced a formidable Japanese naval force in the Pacific, and in the Atlantic Adm. Karl Dönitz launched Operation Drumbeat in early 1942 against seagoing commerce along the East Coast of the United States. It quickly became obvious that interwar politics and policies had delayed the development and fabrication of modern ships and weapons in sufficient numbers to respond as quickly as the two-ocean threat demanded. Surprised and angered, U.S. naval officers did not want laboratory science. They wanted answers with combat applications.

As never before, naval culture had to solicit and welcome unusual allies. Until that time, officers trained to fight a naval war only sought the help of the scientific community in response to unusual circumstances, such as the advent of the U-boat during the Great War. While agencies like the NRL and Hydro regularly worked with the civilian scientific community and reaped great profits for all concerned, naval tradition and training denied scientists a significant role in questions of strategy, tactics, and planning. Thus, Commander Pryor's instinctive decision to depend upon ordnance and ship handling to the exclusion of scientific observations accurately reflected the attitude of his colleagues—officers who were far less familiar than the former occupant of BuShips's sound desk with the opportunities offered by an intimacy with the ocean environment.

In 1942 and 1943, involvement in war opened the eyes of the operational navy even further to the edge oceanography might provide in battle. These same circumstances motivated scientists like Columbus Iselin, Roger Revelle, and Maurice Ewing to command the navy's attention and weave the ocean sciences into the fabric of naval war. At the same time, wartime science-on-demand opened the eyes of many scientists and technicians to the possibilities and profound problems in working for the government. The obstacles and compromises proved too great for some scientists. Others, like Maurice Ewing, could overlook the trials of doing scientific research for the armed forces in exchange for appealing new fields of study, interesting projects, and generous sources of funding for the future.

The Navy Department under Secretary Frank Knox brought its officers into contact with the scientific community outside the naval laboratories in two ways. The department already had a working relationship with Vannevar Bush's OSRD, and cooperation initiated in 1940 was bearing fruit. In evaluating the response from some quarters to the Colpitts Report, Knox decided that a truly productive relationship between civilian science and industry required yet another step. He asked Naval Academy graduate and talented aeronautical engineer Jerome Hunsaker of MIT to evaluate the structure of research activities within the navy. In what ways should the department reorganize in order to take full advantage of civilian and naval scientific talent?

Hunsaker's advice took practical form with General Order 150, which created the Office of the Coordinator of Research and Development (OCRD) on 12 July 1941. At Hunsaker's urging, Knox altered the order on 1 December 1941 to permit an officer rather than a civilian to take charge of this office. The MIT engineer felt strongly that the navy needed a direct way of making its requirements known to the civilian scientific community. Knox realized the wisdom of this given Admiral Bowen's reaction to the Colpitts Report and his strong opposition to civilian influence over naval research. This outlook would no more promote productive relations between the navy and civilian science than the opposite, complete civilian control over wartime research without benefit of naval advice. In recommending an officer, Hunsaker perceived a need for a naval counterpoint to Vannevar Bush. The job required a capable and flexible naval professional who would further enhance the navy's relationship with OSRD. In short, wartime research and development needed a naval officer who understood the culture and could easily work within it. This link would serve both the naval and civilian side of the effort to utilize science for war and would set a valuable precedent for the days when global war no longer plagued humanity.[17]

Lieutenant James Parker, one of roughly a hundred naval reserve officers selected for the coordinator's office by Hunsaker and his assistant, Capt. Lybrand Smith, described their activities as a clearinghouse "for all technical and scientific undertakings of the Navy Department." The first coordinator of research and development, Rear Adm. Julius A. Furer, took charge on 13 December 1941 and immediately launched into the role of communicator and facilitator. He strove to eliminate duplication in the research applied to naval problems, worked to ensure that information reached every scientist and officer who needed to know regardless of bureau boundaries, and sought out foreign sources of information from both allies and adversaries, intelligence sources permitting.

Admiral Furer's office also assumed the role of naval liaison with the OSRD/NDRC. His staff worked with Bush's office to focus scientific effort on the most pressing wartime problems. Captain Smith assumed the same supporting role for Furer and served as the liaison officer between the navy and the NDRC. Smith, an exceptional engineer, held a doctor of science degree from American University in Washington, D.C., and had research experience at the David Taylor Model Basin.[18]

Furer and Smith did not direct scientific programs. The coordinator had the authority to develop an idea or study a problem of concern to the operational forces by giving it priority and visibility. He could also direct it to the appropriate division of the NDRC or the naval laboratories.

On the West Coast: Scripps and UCDWR

As a part of the University of California since 1912, defense research began for Scripps much later than it did for WHOI. On 20 August 1941, an informal board of NDRC representatives and the navy's Point Loma USNRSL met to compose an oceanographic program for the West Coast at UCDWR to accomplish the goals set forth in the Jewett-Tate plan of 18 April.[19]

The group that assembled at Point Loma in August consisted of Capt. Wilbur Ruble as commanding officer of the Radio and Sound Lab, and Lt. (jg) Roger Revelle, USNR, acting as his fellow naval member and committee recorder. Vern Knudsen and Scripps director Harald Sverdrup participated as the NDRC members. These discussions preceded Sverdrup's difficulties over his security clearance. As a Norwegian national, the navy eventually viewed Sverdrup as valuable but vulnerable. He still had family in occupied Norway, and this placed him in a situation the navy soon found unacceptable. For the moment, however, he acted as the leading oceanographer on the West Coast and participated in early discussions on the role Scripps, the UCDWR, and Pacific oceanography would play for the navy and NDRC.

In the months preceding this gathering, the Scripps staff collected a significant amount of ready data for various strategically significant parts of the Pacific Ocean. During Thomas Wayland Vaughan's interwar directorship, the institution conducted a considerable amount of research while working in cooperation with Hydro in the Pacific. As war came uncomfortably close, Roger Revelle and other Scripps scientists assembled material on ocean-bottom sediments, temperature, salinity, SDF soundings, and any other data that could possibly prove useful.

This data collection effort naturally generated a great deal of information, but the size of the Pacific and the legacy of national and individual research in this immense part of the world ocean demonstrated the extent to which the Pacific remained a mystery. According to a memorandum submitted to Captain Ruble on 15 July 1941, the navy lacked essential data on Central America's western coast north of Panama, the Gulf of Alaska, and the China Sea west of the Philippine Islands. When Hydro plotted the available information on Pacific maps, dangerous gaps in scientific knowledge also appeared in the strategically significant waters north of Hawaii along the route eventually taken by the Japanese Pearl Harbor attack force and the region just south of the Aleutians, where the Japanese precipitated the most decisive battle of the Pacific war during their 1942 Midway-Aleutians campaign. Naval authorities responsible for coastal defense did not fare any better, detecting a disturbing shortage of vital data on the San Francisco area, Dutch Harbor, and Manila. In spite of important individual and national efforts to explore this expansive body of water, international cooperative efforts aimed at assembling the personnel and resources to understand the Pacific more comprehensively began only after the Great War. During the 1920s, scientists initiated collaborative research that might have begun filling the void only to find their efforts reduced to an agonizing struggle by the Great Depression. They thus managed only a modest return for a considerable effort initiated barely a decade before Bush created the NDRC.

After examining the available resources and knowledge, a UCDWR committee suggested that West Coast scientists and naval officers emulate the general program followed by WHOI and the USNUSL on the other side of the continent. Available underwater sound information on critical areas of the Pacific proved particularly uneven, a source of naval concern, and a very high priority. They found that "in the limited regions where data are available on underwater sound conditions, some areas are notoriously poor, some are fair and some are good. In major portions of the Pacific containing some of the most probable strategic and tactical operating areas, information on sound conditions is non-existent or too meager to be of value."

As part of its 20 August meeting, the committee promptly adopted a plan of action. The navy and the UCDWR would begin by adding to Hydro pilot charts and weather maps of potentially critical Pacific operating areas information on underwater sound conditions, and, where possible, data on the bottom composition that might affect echo ranging and listening. These charts would serve the navy in planning submarine and ASW assignments, in providing essential information for deploying destroyer escort screens, and in determining the best routing alternatives for merchant convoys and naval task forces. With adequate inshore information for the American mainland and important island groups, the navy could choose or design the most suitable sound and defensive equipment for coastlines and key naval bases. More comprehensive information would also serve the sound schools in training and qualifying submarine and surface sonar operators. The gaps in knowledge essential for the USNRSL and the UCDWR to develop new systems and training services dictated the type of oceanographic research pursued by the scientists and naval officers at Point Loma.

Emphasizing the responsibility shared by the navy and the NDRC at Point Loma, Sverdrup, Ruble, Revelle, and Knudsen insisted on the creation of a comprehensive BT data collection system on the West Coast comparable to the one employed at Woods Hole. In each of the West Coast naval districts a group of officers and civilian scientists or technicians acquired BTs and made sure they found their way on board naval and merchant ships. The UCDWR helped by collecting and processing data and by organizing a program of BT observer instruction in imitation of the one already underway at WHOI. Scripps's Eugene C. LaFond, a Sverdrup student, took charge of the BT data collection and training program.[20]

To supplement BT collection and the publication of this data on charts and in hydrographic intelligence reports, Francis Shepard of the University of Illinois, a frequent researcher at Scripps, contributed his knowledge of sediments and the character of the sea bottom to essential maps and charts. He did prewar research in this field at La Jolla funded by a grant from the Geological Society of America. One of his students, Kenneth O. Emery, later played an important part in providing ocean-bottom composition data for charts used by American submariners in Japanese regional waters.[21]

In addition to the work at Point Loma, some war research also took place at La Jolla. After a prewar stint in the army, Scripps student Walter Munk investigated both the temperature microstructure of the ocean and wave prediction. Munk came to Scripps after his undergraduate years at CalTech to do graduate work in oceanography and geophysics with Harald Sverdrup. In the first weeks after America's entry into the war, this promising student, now a world leader in submarine geophysics, investigated the fine points of the ocean's thermal structure as roughly revealed by the BT. Over a half-century later he recalled: "we worked on underwater sound propagation . . . I remember working on intensity. . . . I did some work in a field that was later on to become very active, namely, the effect of . . . temperature microstructure in the oceans. Our measurements were very coarse, but they showed that the temperature profile is not smooth. There are small thin layers, and I worked on that problem. And later, in a complicated way we became involved in wave prediction, and that's probably the thing that's most interesting. That was actually under [U.S. Army] Air Force auspices."[22]

Naval officials later briefed both Munk and Sverdrup on the general plans for Operation Torch—the Allied invasion of North Africa eventually launched on 8 November 1942. The planners were worried that the height of the breakers on the African shore would overwhelm the landing craft. According to Munk, "they were practicing on some beaches in North Carolina and if the breakers exceeded six feet they had so many . . . broached ships that they went home for the day."

The two Scripps scientists worked with meteorologists to determine the kind of environmental circumstances that might provide the most benign wave conditions for amphibious landings in that region and others. In the process, Munk and his mentor worked out a method of predicting sea swell and surf conditions that significantly contributed to planning the Torch landings in North Africa and the June, 1944, Normandy invasion. Scripps also instituted a program to train naval officers in their sea swell and surf predicting technique so that teams of these trainees and military meteorologists might make on-site evaluations at the request of commanders and their staffs. In the coming year, the submarine and amphibious campaigns in the Pacific would draw heavily on the talent at both Scripps and the UCDWR.[23]

Thus, after the Japanese attacked, prewar West Coast research programs did not entirely vanish, but they certainly yielded to a wide spectrum of navy priorities. A UCDWR oceanographic group, initially led by Scripps's Harald Sverdrup, was committed to oceanographic inquiry in support of the research and development tasks that took center stage at the UCDWR. Besides offshore and harbor surveys that dealt with a variety of seawater and ocean-bottom characteristics, the oceanographic group worked on sound transmission below the surface, with the BT as its primary tool. This demonstrated not only navy priorities, but also the needs of the research and development teams at the Point Loma laboratories as they worked to dramatically improve American sonar systems.

The August, 1941, meeting between Knudsen, Ruble, Sverdrup, and Revelle at Point Loma also produced a proposal for a joint navy and NDRC oceanographic organization patterned after the Naval Aerological Service, to promote interaction and close cooperation between West Coast scientists and serving officers in the fleet. Although the UCDWR's annual report blamed the demise of this idea on the advent of a shooting war on 7 December, it seems more likely that its redundancy proved fatal. The productive relationship between Hydro, WHOI—as the NDRC oceanography laboratory—NDRC

Division 6, and Admiral Furer's office already addressed the need for integration between naval personnel and the scientific community. For West Coast scientists, this organization may have had a disturbing East Coast flavor, but the navy and the NDRC had little interest in creating another agency. Instead they called for completing the job of integration and fulfilling the navy's wartime needs within the system of relationships created by Bush, Jewett, Tate, Iselin, and Furer.[24]

7

The Critical Innovation, 1940–41

Our problem was not to make it necessarily more accurate,
but to make it so that it was ten times more usable, so it could
be used off ships at convoy speeds, so it could go to—so it
could attain reasonable depth.

—ALLYN COLLINS VINE describing the bathythermograph, 27 April
1989

C OLUMBUS ISELIN NOW HAD PROJECTS TO MANAGE AND
new equipment to gather.[1] In a 6 September 1940 letter to Jewett, he
briefly mentioned the fouling project and the possibility of research
money for meteorology, but dwelled at length on a recent trip to the Submarine Signal Company in Boston to discuss sound instrumentation.

Although headquartered on State Street in Boston, Submarine Signal operated a laboratory specializing almost exclusively in underwater sound apparatus at 27 Atlantic Avenue. Submarine Signal had employed Reginald Fessenden during the Great War, sued the navy and Harvey Hayes over the Sonic
Depth Finder in the 1920s, and, with Iselin's visit, sensed something important in the offing. The scientists and engineers at Submarine Signal knew the
WHOI staff had designed something very special and significant. In September, Harold Fay, chairman of the board, and T. R. Madden, the company's
president, welcomed Iselin with open arms. The WHOI director later wrote
to Frank Jewett: "Mr. Fay is terribly interested in the whole scheem [*sic*] and
has turned over to us the full facilities of their shops and engineering experience."[2]

The refinement of the bathythermograph at WHOI lay at the source of Submarine Signal's considerable enthusiasm and Iselin's determination to move ahead on the underwater sound program in the autumn of 1940. Effective ASW and successful submarine operations depended upon an accurate knowledge of ocean temperature and depth that the BT could quickly provide. These two factors, plus a minor role played by salinity, governed the behavior of sound in seawater and thus the performance of sonar in the battle between submarines and ASW vessels. Wise use of this instrument could well provide a decisive advantage.[3]

The development of this critically important device began with an effort in 1936 by MIT's Carl Gustav Rossby to create a prototype device to measure water temperature variation as depth increased. In the midst of groundbreaking Gulf Stream studies, Rossby wanted to examine the effects of wind and current on water movement in the upper ocean layers. Because eddies on the edge of the Gulf Stream brought warm water to the Nantucket region, he sought an accurate way of measuring the movement and characteristics of this water.

Assisted by his South African graduate student, Athelstan Spilhaus, Rossby built an instrument affected in two ways by the natural forces of the ocean. At one end, the device contained a metal bellows mounted inside and parallel to its tubular housing which, as the instrument descended through the ocean, compressed in response to the ever-increasing pressure of depth. The other end contained a bimetal strip that would expand and contract in response to the water temperature. The two metals composing the strip contracted at different rates, causing the strip to move. Rossby attached a stylus to the end of the strip that would inscribe the its motion on a glass microscope slide smoked with skunk oil. As the stylus began to stir, the frame containing the glass slide moved in response to the action of the pressure-activated metal bellows at the other end. The natural characteristics of the ocean thus initiated motion within the instrument that placed a stylus trace on the slide in much the same way that one might connect data points according to the values of X and Y on a graph. The finished product gave the appearance of a small, flat-black glass negative with a clear tracing across its surface.

Only a few months after working with *Semmes* on the discovery of the afternoon effect, Iselin, then WHOI's associate director, convinced Spilhaus to take the BT to sea on the R/V *Atlantis*'s Cruise 70 between 23 and 31 August 1937. Spilhaus and the WHOI team conducted BT tests in the company of a submarine, R. L. Steinburger of the Washington Navy Yard's radio laboratory, and the destroyer *Semmes* to determine the value of the new instrument in submarine detection. Spilhaus later recalled that "sound engineers were attributing failures [to detect submarines] to deficiencies in the sonar equipment, whereas we were trying to convince them that it was the thermal layering of the oceans and the lens-like bending of the sound waves by the thermocline that were responsible for the misses."[4]

The thermocline was that layer of the water column in which the greatest degree of temperature change took place with depth. In some cases the water became warmer, but in most instances the environment became colder as the depth increased. When sound passed through the water, the change in temperature acted like a prism, bending the signal and refocusing it downward or back to the surface. If the sound bent toward the abyss, not only the colder temperatures, but also the pressure of the deep ocean played a role in the course it traveled. With temperature and depth decisively influencing the behavior of sound underwater, the BT could well have a telling effect in ASW or in an attempt to use the ocean environment to shield American submarines from enemy searches.

In October, 1937, only weeks after Spilhaus successfully tested his device, the navy demonstrated its conversion by ordering two of the instruments. At first, Sperry agreed to manufacture the navy's BTs, but a change in the technology and design resulted in the job being given to the Submarine Signal Company. Spilhaus decided to abandon the bimetal technology responsible for the motion of the stylus in favor of a Bourdon tube filled with xylene gas. The Bourdon-tube model provided protection for the stylus, now enclosed within the instrument, from the rush of the water along the BT's cylindrical housing as it descended through the ocean. In the previous design, the bimetal temperature element, which carried the stylus, needed direct contact with the water. In the newer version, an external coil of xylene-filled copper tubing could react to seawater temperature and transmit the required X-axis motion to the stylus attached to the interior-mounted Bourdon tube. In this way, the entire mechanism remained shielded from any rushing water that might induce unwanted vibration. Although Submarine Signal did not at first express enthusiasm at the prospect of manufacturing the BT, it did so to retain its reputation as one of the leading submarine instrument companies and to secure a potentially promising patent. On 10 August 1938, Submarine Signal took out the BT patent in Spilhaus's name, but the company received the rights to the design.[5]

A few months after Spilhaus embarked on the *Atlantis* to test the BT during Cruise 84 to the Gulf Stream in July, 1939, Maurice Ewing moved to Woods Hole with his students Allyn Collins Vine and J. Lamar Worzel. Having the Lehigh physicist and his corps of students at Woods Hole while he planned the underwater sound project for WHOI in 1940 gave Iselin a distinct advantage. The behavior of sound underwater was a critical element in the refraction methodology Ewing adopted from land-based geologists for use at sea. His team already had extensive experience working in Woods Hole and aboard the *Atlantis* with their explosives and free-floating submarine seismograph systems.

Ewing became Iselin's natural ally in soliciting and executing the early sound work for the navy that began with the NDRC sound research that Iselin christened WHOI Project Number 2.[6]

The Rossby-Spilhaus BT became an essential tool for this group—one that would figure prominently both in their future personal research and in their effort to provide the navy with an unprecedented knowledge of the submarine's operating environment. In spite of the instrument's promise, in the autumn of 1940 its construction remained impractical. It did not operate well if the deploying vessel moved at high speed, and it did not descend fast enough through the water to ensure a reading at the desired depth. After the WHOI sound project started with the NDRC's blessing on 1 October 1940, Allyn Vine took the BT in hand with an eye toward making it more functional and durable and thus more useful in both peacetime geophysics and antisubmarine warfare. Before his death in January, 1994, Vine would become a translator of world reputation, with the insights and knowledge of a physicist and a natural talent for engineering. Recalling the BT problem, Vine commented that

> we were looking very seriously at what is the overall principal effect for which temperature was obviously good enough? Then what might it turn out that we would really need to know in much better detail to make salinity corrections and so forth, in case we needed more refined measurements? So after a few months of looking at the data and considering the alternatives and the extra cost, why, it was decided the basic bathythermograph was the kind of thing that made sense to put on ships. Our problem was not to make it necessarily more accurate, but to make it so that it was ten times more usable, so it could be used off ships at convoy speeds, so it could go to—so it could attain reasonable depth. At that time, I believe our first one went 250 feet, then 450 feet, and these could be done at speeds of 15 to 20 knots. We also put in some finesse point features to make it a little bit more precise and so forth. Then we designed it—another thing, we designed it so that they basically threw it over the side with a fine wire on it and let it fall free, and then by the time it got to depth, it got to the end of the wire, and then they could hoist it in. Like most things, it was part of a system. We had a big debate of whether it should be—the bathythermograph should read in the English or metric system. The oceanographers in Washington thought it should be in the metric system, and the oceanographers here at Woods Hole, we felt that the poor kids from Kansas that were dragged into the Navy shouldn't have to learn the metric system while they were seasick so we made the first 200 in English sys-

tem, and by that time the die was cast, and it all went English. As a consequence, however, I was brought up thinking about the ocean temperatures from the top down in English system, and the bottom up in the metric system, and so at a depth of a few hundred meters, I still get vertigo![7]

The BT that Vine and Ewing refined at WHOI in the autumn of 1940 proved truly remarkable. Its rather awkward, oversize, cylindrical, and cagelike appearance gave way to Allyn Vine's streamlined projectile armed with a ten-pound brass weight at the tip and fins at the tail to promote a quick descent through the water. Vine's BT retained the Spilhaus bellows in the nose of the instrument as the pressure element, but he altered the temperature element considerably. He rigged a coil or bulb of very thin copper tubing filled with xylene around the fins to play the role of thermometer with sufficient external exposure to the water. Vine then soldered the end of this copper bulb to a flat Bourdon tube shielded inside the slim cylindrical instrument housing. The tube coiled like a watch spring around an axle passing through the diameter of the housing. Reacting to the temperature change transmitted by the external copper tubing, the xylene expanded within the tube, initiating a horizontal motion at the coil's outer tip, which held the stylus. As in the Spilhaus version, this provided the BT's X-axis movement.

To ensure the proper temperature sensitivity of the Bourdon tube, Ewing and Vine instructed the ship's technicians to make sure that, upon deployment but before descent, the BT housing flooded in the surface layer adjacent to the ship so the temperature elements could acclimate to the ocean environment. Unfortunately, this trapped ambient surface water inside the instrument, causing temperature conflicts upon descent and once again introduced excessive stylus oscillation, distorting the trace on the smoked slide. The vibration problem had earlier prompted Spilhaus to suggest that an internally mounted Bourdon tube should replace the bimetal temperature element of earlier models. Vine had followed these general suggestions, but the difficulties reemerged with Ewing's deployment technique. With the need for internal flooding, some of the oscillation characteristics remained, as well as the nagging temperature differences between the residual surface water inside the BT and the external seawater the scientists needed to measure.

In his 1940 work on the BT, Vine introduced one of his "finesse point features" to solve both remaining problems. He combined the old bimetallic strip technology with the advantages of the internally mounted Bourdon tube. Working in the WHOI machine shop, Vine attached a coiled bimetal strip to the axle holding in place the Bourdon tube attached to the external copper tubing. He also soldered the arm that held the stylus at its tip to this strip. Instead of permitting it to react to the water flooding the instrument housing in the same way as the Bourdon tube, Vine mounted his bimetal strip to ensure that it applied force in the opposite direction. Both the tube and the strip would react to the internal water in the same way, but Vine's design had them mounted on the same axle working against one another. This virtually eliminated the oscillation of the stylus arm as water passed through the housing and completely nullified the effect of the internal water temperature. The internal water would simply initiate two motions that would cancel each other. Only the temperature conveyed from the external coil of copper tubing to the Bourdon tube would actually inaugurate a restrained motion that would rotate the axle that held the tube and strip, moving the stylus arm and providing a trace on the smoked-glass slide. This mechanism provided an accurate external temperature measurement to complement the pressure effect provided by the bellows at the instrument's opposite end. The bimetal coiled strip acted as a reverse response compensator to eliminate instability and error.[8]

Ewing, Vine, and Worzel manufactured the first seventy-five instruments themselves in the Woods Hole machine shop and demonstrated their value in the underwater sound program. In recalling the simplicity of the Vine design, Ewing noted in a letter to Capt. E. John Long more than twenty years later that the Vine BT was not simply more accurate but more durable and easier to use: "our device could be handled like a sounding lid is handled. It could be put into the water on a thin line, allowed to fall freely by paying out lines to make up for the motion of the ship, and then when the brake is applied, the towing forces would cause the instruments to rise almost to the surface and the reeling in of the wire simply brought the instrument more or less along the surface to the ship."[9]

The new BT descended to the desired depth more quickly thanks to its sleek hydrodynamic shape, and the elegance and simplicity of Vine's design and engineering guaranteed unprecedented accuracy and low production costs. Ewing estimated that a team of builders would cost $1,100 per month and supplies and production overhead would cost another $600. When he reached the bottom line, the navy paid only $70 for each BT, which required a total assembly time of sixteen hours per instrument. Woods Hole began using the instrument immediately. By March, 1941, Vine and company began shipping BTs out to Sverdrup at Scripps to begin the West Coast program.[10] In unpublished memoir notes that Iselin prepared long after the war, the WHOI director recalled the retrospective significance placed on the institution's refinement of the BT when the war ended. Despite the welcome praise lavished on WHOI, he knew well where the credit lay: "I received the Medal of Merit from Harry Truman for this development; it should have gone to Allyn Vine."[11]

Armed with the new BT and WHOI's commission from the navy and the NDRC, Iselin and Ewing began writing a report on the transmission of sound underwater that the navy needed at its underwater sound schools. They finished the preliminary version in less than two months. Iselin also entered into discussions with the navy to continue the research program at sea. On 11 December 1940 he reported to Jewett on a meeting in Washington, D.C., held just a few days earlier. In consultation with Harvey Hayes, R. L. Steinburger, and Lieutenant Commander Pryor at BuShips's "sound desk," Iselin placed the final touches on research plans for the *Atlantis* and USS *Semmes* for the next few months. The WHOI director brought copies of the Iselin-Ewing preliminary report on underwater sound with him to the gathering. The first two copies went to the NRL, Pryor and Jewett received copies, and Woods Hole provided others for distribution. With Iselin at the helm and receiving support from the navy and the NDRC, the underwater sound program proceeded rapidly despite America's neutrality.[12]

From 1940 onward, work with the BT took priority at Woods Hole. Robert W. King of Jewett's Bell Laboratories initially directed the WHOI effort in submarine detection by sound, but a subtle change in emphasis at Woods Hole set this work on a different track.[13] King used explosive charges and the characteristics of the water to test the feasibility of using sound to triangulate on the hidden submarine. While this may have shown some promise, the efforts of Iselin, Ewing, Vine, Worzel, and others with the BT sought to employ the properties of the water as the primary ASW weapon. With the BT providing a profile of the water column in the search area, the warships hunting a U-boat could better estimate the options open to the German commander. This would focus the search sufficiently to permit effective echolocation using hull-mounted transducers rather than explosives as the sound source. The difference was subtle, but the former approach made use of more passive techniques until the final search phase and filled an already noisy ocean with fewer extraneous sounds.

Thus, by early 1941, the preferred WHOI approach to ASW required placing BTs on as many surface ships—merchant, naval, or Coast Guard—as possible and training officers in the instrument's proper use. Iselin started his appeal along these lines even before the navy's ASW organization took shape in the summer of 1941. He managed to get BTs on board destroyers, destroyer escorts, and some navy transports by the spring of 1941.

Back at Woods Hole, Vine and Ewing kept refining the instrument. They proceeded to develop a circular slide rule to speed interpretation of the BT readings. Experienced BT readers could determine the effective sonar range in search area water without any calculation. The indications of temperature and depth supplied by the BT provided sufficient data. Inexperienced operators had to employ a system of calculations that would result in a sound-ray diagram—

a paper rendition of an image similar to that observed on an oscilloscope—that would provide an idea of the current effective range of the sound equipment on board. Even though the calculations did not take too much time, many officers expressed their desire for a quicker alternative. The circular slide rule spared them the calculations. Like the BT, WHOI technicians fabricated these precision instruments in the Woods Hole machine shop.

The WHOI team employed the BT to discover a few revealing and extremely valuable regularities in the undersea environment. For example, a natural or mechanical sound introduced into the ocean will always travel toward the region of lower velocity. While this hardly seems tactically important on first hearing, more careful consideration reveals this seemingly modest statement as truly powerful knowledge. Sound bends as it travels into the depths in response to both temperature and pressure. Once a signal reaches six hundred feet, for example, its speed has increased by eleven feet per second over its surface rate. This effect derives from pressure alone. A consideration of temperature reveals an even greater performance variation. Changing the temperature just ten degrees will alter the velocity of the sound signal five times as much as the pressure of the ocean at six hundred feet.

These barely interesting observable facts became far more significant when BT data revealed the combined effect of pressure and temperature. In a very common situation, where temperature decreased with depth, a phenomenon described scientifically as a negative gradient, the signals from navy sonar curved sharply downward immediately after transmission, leaving "shadow zones" both inside and outside the broad arc made by the sound path. While the angle of the curve and the velocity of the sound varied according to water conditions measurable by the BT, the consequences remained the same. Without a BT profile of the immediate area, a U-boat or Japanese submarine might move into the inner or outer shadow zone and go completely undetected with possibly tragic consequences.

Going one step farther, the BT and the accompanying circular slide rule provided the data and calculations necessary to approximate what became known as "assured range" (AR).[14] Assured range gave the commanding officer of an ASW ship the range between his sonar projector and the beginning of the outer shadow zone. Within this sector he could, with some certainty, discover the location of an enemy submarine by active echo ranging. A BuShips publication from June, 1942, described AR as the "maximum range for a target located at depth where the outer shadow zone is closest to the [sonar] projector (the best depth for avoiding detection)."[15]

Lest this sound too easy or formulaic, as one might expect, varying ocean conditions and different types of shipboard sound equipment changed the AR and made the estimation process difficult. If the ASW vessel's sound projector did not point in the direction of expected contacts, the likelihood of detection naturally declined. A momentary loss of contact would also ruin the prospect of calculating an AR. In addition, nature played havoc with man's best efforts. In heavy seas, a sonar operator might find it difficult or impossible to keep the sound projector in a fixed position. When searching in shallow water, excessive reverberation easily masked a true echo, giving the submerged enemy a distinct advantage.[16] Save for the possibility of defective equipment and ineffective operating procedures, these obstacles to calculating a true AR usually resulted in estimated ranges less than the actual assured range. In rare cases, surface vessels detected submarines beyond the calculated AR if the latter slipped to a depth that placed them in the outer reaches of the sonar curve. Taking up position just beyond the AR and at the same depth as the surface ship's sonar projector provided the safest place for a submarine to hide and presented the greatest threat to ships on the surface. Solutions to the detection and evasion problem and their tactical implementation remained a challenging dilemma because applied science still had to contend with nature's capriciousness. The BT helped the navy cope more effectively with this dilemma by providing an important tool that improved the ASW officer's chances of finding his quarry and promised the submariner better ways to evade a pursuer.[17]

The value of experienced BT observers rose concurrently with the significance of the BT to undersea warfare. If observers executed their task in a professional, well-informed manner, then those who commanded the ships would accept them, value the instrument more highly, destroy U-boats in greater numbers, and increase the general knowledge of the oceans in the process. In the months before Pearl Harbor, Columbus Iselin also hoped that warships in the eastern Atlantic would contribute BT readings to permit comprehensive charts and publications for use by shipmasters across the entire ocean.

In the summer of 1941, as the Jewett-Tate ASW network went to work, Iselin finally convinced the navy to send freshly minted ensigns to Woods Hole for training as BT observers. The first group arrived on 15 August and trained through mid-September. Two of the graduates were sent to the sound schools in San Diego and Key West, and the others were assigned to ships.

With these developments, Iselin knew that he and his colleagues had accomplished several important goals at once. Although by no means as proficient as scientists, the new BT observers would still help the fleet gain experience and knowledge by accumulating valuable data during Atlantic patrols and escort duty. His persuasive arguments, in this case illustrated by the ingenious practicality of Allyn Vine's BT, helped the WHOI director break down naval cultural barriers and initiate meaningful dialogue. As he noted in the spring of 1942: "a more immediate and important consequence of any such plan whereby bathythermographs might be put into use on patrol vessels would be the opportunity for commanders to gain confidence in this method as a means for determining the effective echo range. At present only a few officers understand the practical potentialities of this instrument and its is believed that the technique of its use is now sufficiently developed so that with proper supervision it can rapidly become a most effective anti-submarine tool."[18]

Oceanographers at both Woods Hole and the UCDWR spent many hours teaching navy ensigns the art of reading a BT slide and preserving it for future use. Each instrument needed careful and individual treatment because each possessed its own eccentricities. After the BT's emerged from the manufacturing shop at Woods Hole, the Bristol Company, or Submarine Signal Company, technicians calibrated each one to evaluate its performance and to establish parameters for its use. Although they all worked on the same principles, the variety of manufacturers, the variation in material quality, and the fact that no two instruments performed in exactly the same way required individual calibration. As a consequence, the scientists responsible for instruction and data collection treated every BT as unique. All of the instruments received identification numbers and each came packaged with its own grid. The latter provided the graph-like backdrop that converted a seemingly random trace on a smoked slide into plotted data and usable knowledge.

In learning to use the BT, naval officers also learned basic science, the significance of the device, how to interpret the slide trace, and the expanding place of naval oceanography in warfare, as well as the language spoken by the initiated. Following instructions formulated by Ewing and Vine, a ship would tow a lowered BT for a few moments to acclimate the temperature elements to the water before releasing the brake on the winch to permit the device to drop to the desired depth. This towing phase always placed a horizontal cap, very much like the cross-member on a capital T, at the top of the slide trace. When they placed BT slides against a graph to evaluate the temperature and depth information offered by the trace, the ensigns quickly learned to measure depth from this horizontal cap, whether or not this stylus stroke fell exactly on the grid's zero line. They also had to allow for wide, funnel-shaped traces caused, in rare cases, by stylus vibration induced by ship speeds in excess of twenty knots and the speed with which the BT ascended during retrieval. The pointed end of the funnel would appear at depth and the trace would broaden as the instrument approached the surface. In these cases, the BT reader would have to estimate a line running through the center of the cone equidistant at all points from the outer edges of the trace and use that as his authority on temperature and depth in that location.

During instruction at WHOI or the UCDWR, scientists like Woods Hole's Dean Bumpus introduced naval officers to the importance of isothermal water and BT readings that indicated negative and positive gradients. If the BT trace extended directly downward, parallel to the lines on the left side of the grid indicating depth, then oceanographers called the water through which the ship passed isothermal, suggesting a roughly uniform temperature. This isothermal trace might extend down hundreds of feet or comprise only the first fifty to one hundred feet of the water column through which the BT passed on its way to the desired maximum depth.

Isothermal waters promised minimal refraction, good hunting for ASW vessels, and nightmares for submarine commanders. The latter much preferred the other, more common scenarios revealed by the BT. If, upon retrieval, the BT slide trace sloped to the left indicating a negative gradient, the temperature of the water fell with depth. In this case, refraction would provide the possibility of shadow zones. If the opposite occurred and the trace sloped for a time to the right before swinging left in deeper water, then for a time the temperature rose with depth, betraying a positive gradient. In these latter circumstances the temperature effect and the upward bending of the sonar sound signal provided unusual opportunities for the submarine to exploit water conditions to elude its pursuers.

The BT instructors gave composite names more often identified with wireless radio communication to these scenarios. "Charlie," for example, was the code name for a class of BT patterns that showed a slightly negative gradient continuing well below thirty feet before the trace displayed a sharp bend to the left, indicating the thermocline and a dramatic drop in temperature. Another class of BT readings, called "Mike," exhibited a slight negative gradient near the surface and then a relatively straight trace drop indicating an isothermal layer extending past a depth of more than fifty feet.

At every lowering, the BT operator deployed the instrument, retrieved it after a brief stay at the desired depth, and then removed the slide. After rinsing the slide in fresh water, he fixed the trace by dipping the small glass rectangle into a special lacquer, draining off the excess, and allowing it to dry. A trained officer immediately examined the slide against the proper grid for the information it had to offer. Thereafter, the precious piece of lacquered glass went into a small, wooden storage box along with a record of the date, the BT number, and the serial assigned to that particular reading. The slides and information promptly found their way back to Woods Hole or the UCDWR for further analysis and storage. As information accumulated, the oceanographic community rapidly found itself in possession of a growing and valuable database on those ocean characteristics most critical to both wartime naval oceanography and the civilian peacetime interests of the participating scientists.

Although the composite readings helped to simplify the total picture, naval officers in BT training had to study many natural variables as well as information in naval publications that would offer them a better appreciation of sound conditions at sea. Once they returned to the fleet, these officers helped ensure a steady accumulation of BT data and an unprecedented level of knowledge that enabled scientists to appreciate the ocean as both an ecosystem and a marine battlefield. From Pearl Harbor through the early spring of 1943, WHOI trained roughly thirty officers in these skills. Two of Iselin's field scientists, R. J. McCurdy and Martin Pollak, conducted classes in the Caribbean area. Even though all of the trained officers did not remain in positions that permitted them to continue working with oceanographic data, a sufficient number did to guarantee the regular accumulation of BT data. Under the direction of Eugene LaFond, a similar program took shape at Scripps and the UCDWR employing BTs supplied by Woods Hole and East Coast manufacturers. All the while, Iselin urged Admiral Furer, the navy's research coordinator, to persuade an often reluctant Bureau of Personnel to support this program. With BT production accelerating, and important tactical applications for the data gathered, the fleet needed more officers familiar with the instrument's significance and operation.

Ever the salesman and entrepreneur, Iselin, in a letter drafted in early 1943, noted the many important results of effectively employing BT information. Antisubmarine warfare searches became more precise with the possibility of determining AR. In addition, the BT gave clues to the depth of an enemy submarine, knowledge of the water conditions helped the commanding officer adjust the "ping" emitted by his sonar to the most effective range, and an informed officer would naturally make better use of his crewmen by keeping them alert in the poorest listening conditions.

Iselin also felt that BT information should influence the performance of convoy escort duty. With BT data in hand, convoy planners could give greater weight to any water conditions, seasons, and latitudes that would likely give enemy submarines environmental advantages. With a heightened appreciation of these factors, escort groups could arrange their convoy coverage so each warship might determine the assured range given the region and weather conditions. If circumstances permitted, escort vessels had the potential for complementing each other with joint and overlapping coverage of the underwater approaches to their convoy. In addition, with the BT, an escort group commander would know whether active sonar pinging might betray the convoy by revealing its presence at some distance. Given the German preference for operating in packs, it paid to keep the number of adversaries to a minimum.

Lastly, the WHOI director reminded Furer that BT data also revealed the extent of the afternoon effect as well as the depth of the thermocline. An ASW officer easily recognized the thermocline in any water column when the BT slide trace shifted sharply to the left, as they all did at some point. In that particular ocean layer the temperature changed most dramatically with depth.

There, a friendly submarine or threatening U-boat might silently balance on a layer of deeper, colder, denser water with motors off while trying to escape a pursuer or seize an opportunity. Whether ally or adversary, the submarine commander also had the chance to take advantage of the pronounced refraction effects of this layer to escape detection below the thermocline while he plotted his next move.

While he clearly demonstrated the strategic and tactical value of both the BT and the wartime sound research programs in his draft letter to Furer, Iselin realized that the admiral had long ago come to appreciate the naval significance of the ocean sciences. Nonetheless, he penned this oceanographic litany to one of the converted and said the same things to anyone else who would listen. A brutal sea daily taught naval officers faced with an aggressive enemy the significance of oceanography. Those who rarely probed the character of the navy's natural habitat before the war, minute by minute learned its significance in desperate and costly battles. In cooperation with oceanographers, naval officers could now meet these challenges with greater readiness and confidence. For his part, Iselin repeatedly made his point simply and clearly to everyone he encountered. The ocean would serve neither side in the war. It would merely treat more kindly those who knew it best.[19]

Submarine Bathythermograph

Barely more than one year after Allyn Vine brought his engineering expertise to the Rossby-Spilhaus BT, the commander of submarines in the Atlantic Fleet (ComSubLant), Rear Adm. Richard S. Edwards, had requested that Woods Hole adapt the BT for submarine use. The surface-deployed instrument had quickly proved its worth in ASW, and Edwards wanted the same principles and data working for American submariners. He wondered in his correspondence to Iselin if Ewing, Vine, McCurdy, and the Lehigh group at WHOI could adapt the instrument for prosubmarine purposes. The admiral's request, dated 15 October 1941, had no sooner landed on Iselin's desk at Woods Hole than Ewing, McCurdy, and Vine had formulated a prototype. McCurdy had already spent some time experimenting with a different BT configuration in an effort to create a basic temperature-depth recorder for submarines. By the second week in November, the Lehigh team had Instrument Number 1 installed on board the submarine S-20, and they conducted tests until the end of the month. With the S-20 operating out of New London and winter near at hand, Ewing and his people had only a few weeks for testing their submarine bathythermograph (SBT). The cold soon made the water and shipboard conditions unsuitable, and the technicians at New London removed the SBT from the submarine until the conditions for evaluation improved.[20]

Less than a month later the scene shifted to Key West and warmer climes, demonstrating both the navy's interest in the SBT and the value placed on the benefits the instrument might provide. On 9 January 1942, Instrument Number 2 began its journey to Key West destined for a place in the control room of the submarine R-20. Navy technicians effected the installation on 20 January, but not without significant difficulty. Both this second variation of the SBT and the third iteration, shipped to Key West on 2 February, arrived damaged. Although installed in the R-20, SBT 2 never went through testing. The shipping damage caused a malfunction that permitted excessive hysteresis. This condition produced an objectional delay between thermal stimulation of the instrument and the motion that actually recorded the SBT trace. Technicians at Key West repaired Instrument Number 3 by 18 February and went through a series of satisfactory trials aboard the R-14. In a series of five designs developed in response to Admiral Edward's request, SBTs 1, 2, and finally 5 became the operational prototypes destined for deployment in the submarine fleet.

Totally different in appearance, the early BTs and SBTs employed the same basic technology to measure water temperature and record depth. The course of the temperature component of the SBT began on the outside of the hull. Mounted at first on the conning tower and later low on the hull just above the keel, cylindrical frames welded at right angles to the conning tower exposed the xylene-filled copper tube coiled about each to seawater as the submarine descended. The blister covering the frames allowed the water to flow freely through. Reacting to the ocean temperature as the boat plunged, the warmth detected by the copper tube traveled its length through a packing box and valve in the submarine hull wall and into the SBT mechanism mounted on the overhead in the control room next to the Fathometer. The tube passed into the device from the top at a right angle to the overhead and transferred its temperature to a Bourdon tube, coiled also at this same right angle. As it gently moved in a clockwise rotation, simulating an unwinding movement, the flat Bourdon tube imparted a horizontal motion to two slim shafts attached to its coil. At the other end of the shafts, a single pen, the SBT version of the BT stylus, also moved in the prescribed horizontal fashion. This system provided the X-axis motion.

As with the BT, the ocean pressure contributed the vertical Y-axis action. Another slender copper tube, attached to the pressure line on the depth gauge, transmitted water pressure to another Bourdon tube mounted in front of the first and parallel to the overhead. Its gentle unwinding action in response to the water pressure transmitted through the copper tube initiated vertical movement in a rectangular metal frame. At the outer end of the frame, facing and parallel to the pen, Ewing, McCurdy, and Vine mounted a lightweight metal box that acted as a cardholder. Instead of using smoked-glass slides, the

SBT employed cards already inscribed with the BT temperature-depth graph. Affixed to the face of the box, the card presented a perfectly aligned surface ready to receive the pen inscriptions. As the pen recorded the horizontal motion caused by temperature change, the pressure of depth caused the second Bourdon tube to move the cardholder vertically. The pen thus traced a graph similar to the one recorded on the BT's smoked slide.[21]

Lieutenant Commander Rawson Bennett's BuShips sonar design branch (Code 940) acted as the SBT procurement agent for the navy with Bristol as prime contractor. Lieutenant Butler King Couper, a WHOI-trained BT observer, acted as Bennett's manager for the SBT production process, and Iselin asked Allyn Vine to step in as the WHOI liaison. The SBT proved to be mechanically durable and reliable from 1942 through to the end of the war. Although Admiral Edwards's original request for an SBT simply reflected a desire to make temperature and pressure data available to the submariners, the SBT's utility quickly surpassed ComSubLant's ambitions. On the surface, the problem belonged to the hunter. The ASW officer had to find the submerged enemy and the BT provided valuable insights. When seen from the submariner's perspective, the SBT had even greater significance. The object of the hunt naturally knew its own location and, with an effective hydrophone listening system or passive sonar, very possibly that of his pursuer as well. Equipped with an SBT, the evading submarine would immediately discover the character of the water as it dove. Armed with information on temperature and depth, the crew could then employ the circular refraction slide rule very early to determine the behavior of sound in the submarine's vicinity. Thus, while their pursuer frantically tried to determine its location, the submariners knew the approximate capabilities of their adversary's sonar, his assured range, and the probable location of lifesaving shadow zones. In addition, the SBT trace would illustrate the position of the thermocline and the colder water of increasing density below it. The submarine might thus hide below the thermocline's swiftly changing temperature or balance on a density layer without propulsion sounds or the pumping and blowing cacophony that accompanied ballast adjustments.[22]

Constantly on the lookout for further capabilities made possible by the SBT, Allyn Vine recalled both how the submariners worked the SBT into their daily shipboard routine and the origin of further refinements that made the device a truly potent ally. His comments reflect months of riding submarines in the Pacific and the Atlantic teaching basic science, engineering, tactical applications, and himself learning from the trials faced by his experienced students:

> The biggest difference, of course, was that for the surface vessel it [the BT] was something that you threw over the side and made measurements. The submarine itself dove, so you just strapped the

temperature measuring device under the hull like an external thermometer in a house, and you dove the house. And in normal operating procedures, most submarines, after they got these, most
submarines would make a dive a day, sometimes more than that,
to keep track of what the temperature gradients were so that if they
had to go into action suddenly, they would have an idea of what the
sound conditions might be, how deep they might have to go to get
underneath the layer to sneak away and things like this.

The first ones were mounted up on the conning tower, . . . about
as high up as we could get them, because originally the [submarine]
bathythermograph was put on for sonar purposes. Then when
we . . . were asked to look into the diving characteristics and see how
the measurement gradients affected the diving of the submarine and
so forth, we spent quite a bit of time trying to make densitometers.
This was mostly under [Alfred] Redfield [the WHOI biologist] and
so forth, of half a dozen classical ways of making densitometers. It
usually turned out that the flow passed in the opening would produce Bernoulli effects larger than the difference in density that you
were dealing with. Because we had to do this on a man-of-war and
keep everything small and simple, it began to appear that was not a
very fruitful way of doing it. We already had the bathythermograph
aboard, and because salinity gradients were usually so small on the
high seas that it appeared foolish to put another . . . specialized instrument aboard the submarine when we could just use the [subma-
rine] bathythermograph.[23]

Vine had also worked with Alfred Redfield at Woods Hole on the effect of
density on submarine buoyancy. Hence his quick recognition that careful use
of the SBT precluded the need for yet another device to measure density.

As the navy and civilian science continued to evaluate the SBT's performance, the true beauty of the instrument began to emerge. Its great utility did not
end with extremely valuable temperature, depth, and density data. In the SBT,
submariners also found an important tool that effected buoyancy, ship handling,
and quieting. Balance, trim, and submerged stability kept constant guaranteed
the kind of performance that every submarine commander required. When the
vessel submerged, the ocean pressure changed the submarine's buoyancy by
slightly compressing its hull. In the process, with greater depth, both the ocean
and the submarine became denser. At a certain point these two effects canceled
each other out, but above and below that point, maintaining depth and trim or
even diving deeper became a problem in control due to excess buoyancy or the
absence of a necessary buoyant quality in the boat.

In a remarkable flash of insight, Ewing and Vine devised an elegantly simple way of using the SBT to address the buoyancy problem and in one stroke they gave submariners their solution and eliminated the need for yet another device to measure hull compression. Vine remembered well that late night in the lab at Woods Hole: "Then one of those marvelous, small, coincidental miracles that happen once in a while . . . Maurice Ewing and I were doing this. I remember the night we figured this one out very clearly. On the bathythermograph card, which recorded temperature versus depth, it turned out that if we drew some sloping lines on there that indicated those changes in depth and temperature that compensated for the compression of the hull, such lines could be drawn. By drawing them the proper distance apart and sizing them to the [individual] submarine, we drew lines—just merely drew lines that represented one ton of ballast, and overprinted the cards, and we had the machine already on the submarine, already installed, and it was easy to understand. It was a great piece of luck."[24]

These isoballast lines, as they were called, sloped downward from right to left and the distance between each represented two thousand pounds of ballast water. The value of these modest strokes on the SBT graph card becomes clear in a hypothetical battle scenario. For example, after diving through a surface temperature of seventy-four degrees Fahrenheit, a commanding officer might decide to take up position at approximately 200 feet to evade his pursuers in a water column with a thermocline occurring between 150 and 200 feet. Counting the six isoballast lines across the SBT card from the line roughly congruent with the point of submergence to the line crossing the desired destination would lead the diving officer to flood six tons of water as the descent progressed. Upon arrival at the preferred location, the submarine would have neutral buoyancy and could balance in relative silence with a minimum of machinery noise. Thus, in the time it took for an evasive dive and quick layer and position calculation, the submarine would have readied itself physically to take the best advantage of the environment in that particular location.

With sufficient training and practice, a crew could obtain a multiplicity of services from the SBT. To ensure that the submarine fleet received proper instruction in the instrument's potential, a great many oceanographers—including Allyn Vine, Dean Bumpus, William Schevill, and Alfred Redfield—spent time in New London, Pearl Harbor, and other locations teaching, advising, and actually going to sea with the submarines. Admiral Edwards thus received far more in response to his request than he ever expected.[25]

As the war progressed, scientists and technicians constantly sought to improve oceanographic instruments in response to defense requirements. The BT itself underwent refinements at the hands of Allyn Vine and another WHOI instrument specialist, Ray Deysher. Maurice Ewing volunteered to rework the

BT for 500 feet at the navy's request. However, he warned Bennett at BuShips that the necessary alteration of the device for a 450–500 foot depth would demand the compression of data card information and reduced precision. Woods Hole could manage the transformation with the existing device if the bureau wanted only macromeasurements such as the approximate location of the thermocline. As a result of Ewing's cooperation and because of his sound advice, Bennett then and one year later recommended against the alteration because of the loss of precision. Oceanographers had clearly found a recognized place in the national defense. By 1942–43, the level of mutual trust had improved and effective cooperation brought results in battle and on the home front.

From the beginning, Maurice Ewing was the driving force behind the BT and most of the pioneering research in underwater sound. By any standard he cut an impressive professional figure and demanded the attention of the navy and his peers. While certainly an insightful scientist and leader, Doc's occasionally abrasive manner and relentless work habits did not suit everyone. After working with Ewing during the photography research on board the *Gentian,* physical oceanographer Gordon Riley felt that getting to know Ewing better proved a most memorable part of the experience. While not uncritical of Ewing's manner and prejudices, Riley's insights provide an interesting additional perspective on a remarkable and occasionally disturbing personality:

> Our job on the *Gentian* was finished around the end of August [1943]. One of the biggest dividends of that summer was the opportunity to get to know Maurice better. My respect and liking for him had grown, despite the fact that I saw foibles and flaws that I had not previously been quite so aware of. He had some of the attitudes as well as the speech of a Texas boy—racist and chauvinistic attitudes that I found a little uncomfortable, or sometimes just amusing. . . . [Furthermore] He was not only a workaholic but liked to be seen as one. I remember one time he came to an evening seminar at Woods Hole dressed in shabby and dirty work clothes, apologizing as he got up to speak, saying he had been working in the shop until the last minute and didn't have time to change. . . .
>
> Like a good many other brilliant and egotistic people—and some who aren't so brilliant—he had to be the boss. There were some very good people who chose to stay with him and do his bidding, but some of the best and most independent ones drifted away. As far as his professional life was concerned, that was probably the most serious flaw. And having said the worst things that I can about Maurice, I'll say again that there are few people in the world whom I liked and admired more.[26]

Lamar Worzel, one of the few students to spend an entire career working with his mentor, also readily recalled that while the work always proved interesting and challenging, it could quickly turn into a trial: "he was a great person to work with most of the time, but he pressed you to the limits of your endurance. He would press you to the point where you were just about ready to call it quits and say, 'That's it. No more. I quit. I'm through,' and he'd back off and he'd be just as nice as pie. You'd change your attitude and say, 'All right, we'll go on from here.'"[27]

Time and experience finally increased the appreciation of frontline officers for the work and contribution of the ocean scientists who had officially labored on both coasts since 1940. Most scientists and technicians who worked with the fleet felt less the isolation and exclusion that made Iselin uncomfortable at Key West in March, 1941. Discovering in combat the value of oceanography's contribution made believers of many officers, especially submarine commanders. In more than one instance, the SBT, for example, gave submarines and their crews a chance to come home after a particularly harrowing encounter. As Rawson Bennett wrote to the commander of the Atlantic submarine force in 1943: "One of the reasons for the favorable acceptance of the submarine bathythermograph is the widespread experience of submarine commanders that a submarine below a sharp density layer due to temperature gradients is relatively safe from detection by echo-ranging or listening."[28]

Allyn Vine echoed Bennett's observations and provided a few of his own on the growing wartime relationship between operator and scientist:

> [A] true operator is responsible for everything. He knows your unusual circumstances. He knows that some of his buddies who were just as smart as he was got sunk, for reasons he's not quite sure of. These are great helps to communication. [For example, Commander Lawson P. ["Red"] Ramage didn't tend to talk very much. He tended to be quite caustic. . . . I remember one of my introductions to education and working with the Navy, [William] Schevill [of WHOI] and I went up to Portsmouth when . . . [Ramage] was taking over being skipper of the *Parche,* and we were talking to a roomful of their commanding officers, officers, and so forth in that squadron. At one stage, "Red" couldn't stand it any longer, and he got out of his seat, and kind of halfrising, said, "Are you trying to tell me that I should evade shallow?" Interesting. "I've said only under some circumstances, sir, that if Mother Nature says that you are acoustically the safest near the surface, it's maybe better to stay there where you can remain in a more offensive position. But as I understand it, sir, it is your option, not mine." The point about Ramage, which is so

typical, he went right to the core of the problem. But the point was, he was typical of those people who are extremely interested in anything that will make their boat operate better, . . . while he might almost look like he was pooh-poohing the bathythermograph, he saw to it that his officers who were responsible for that spent a lot of time with me. Then, of course, in later years, he spent a lot of time in the R&D community, and he was again very typical of the kind of person who is . . . extremely supportive when he thought it was something that was really good for the Navy. I don't think it would be appropriate to call "Red" Ramage typical. But in this regard, he was more than typical. But there are a great many of them, an unusually high percentage of them. You had the privilege of riding many submarines, and many you'd get off and you just knew that they were draining you of every piece of information you had that might help them.[Lieutenant Commander] George Street [commanding officer of the *Tirante*] was another one of those kind. In his submarine, you just felt that guy was taking everything you knew that might help him out of your head and tucking into his.[29]

By 1943, oceanography had clearly demonstrated to the frontline officer that knowledge of the environment could mean the difference between victory and defeat, life and death. As Allyn Vine knew all too well, you ignored Mother Nature at your own risk.

8

Operational Applications, 1942–43

Are you trying to tell me that I should evade shallow? . . .
I've said only under some circumstances, sir, that if Mother
Nature says that you are acoustically the safest near the
surface, it's maybe better to stay there where you can remain
in a more offensive position. But as I understand it, sir, it is
your option, not mine.

—ALLYN VINE's response to a provocative question asked by
LT. COMDR. LAWSON RAMAGE of *Parche*, summer, 1943

THE ACTUAL TACTICAL APPLICATIONS OF THE BT AND SBT
only scratched the surface of the opportunities these instruments provided. A wide variety of useful information attributable to bathythermograph observations spilled out of the navy and the NDRC laboratories into
the waiting hands of ship and submarine commanders. Many of these aids,
mostly rendered in print, represented the latest appraisal of the temperature
and pressure data collected from warships and support vessels all over the world
by junior officers and technicians trained in the WHOI program or by Eugene
LaFond at the UCDWR. Periodic observation summaries appeared in mimeo-
graphed copies for distribution. These would include analyses made possible
by BT and SBT data on temperatures, diurnal warming, wind effects, refraction, mixed layers, reflection of sound both to the surface and back down to the
depths—each of these according to seasonal and geographic variation. Environmental conditions often changed dramatically, and operators above and below the surface needed sufficient information to permit them to adapt tactically.

Woods Hole cooperated closely with the UCDWR in both the BT/SBT
training programs and the efforts to translate scientific data and conclusions

into useful publications, which might progress well beyond affecting sonar performance to influencing strategy and tactics. Iselin frequently corresponded with Richard Fleming of the UCDWR oceanographic section concerning BTs sent to California from both Woods Hole and East Coast manufacturers to make sure they both agreed on use, deployment, critical training issues, and problems presented by flawed BTs, missing grids, calibration difficulties, or collection and analysis practices.

All of this coordination became essential as both Iselin and Fleming realized the effect the veritable avalanche of wartime BT/SBT information would have on oceanography during and after the conflict. The systematic collection of basic data exponentially improved oceanography's contribution to offensive submarine strategies and ASW. Thomas Wayland Vaughan and Admiral Gherardi in their wildest interwar dreams would never have envisioned the irregular but steady accumulation of substantive oceanographic data from locations across the world ocean by those interested in bringing the war to a successful conclusion for the Allied powers.

The Navy Department also recognized the need for a naval officer to coordinate the collection operation within the fleet. Thus, Lt. Comdr. Rawson Bennett's Code 940 in BuShips's Sonar Design Section received additional personnel to work with Woods Hole and the UCDWR in managing the BT/SBT data collection effort. Tired of training radar operators at the Point Loma U.S. Navy Radio and Sound Laboratory (USNRSL), Lt. Roger Revelle had earlier begged his friend Bennett for a chance to move to Washington and a new, more interesting task. In early 1942, Revelle's patron helped him move to Hydro, then under the command of Rear Adm. George S. Bryan, to supervise the BT data collection program, act as the navy's oceanographic point man in Washington, and perform hydrographic duties related to Hydro's survey program.[1]

Charts and Sailing Directions

Composing charts and sailing directions from the data produced by the BT/SBT projects and other sources proved one of the most meaningful and practical services rendered to the operational forces by naval and civilian science. Hydro and naval contractors designed these publications for distribution to ship commanders before deployment, and their content reflected the oceanographic and meteorological characteristics of the theater of operations discussed in the ship's orders. Some of these operational aids summarized BT observations for the area in question during a particular season. Others outlined the results of hydrographic surveys, and still others, like the submarine supplements to sailing directions, provided a synopsis of all factors influencing sound transmission and echo ranging. In addition to the significance of bottom composition for sonar performance, these published aids also discussed density and the natural influences on ballasting, buoyancy, and ship handling for submarines. Thus, with a relatively modest number of naval pamphlets on hand, a surface or submarine commander had at his disposal all of the oceanographic and meteorological information available for his patrol sector.

This activity, even more than the BT, helped oceanographic entrepreneurs translate into a familiar language the importance of oceanography for the American submariner and surface officer. Building on the foundation of interwar Hydrographic Office maps, scientists like Ewing, Vine, Iselin, Revelle, Redfield, LaFond, Francis Shepard, and Kenneth Emery added information harvested daily from international publications and instruments on warships, support ships, and specialized vessels. The charts thus reflected BT/SBT readings, more accurate statistics on bottom contours and currents, the regularity and disturbing effects of waves and swells, comments on the formations and the soil and rock composition of the ocean floor, and information on the effects of marine life on combat conditions. From his position at Hydro, Roger Revelle assumed the job of coordinating the information for these charts and tactical publications. He set general standards of quality and expression down to the kind of legend symbols and units of measure employed.[2] Interested in advancing, and craving more engaging work, Revelle in late 1942 again appealed to Bennett and received a long-term temporary active duty (TAD) assignment to BuShips Code 940 effective 22 December 1942. He worked under Comdr. Jacob Meyers in 940D and, in essence, served as the bureau's oceanographer.

To assume some responsibility for the duties he left behind, such as coordinating data collection and composing the environmental aids at Hydro, Revelle found a truly remarkable replacement. As he recalled nearly forty years later, he did not choose Mary Sears, she chose him. Sears, a WHOI biologist and Radcliffe graduate, had just completed the training course for female officers and asked Revelle for suggestions on the best billet for her skills. Sears would spend her postwar career at Woods Hole and become the founding editor of the well-respected journal *Deep Sea Research*. In 1943, as she drew her five-foot-three-inch frame to attention before the six-foot-four Revelle, the disparity in their height made a lasting impression and she later described the Scripps oceanographer as an intelligent but physically slow and lumbering giant. She had to walk very fast to keep pace with his long strides and found him altogether amazing: "I think as an administrator, he seemed to do it just effortlessly. People just sort of fell into place. . . . He looked at you as if he looked right through you, and that might discombobulate you, but he was an amazing man, because I think he remembered everything he'd ever heard."[3]

Revelle helped arrange for Sears to work at Hydro in Suitland, Maryland, as leader of a unit charged with drafting many of the environmental aids the navy's worldwide data-collection network made possible. She spent the war at Hydro contributing to and improving the most significant printed oceanographic and scientific intelligence aids published for the operating forces.[4]

Just as Revelle left Hydro for BuShips, the navy's role in supplying critical environmental information and analysis increased dramatically. Early in 1943, Admiral Furer's office began working with the army in a Joint Meteorological Committee (JMC) that had as one of its subcommittees a group on oceanography. Working with Lt. Comdr. D. W. Knoll, Revelle joined the group as participant and recording secretary, and their subcommittee set about establishing a function for itself and setting priorities. Admirals Furer and Bryan obviously hoped to use this opportunity to extend naval authority in this area of research and to define more precisely their wartime roles as oceanographic practitioners and intermediaries. One of the subcommittee's functions was to maintain a list of both available research personnel and activities interested in oceanographic information. The JMC also asked the oceanography subcommittee to keep a bibliography of oceanographic reports from the army, navy, NDRC, and all private research contractors, as well as those available from foreign countries. Specific authorization permitted the subcommittee to recommend projects addressing specific needs defined by army and navy activities to pinpoint the facilities that might undertake this work, and to assign priorities according to the most pressing wartime needs. In this way, Revelle found it possible to direct the army's questions on wave and surf forecasting to Sverdrup and Walter Munk at Scripps.

The subcommittee soon became a completely naval affair. Over the next six months, the Army Air Forces was reorganized and rethought its role in the joint oceanography subcommittee. Given navy experience in this field, the army decided to surrender all oceanographic services in the summer of 1943. While this move created some difficulty in realigning contractors from army to navy service, from 1943 on the navy had complete control of oceanographic activity within the military.[5]

A Pebble Tossed in a Pool

Like a pebble tossed in a tranquil pool, the pioneering work of Ewing and Iselin on underwater sound transmission and Vine's refinement of the BT caused a ripple effect that touched every aspect of the naval war between 1940 and 1945. Work at the UCDWR, Scripps, and Woods Hole built on the fundamental temperature and pressure data provided by the BT and the perspective offered by the Iselin-Ewing analysis of the behavior of sound in the ocean. These tools provided new ASW techniques, innovative submarine evasion tactics, submarine quieting, a new understanding of submarine buoyancy, sonar requirements and advanced designs, data on ocean-bottom composition, maps and charts making all of this information accessible, the significance of the deep-sea sound channel, and studies of both the ocean's microstructure and wakes.

In all of this, as mentor or colleague, Maurice Ewing made the critical contribution. His everyday vitality easily matched the intensity of mobilization, and his personality and work habits influenced everyone in wartime oceanography. The Lehigh physicist's leadership in the study of underwater sound and the importance of that field to the war effort made his group—which included Vine, Worzel, McCurdy, and Albert Crary—a critical component of the naval war in both the Atlantic and Pacific.

For the navy, the intellectual diversity, energy, and experience of the members of Ewing's Lehigh group made them the perfect translators of oceanography's valuable mysteries. Ewing, Vine, and Worzel had extensive experience at sea and worked in the embryonic fields of oceanography and geophysics, which demanded seminal theoretical work as well as a gift for instrument design and engineering. Embryonic in its own right, undersea warfare required an application of the same kind of fundamental experimentation, analysis, and engineering. Ewing, Vine, and the others thus felt comfortable in their wartime tasks and found the demands very familiar.

The underwater sound effort sponsored by the navy and the NDRC during the war did not focus activity exclusively at Woods Hole. Nor did the Lehigh group monopolize the assignments and research. The Columbia University Division of War Research (CUDWR) at its Fort Trumbell New London laboratory collected salinity measurements, bottom samples, and basic oceanographic data—including BT readings—along the New England and Middle Atlantic coast. Often, the Columbia lab worked in conjunction with the UCDWR on both coasts gathering data that would support their respective research and development projects in sound detection instruments and systems.

The UCDWR, of course, had its own Oceanography Division, staffed largely by Scripps scientists. Sverdrup initially directed this group, but in the scare following Pearl Harbor, those with Asian racial characteristics and others with deep roots in Axis-occupied countries such as Norway, found the U.S. government a reluctant sponsor. Afraid that the Scripps director could prove vulnerable given his Norwegian citizenship, the Navy Department withdrew his security clearance and confined his work mostly to unclassified, less sensitive projects. In the annual UCDWR report composed by staff secretary John M. Adams, the university noted with obvious mixed feelings that Sverdrup no longer led its wartime oceanography program: "The loss of Dr. H. U. Sverdrup of the Scripps Institution of Oceanography as a part-time consultant might be especially noted. His unique knowledge of the physics of the sea was largely responsible for the sound basis upon which the oceanographic program was established. Dr. Sverdrup's services were relinquished in April 1942, because his access to confidential material was unacceptable to naval authorities." Richard Fleming, who received his doctorate in marine chemistry from Berkeley after completing his research at Scripps, immediately replaced Sverdrup as chief of the UCDWR Oceanography Division.[6]

In addition to studies of underwater sound attenuation and reverberation conducted by various sections of the UCDWR in support of sonar development at the university laboratory, Fleming's group studied the characteristics of the ocean that promoted or inhibited sound transmission. They accumulated sound-ranging data with the BT, focusing particularly on horizontal gradients. The current training informed submarine officers of the vertical temperature variation with depth without giving sufficient time to the finer, short-range horizontal variations in temperature and density. The UCDWR team examined ASW surface ship detection problems possibly attributable to the environment at ranges of roughly two thousand yards or less. Munk's microstructure work at Scripps and later with Fleming fit well into this examination of the finer points of ocean temperature distribution and its affect on short-range antisubmarine work. In 1942, Fleming also presided over investigations of surface reflection of sound and ocean turbulence. As the laboratories converted raw data into concepts the frontline navy needed to know, he carefully coordinated teaching methods and warfare techniques with Iselin and Ewing at Woods Hole to ensure a uniform and accurate curriculum for naval trainees. All through 1942–43, the oceanographic group supplemented these primary efforts with work on marine life, harbor conditions, bottom photography and sampling, and, for a short time, collecting oceanographic data published by the enemy during the interwar period on the Pacific Ocean and the Japanese home islands.[7]

Although it cooperated with the other laboratories and the strong oceanography program at the UCDWR, Woods Hole remained the NDRC's premier oceanographic research center. Iselin's institution served in most cases as the source of the theory, experimentation, and instrument development in the ocean sciences that advanced the technical systems development ongoing at Columbia's New London laboratory and the UCDWR. The Ewing-Iselin analysis of sound behavior in seawater provided the foundation for numerous wartime projects in every aspect of geophysics and physical oceanography. In just as many instances, the Allyn Vine variation on the Rossby-Spilhaus BT provided the data to understand the ocean as a battle environment and to provide the basis for instruments and publications used in combat by naval officers.

Sonar and Sound

By 1944, moving systems hardware from applied research to shipboard installation became almost as commonplace as the preparation of printed oceanographic intelligence and regular data collection. This proved especially true in the area of underwater sound. The NDRC's Division 6 concentrated its acoustics program at Columbia University's New London laboratory and the UCDWR. This effort had already begun to tie sound behavior underwater to the development of hydrophones and transducers, as well as entire suites of ASW detection and prosubmarine combat systems. The work of J. Kneeland Nunan and William B. Snow at the CUDWR in New London demonstrated well the translation of oceanographic concepts and knowledge into more effective sonar designs. This proved true especially in the development of the QBF type, an echo-sounding and listening system based upon Rochelle salt crystal projector technology and manufactured in the hundreds after 1943 by Western Electric.[8]

Probably the most comprehensive effort made by oceanographers during the war, the underwater sound program reached its stride in the last two years of the conflict, applying considerable talent and resources to investigate virtually every aspect of ASW and prosubmarine problems. Despite the persistent press of events, the navy now had scientists working full time on aspects of the underwater war many did not even imagine when the European conflict began in 1939. Since that time, Allied fortunes had changed in the Pacific, and the effective, coordinated German onslaught against the transatlantic convoys had begun to wane. Insights provided by Allied mastery of the German Enigma cipher machine and effective air and surface ASW strategies aided by the BT and a greater appreciation of the underwater environment reduced the German U-boat force to a fraction of its former numbers and effectiveness.

Heeding the hard-won lessons of the underwater war, the NDRC, MIT, the NRL, and the UCDWR worked closely with the submarine and ASW communities to produce a series of effective sonic decoys or noisemakers. The researchers employed transducers, ammonia discharge, and carbon dioxide to simulate submarine and propeller sounds hoping to attract weapons like the German T-3 acoustic torpedo or enemy ASW sonar sets.[9] Although devices like MIT's FXP Mark II and III developed by the spring of 1944 failed in their initial effort to achieve the desired sounds and acceptable decibel level, and too often ran into towing problems, remarkable success crowned their efforts later that year.[10] By early 1945, the navy had effective towed decoys, better towing techniques, and self-propelled submarine decoys designed for blue-water deception or coastal and harbor work. The UCDWR successfully developed the latter, called NAD beacons. The submarine force employed the NAD-6 and NAD-10 as ejectable underway countermeasures to enhance evasion tactics.[11]

Yet another aspect of underwater sound's importance arose with research into ship and submarine wakes. The mass of bubbles caused by propeller and rudder motion, as well as the complex water patterns generated by the force of the propulsion system, presented considerable challenges. Wake noise not only provided unwanted indications of submarine and ship positions, but it also almost completely masked a ship's echo. Acoustical torpedo development, propeller design, the effectiveness of self-propelled decoys, and the more advanced versions of sonar then under development all depended on an understanding of this phenomenon. In the process of wartime research at Woods Hole, General Electric, the UCDWR, and the navy's Point Loma radio and sound laboratory, oceanographers and physicists discovered the considerable varieties and intensities of sound made by wake bubbles. In addition, the examination of depth as a factor in sound behavior also revealed that the accumulation of bubbles and air in the immediate realm of the propeller also caused vapor cavities that emitted distinctive sounds when they collapsed. This common occurrence became the dreaded "cavitation" that later submariners would avoid at all costs, lest they compromise their location and their lives. Scientists exploring sonar detection methods discovered that the depth of the propellers significantly affected cavitation. Deep-ocean pressure compressed the bubbles and the vapor cavity, reducing the intensity and frequency of cavitation noise at depths from 250 to 450 feet—the maximum operating depth of early postwar American submarines. This permitted considerable strides in evasion and countermeasures programs.

At Roger Revelle's request, the USNRSL at Point Loma also explored the vertical effect of submarine wakes. For years, scientists and engineers working with propulsion systems and hydrodynamics studied the familiar horizontal water distribution pattern at the stern of surfaced submarines and conventional vessels. By war's end, ongoing horizontal wake projects discovered a way of making minor adjustments to surface-ship sonar to permit the detection of most vessels generating wakes within sonar range of the searching vessel. If the horizontal effect seemed familiar, what about the wake distribution in the third dimension between the moving submarine and the surface? In 1944, research into the vertical phenomenon by navy scientists on the West Coast uncovered what has become known as the surface scar. Tests at sea demonstrated that a submarine moving at periscope depth with a 5.5-knot pace left wake scars at the surface forty-eight feet above the screws! At this point the scar became visible acoustically and, at times, visually. This presented both an advantage and a handicap, depending on one's viewpoint. For American ASW personnel submerged, on the surface, and especially in the air, it provided yet another way of making a submarine's future uncertain. From the prosubmarine perspective, understanding the nature and extent of the scar could mean the difference between survival and certain death. In postwar-era investigations, thermal wake searches complemented sonic detection methods in a quest to understand and explore the surface scar. As in virtually every other aspect of naval warfare, the submarine forced the navy to think and explore in three dimensions.[12]

Sometimes the navy's understanding of sound transmission in seawater revealed the limitations and dangers inherent in using this form of detection. As the conflict progressed, submarine commanders discovered that a single ping from an active echo sounder to confirm position and range dramatically increased the likelihood of their torpedoes hitting the target. Unfortunately, introducing a sound into the ocean for this purpose or any other might provide an ASW vessel with a way of tracing the attacking boat. The navy needed a way of limiting the detectability of the active sonar signal source. The problem called for developing a "secure" echo-sounding device.

After renowned theoretical physicist Carl Eckart of Scripps devised a feasible concept for a secure echo sounder, the navy moved quickly to push the innovation into prototype and production with help from the scientists and engineers at the UCDWR.[13] By June, 1944, the UCDWR had successfully installed and operated its Secure Echo Sounding Equipment (SESE) in a submarine at the urgent request of the Pacific submarine force commander (ComSubPac), who granted it very high priority. A submarine commander usually employed both his Fathometer and echo sounder to keep his boat off the bottom, especially in shallow water, all the while fearing that active sonar might bring detection. However, tracing the origin of the echo-sounding signal became impossible within 250 yards of the point of origin when using the SESE. This made a fix on an actively ranging American submarine very difficult and a successful escape more likely. Furthermore, this type of sonar could well prove useful in shallow, poorly charted enemy waters.

Scientists at the UCDWR first installed the SESE on the USS *Spadefish* (SS-411) at San Diego, and the system went through a series of tests during the voyage to Pearl Harbor. The system had performed so well that the submarine retained the equipment on its war patrol for further testing and evaluation. Technicians at the UCDWR wanted to evaluate any problems in the system's combat performance. Human factors and ease of use were important considerations from the outset. The final combat version of the SESE went into the USS *Spot* (SS-413) at the end of July, 1944. Thereafter, the university considered the project concluded turned it over for production.

The commander of SS-413 wrote to the director of the UCDWR in early February, 1945, to confirm the laboratory's positive estimate of the system and to praise the staff for its development. He felt that "Susie," as he called the SESE gear, "was of even greater value than had been anticipated. The safety with which SPOT navigated shallow, poorly charted waters is evidence of the valuable contribution rendered by SUSIE to the success of SPOT's first war patrol. There were frequent instances that prohibited the use of a standard 24 kc fathometer, yet soundings were taken with greater, easier accuracy than could otherwise have been done, and with complete security." The *Spot*'s crew discovered that the new SESE installation provided accuracy within one-half fathom up to 180 fathoms and within six inches in depths of less than forty fathoms.[14]

Effective shallow-water sonar became important for a wide variety of tasks. As American submarines and surface ships closed on the Japanese home islands and western Pacific possessions, the ranges became shorter and the operational requirements for attack or reconnaissance at close quarters became more frequent. In these cases, the nature of the bottom assumed critical importance. Its composition of hard-reflecting stone and rock or yielding mud and ooze took on great significance for a crew trying to mask their submarine's presence and safely navigate submerged without constantly resorting to use of active sonar or the Fathometer. All during the war, these ocean-bottom characteristics usually appeared in Hydro's publications for regional operations, especially the submarine sailing directions. In this regard, the work of Francis Shepard and Kenneth O. Emery on bottom configuration in the South Pacific and in the seas around Japan proved particularly significant. At the beginning of 1944, the UCDWR had just placed the final touches on a series of five bottom-sediment charts of the Southeast Asia region for publication and distribution through Hydro.[15]

The aforementioned effort to promote cooperation between Scripps and WHOI to achieve comprehensive knowledge that neither could accomplish alone extended naturally to these bottom studies and Hydro publications. Frequent faculty exchanges between the two oceanographic institutions provided additional hands, expertise, and an intimacy with conditions in other areas of the world.[16] In one case, during April and May, 1944, William Schevill of Woods Hole spent time at Scripps working on the summer supplement to Hydro publications 122, 123, and 124 detailing ocean conditions in Japanese waters.[17]

Schevill and other scientists felt acutely any delays in producing material for Hydro because they provided truly critical information for the survival of American submarines in enemy waters. A submarine officer would find commentary on surface and subsurface climates for oceanic areas, the nature and influence of currents, the locations of storm areas, submerged shelves, the impact of river runoff, and information on the contrasts between the windward and leeward sides of important islands. The supplements also included complete BT records of the area, as well as data on marine life, the ocean bottom, and the influence of landmasses on adjacent waters. Appended to these critical publications, a captain would find extracts from recent submarine war patrol reports detailing the experiences of his colleagues with surface currents, bottom sediments, buoyancy problems due to unusual salinity gradients, depth of the thermocline, and estimates of assured ranges that enemy ASW vessels might enjoy. In all of this, sonar played a critical role by providing practical experience of the ocean environment as well as vital data.[18]

Beyond the need to appreciate the terrain under the keel, sonar development could not ignore shallow water because of the importance of locating small objects, including offshore obstacles and shallow and deepwater mines. Most of these devices emerged from insights gained in the development of surface ships and submarine sonar. The NRL worked on the problem within the navy and the Radio Corporation of America (RCA), Harvard's underwater sound laboratory, and the UCDWR addressed the need via NDRC contracts. Engineers at RCA developed a portable obstacle or small-object detector that a diver could carry.[19] Harvard and the UCDWR also adapted their respective sonar specialties to the task. Harvard worked with echo-ranging sonar capability, and the UCDWR with the new frequency-modulated (FM) sonar technology. The latter produced one of the more successful candidates: the UCDWR's FM Subsight, derived from an effort to improve the efficacy of Mousetrap, the forward-throwing ASW weapon.[20] Mousetrap addressed the need to place ordnance more effectively on a sonically detected target. An ASW vessel normally would have to pass over a contact to roll depth charges off its stern or propel them explosively to a location parallel to the ship's track. In each case, the hunter had to achieve a position ahead of the submerged contact and in doing so ran the risk of losing sonar contact or missing an evasive maneuver by the submarine. Mousetrap launched a quick-sinking array of depth charges ahead of the ship's position, thus placing multiple explosives near the sonar contact in a timely fashion. Some of the first modestly successful tests on multipurpose, small-object sonar took place at the Naval Mine Warfare Test Station at Solomons, Maryland, in the spring of 1944. All of the laboratories found difficulties with portability, range, and the interference caused by natural events like waves and surf, but none wavered from the task of developing a locator and perfecting it before the war's end. The army even helped evaluate an Underwater Sound Direction and Range System (USDAR) designed for use in waters just offshore. The successful models varied in nature, size, and style according to the laboratory and the most immediate application for the device in question. Small-object locators served many purposes and therefore required many approaches and configurations.[21]

One of the most significant prosubmarine developments of the war emerged from this necessity to detect obstacles, mines, and other small objects.

Due to its precision and excellent image resolution, the UCDWR's FM sonar graduated late in the war from a secondary detection tool to a primary sonar system for American submarines. Traditional active echo-ranging sonar employed discrete pulses of the same wavelength. The operator had to wait for each return echo before initiating another so as not to confuse the receiver. Frequency-modulated detectors projected a continual stream of signals in two distinct frequencies. The first projected would not interfere with the path of the second. However, when the echo from the first collided with the second sound pulse, the event would produce a beat whose frequency provided a measure of the target range. With a constant series of impulses hitting the target and their revealing beats occurring on the echo return, the on-board cathode ray tube display could provide a much better manifestation of the detected object. In graphic performance representations, the projection paths of the two different active signals gave the appearance of a saw-blade or houndstooth pattern.

Early developmental work on the UCDWR FM sonar took a variety of forms. The prototype apparatus underwent tests on landing craft and other surface ships. It sought irregular bottom formations, obstacles, and, above all, the location of mines. Although the first practical sets did not appear until 1944 and the range for finding mines or anything else proved limited, the precision of the device and the excellent images it offered made it a continual high priority. The principle first found useful application in an experimental device called the Echoscope, developed at the UCDWR in the spring of 1942. By the autumn of that year, applications in the small-object locator and mine countermeasures program had further advanced the research. In addition, the device's precision led to experiments with it as a range-finding device for Mousetrap. Its development as a deployable ship sonar system began in earnest in early 1943, and by February an experimental model was being employed in San Diego Harbor.

The harbor installation gave scientists and technicians at the UCDWR sufficient experience to develop a twenty-channel model from the ten range channels available in San Diego. After tests of the still experimental system conducted aboard the USS *Semmes* in terrible weather off New London in the winter of 1943–44, the UCDWR took the suggestion of Submarine Squadron 45 in San Diego and procured one FM sonar set for the *Spadefish* to conduct preliminary submarine tests.

Research and development at the UCDWR gradually increased the anti-ship range of FM sonar to between one thousand and fourteen hundred yards, with excellent detection results at much shorter ranges for moored and drifting mines. After successful experiments aboard the *Spadefish*, Vice Adm. Charles Lockwood, the ComSubPac, eagerly requested further submarine installations. Thereafter, both the UCDWR and Western Electric in Hollywood, California, manufactured the equipment with constant attention and encouragement from the Pacific submarine force. The admiral desperately wanted to get as many FM sonar–equipped boats into the war zone as possible. With the decrease in the number of available enemy targets, many of Lockwood's men had to seek out the Japanese in hostile harbors, inlets, and along enemy coastlines with the attendant danger of minefields. The FM sonar made navigating these hazards possible and the risks acceptable. The Bureau of Ships eventually christened the new device with the sonar designator QLA.

With the successful performance of the QLA system in the *Spadefish* after its 20 October 1944 San Diego installation, Lockwood immediately authorized others, starting with the *Tinosa* (SS-283) in November, *Bowfin* (SS-287) in December, and *Stickleback* (SS-415) the following June. The navy and the UCDWR managed to deploy twenty sets before the war with Japan ended. These systems performed well on patrol and became the sonar mainstay of the early postwar submarine force.[22]

As the conflict ended, any early wartime imperative to sustain a regular dialogue between the navy and the ocean sciences quickly relaxed. The war offered a reason to initiate the process, but the war ended before practical experience could instill a sense of shared goals and standards. For those still locked exclusively within the naval culture during the last two years of the conflict, the achievements of naval oceanography became part of a "black box," to use a term of recent vintage. Despite the careful and energetic translating by Ewing, Revelle, Vine, and others, it appeared that only a modest number of line officers acquired an appreciation of the ocean environment.[23] They remained completely focused on those weapons and systems placed at their immediate disposal by those who understood the medium. Although black-box issues remained science's province, responsibility for using the weaponry rested with the warrior. Wartime experiences persuaded only a minority of naval officers to view the ocean as an essential tool.

Training and translating thus remained paramount as the war drew to a close. Regardless of Roger Revelle's early impatience with a navy teaching assignment, training retained a high naval priority. Without it, the instruments would not produce the desired results. Lacking the technical training and education in the fundamentals of underwater sound transmission, naval officers could not fully appreciate the meaning of the data and its significance in battle. The UCDWR cooperated closely with the USNRSL and the West Coast Sound School's training squadron in San Diego to promote expertise, understanding, and proficient use of the BT and SBT. Iselin urged the navy to take advantage of the BT school at Woods Hole and asked two members of his staff— Pollak and McCurdy—to remain on station in Trinidad and Miami to address the educational needs of those facing underwater sound challenges in the Caribbean. Despite the U-boat threat, Iselin even kept *Atlantis* at sea in that region, gathering BT data, studying the afternoon effect, and performing hydrographic stations to keep the information flowing and collection skills sharp. Captain F. S. McMurray's orders from the WHOI director strongly suggested both caution en route, and sailing close to the coast when possible. Iselin told Admiral Furer that he sent McMurray south in January, 1942, to avoid submarines and inclement weather. Once *Atlantis* moved into warmer waters, the WHOI director was "inclined to advise the captain not to pay too much attention to submarine activity. Should the ship be sunk in fine warm weather, the crew would have an excellent chance in the boats." In his response, Furer assured Iselin that his instincts would probably prove correct. The admiral did not believe that a U-boat would waste a torpedo on a prize as small as *Atlantis*. Doubtless, Captain McMurray found his assessment comforting.[24]

9

Unfinished Dialogue, 1942–45

[T]here is no longer any feeling on my part that the Navy is failing to take advantage of the physics and oceanography of underwater sound transmission.

—COLUMBUS O'DONNELL ISELIN to REAR ADM. JULIUS A. FURER, n.d., 1943

NEITHER THE NAVY NOR THE WARTIME OCEANOGRAPHIC community forgot the importance of knowing their enemy. Rather, they expanded that well-known dictum to include knowledge of their allies. All during the global struggle, Americans engaged in applied naval oceanography monitored the progress made by the Germans in that field and exchanged information and personnel with the United Kingdom to enhance Allied capabilities at sea.

By virtue of their extensive interwar oceanographic efforts, the Germans had an excellent appreciation of ocean layering, thermal characteristics, and the importance and character of deep and shallow currents in the Atlantic. With over two years combat experience with U-boats between 1939 and America's entry into the conflict, the potential existed for a dialogue between the German oceanographic community and the U-boat force that would have provided Dönitz's submarine commanders with an effective working knowledge of the ocean environment in the Atlantic. Many wondered if a discourse similar to that between the navy and the oceanographic community in the United States already existed in Nazi Germany.

In early 1942, a worried American naval attaché wrote from London that U-boats customarily slipped away from ASW vessels armed with ASDIC[1] by some planned method of evasion. He wondered if ocean characteristics familiar to the Germans often played a role in confounding the Royal Navy's search methods. As Operation Drumbeat unfolded along the American coast, neither the U.S. Navy nor the British knew of a German device similar to the BT, which would have signaled constructive collaboration between German science and the Kriegsmarine. Had the U-boat force perhaps achieved a working familiarity with ocean layering? Did their own very able oceanographers help German submariners appreciate the effects of temperature and pressure on sound behavior?

In that winter after Pearl Harbor, the Allies knew only that the U-boat's natural medium played an important part in the ASW problem in the North Atlantic. Experience at sea also led Americans and their allies to suspect, for example, that U-boat commanders exploited the sharp density layers caused by the confluence of fresh and saltwater at the mouth of the Saint Lawrence River to rest silently and wait for the emerging cargo ships en route to the United Kingdom. In addition, captured charts of subsurface characteristics revealed that German naval training included instruction in the usefulness of the ocean's density in its upper layers.

To their surprise, Allied navies and scientists gradually discovered that Germany's application of oceanographic knowledge to undersea warfare extended only as far as their experience at sea and the general information recorded on those charts. In early February, 1943, Lt. Comdr. J. B. Knight, commanding officer of the New London laboratory, sent two captured German publications to Columbus Iselin for his evaluation and comment. Knight specifically wanted to know if Iselin thought this enemy literature accurately reflected current German naval potential in applied oceanography. The poor quality of the first publication, a report on ocean currents, led Iselin to conclude that while the German navy's Hydrographic Office and Deutsche Seewarte, might have sponsored this work, it did not bear the mark of the best that country had to offer. According to the WHOI director, the Institut für Meereskunde in Berlin thoroughly knew its business and would never have submitted such an error-filled document to the fleet. The authors clearly had little knowledge of the nature of currents and their effect on subsurface operations.

In contrast, he found that the second report exhibited greater knowledge and professionalism, leading Iselin to suggest that its date of composition might provide important insights into the current state of German naval capability. This "Collection of Intelligence and Observations Regarding Enemy Position Finding Apparatus" impressed Iselin as a study that must have appeared at least five years before the war began. It elicited some interesting comments from the WHOI director on possible German naval oceanographic capabilities as compared to those then available to Americans. He argued that the publication showed its age, for he felt "quite certain that of more recent years the Germans have had a better grasp of the oceanography of submarine detection. If on the contrary it is relatively new, then I have overestimated their abilities. . . . You will have noticed that the Germans who wrote this report were confused on two important points: 1) they did not understand the significance of the depth of the mixed layer at the surface and 2) they apparently had no idea of the principle of a sound shadow zone. It is inconceivable to me that this is still the case."[2]

Iselin's respect for interwar German ocean science led him to expect more from the Germans than they accomplished during the war. As part of the U.S. Naval Technical Mission in Europe commissioned by the navy in the autumn of 1944, WHOI staff member and naval technician William Schevill evaluated the capability of the German navy to use the environment to its advantage. The excellence of their scientific community offered great potential, but they addressed such critical problems as underwater sound far too late for oceanography to have a significant effect on the U-boat force's operations. Although German scientists realized the importance of temperature and pressure for U-boat survival and evasion, a significant relationship did not develop between German line officers and scientists until 1944. Germany's translators came forth much later than did Iselin, Vine, Ewing, and Revelle in the United States. In Germany, oceanographers and naval officers did not initiate and focus the process of scientific mobilization as promptly and clearly as they did in the Allied countries. As Schevill indicated in his September, 1945, report: "It was not until late summer of 1944 that any significant liaison developed between the Seewarte scientists and the seagoing navy. Professor [H.] Thorade had been doing what missionary work he could, but only after Oberleutnant z. See d. Reserve Fritz Kallipke, a U-boat commander, called on the Seewarte with some ideas about ocean layers, was much progress made. On Kallipke's only submarine patrol, during which he was stationed some 70 miles west of Stavanger for two weeks in Mid-June 1944, he had been much impressed by the strong density layering in the upper hundred feet and its effect on sound conditions. The first step was to push the development of a thermometer suitable for acoustic use on submarines."[3]

Although Schevill discovered that the Germans had developed a new ocean thermometer, it did not compare to the BT, nor did the Kriegsmarine develop doctrine based upon the possibilities offered by the new device. Although Seewarte scientist Karl Kalle, asserted in a 1942 publication that the record provided by the new thermometer resembled that produced by the prewar BT, he boasted that it was "not so crude and inaccurate as the Spilhaus bathythermograph that was described in America before the war." The Seewarte obviously had no idea that Allyn Vine had dramatically refined the Rossby-Spilhaus BT, that Woods Hole had generated a submarine BT, or that American oceanographers had invested four years investigating all of the major factors influencing sound transmission in seawater.

Schevill also noted that the new thermometer offered the observer no simultaneous view of a depth gauge from his station in the U-boat. A submarine's crew thus did not have an opportunity to benefit from the quick correlation of temperature and depth that might offer shadow zone evasion alternatives. Indeed, the WHOI technician found that the German U-boat operational literature completely lacked an appreciation of reflection, scattering, sound absorption, and the influence on sound transmission of the ocean's bottom composition and marine life. More significantly, save for a course given to seven meteorologists in Ahrensburg in December, 1944, the Kriegsmarine developed no training program for its personnel. Indeed, in Schevill's estimation, "it was too little and too late. Few operating navy men appear to have been taught, and only one operational installation [of the thermometer] in a U-boat has been found." Iselin had certainly overestimated the effect of German oceanography on operational performance. The talent and potential existed, but not the willingness and insight to bring scientist and operator together at an early stage in the war.[4]

American oceanographers did not have to estimate the progress made by their British colleagues during the Battle of the Atlantic. From the very beginning, the United States and Great Britain freely exchanged information and expertise, sustaining a dialogue on the problems of applied naval oceanography that long outlived the conflict. Initial contacts made in Washington, D.C., by Sir Henry Tizard during the British technical mission visiting in September, 1940, quickly bore fruit. Tizard, rector of the Imperial College of Science and Technology and newly appointed research director at the Ministry of Aircraft Production, offered complete British cooperation in exchange for an open dialogue with his American military and scientific counterparts.

This established the groundwork for a relatively free flow of authorized visitors between both countries before and during American participation in the conflict. The first official mission to Britain led by the NDRC's James Conant in February, 1941, set up a permanent office at the American embassy in London. Thereafter, scientists and technical personnel certified by the NDRC visited the United Kingdom on missions usually lasting from four to six weeks. William Schevill made one of these journeys before Pearl Harbor and another in the summer of 1943 in addition to his work during the last two years of the war with the U.S. Naval Technical Mission in Europe.

Becoming an experienced oceanographic traveler during the course of the war, Schevill brought extensive experience to his tasks. This tall, lean, popular scientist with a refined manner and dry wit spent many years working on both collections and a master's degree under Henry Bryant Bigelow at Harvard's Museum of Comparative Zoology. Occasionally unsure of his own ability to completely satisfy the navy's needs, Schevill donned the Brooks Brothers naval technician's uniform he acquired in Boston and began to establish productive ties in Great Britain that served both nations well in the ensuing decades.

Schevill soon overcame his feeling that someone of greater personal authority might better play his role. In later years, in between spells of ground-breaking work on underwater mammal communication, he would tell great stories about his unfortunate technician's uniform with the embroidered eagle on the pocket holding a wrench in its claws instead of the customary laurel branch and arrows. He often had to explain the "stilson bird" to his amused British colleagues.

It quickly proved difficult for any visitor, even with weeks at his disposal, to cover the broad array of commands and activities involved in each critical scientific specialty. During 1941 and 1942, both navies cooperated in devising a method to simplify communication and education aimed at providing visitors with a more complete appreciation of research and achievement on each side. Thus, beyond the particular interests of Schevill and Woods Hole in Admiralty charts and underwater sound, scientists and defense officials could fulfill much more ambitious requests in a relatively short period.[5]

The quick response led to a vigorous dialogue between the Allies. At the beginning of the war the British developed ASDIC far ahead of American sonar. They made considerable headway in detecting both large and small submerged objects using sound projection techniques. As the navy and the NDRC and their participating scientific personnel made rapid progress in underwater sound and ASW after 1940, the relationship between the future wartime allies became collegial, competitive, and often spirited. In one case, a group of scientists with the CUDWR, led by Lyman Spitzer, Jr., questioned the British inclination to de-emphasize the role of refraction in the U-boat detection problem.[6] The Columbia report sparked considerable debate between the operationally experienced British and the American newcomers. At stake was an accurate procedure for determining the assured range of surface-ship sonar under a variety of conditions. As weeks passed, growing American experience, additional data carefully collected by both allies, and a continued dialogue soon focused attention on the relative ranges under consideration. The British concerned themselves mostly with ranges over two thousand yards, whereas the Americans concentrated on much closer relative positions between U-boats and ASW vessels. In the end, Allied oceanographers more precisely determined the relative importance of refraction effects according to the ASW tactics and techniques employed by each national navy. In doing so, they realized that the debate itself would continue, extending knowledge, refining methodology, and responding to the requirements of their respective armed forces. As late as April, 1944, Spitzer continued in this spirit, initiating a dialogue on the role of attenuation in determining range and location.[7]

With his repeated visits to Britain throughout the period from 1940 through 1945, William Schevill's extensive comments on contacts, personnel, joint efforts, and developing friendships revealed the extent of wartime coop-eration in oceanography and some of its postwar possibilities. During a 1943 trip, Schevill pursued an ambitious schedule, which brought him in contact with personnel critical to the Royal Navy's wartime oceanography effort. After arriving in London on 22 August, he proceeded to Bath and worked in the Ad-miralty chart repository with curator F. C. MacKenzie. Woods Hole and Bu-Ships anxiously awaited Schevill's survey of British charts because hand nota-tions written on the maps before the war revealed the bottom composition in various portions of the Atlantic. With wartime advances in underwater sound research, the British and their American counterparts now valued this infor-mation for the critical insights it could provide into sound behavior underwa-ter, especially in relatively shallow areas. By 18 October, Schevill had dispatched 505 chart photographs to the bureau. In addition, he recorded notes taken from 175 others on forty-two spare copies of Admiralty charts, which he also sent to Washington. Many of these handwritten comments permitted the navy and American oceanographers to compare his observations with data on existing and readily available prewar American, British, and Japanese charts.

Schevill owed his success in 1943 to the complete cooperation of Vice Adm. Sir John A. Edgell, the Royal Navy's hydrographer, and to J. N. Carruthers, who guided the WHOI technician through the complexities of wartime Britain. These included both finding J. D. H. Wiseman's laboratory at the Ad-miralty to view the few Atlantic Ocean bottom samples that the British pos-sessed and gaining access to submarine bases for the important reflections of undersea veterans. Schevill and his naval and scientific superiors valued the ob-servations of experienced submariners on the ocean bottom in the shallower regions of the Atlantic war zone. Schevill sent these frontline reflections and observations on the naval applications of oceanography by British civilian specialists, such as George E. R. Deacon of His Majesty's Anti-Submarine Ex-perimental Establishment at Fairlie, to Roger Revelle at BuShips. Deacon later became a postwar leader in British oceanography and a close friend of the pre-mier American oceanographers emerging from World War II.

Before leaving England on 26 October, Schevill met with many British sci-entists and naval officers. In the process he helped coordinate Allied under-water sound work, instrument development, education and instruction, beach studies, tidal action and currents, and many other aspects of the ocean envi-

ronment that would bring success in undersea warfare. Schevill did not monopolize the trips to and from the United Kingdom. Others, like W. B. Snow
of the navy's New London laboratory, also called upon British ASW experts to
compare notes. However, few activities focused so completely on oceanography or proved so informative and directly applicable to the needs of American research in applied ocean sciences than those pursued by Schevill.[8]

Combat: A Considerable Stimulus

Although nearly all of the oceanographic research conducted at Woods
Hole, Point Loma, La Jolla, and New London related directly to naval warfare,
every step in the process also advanced the basic understanding of the ocean environment. Every oceanographer, physicist, and biologist—indeed, every scientist—made significant discoveries or compiled data that would provide rich
rewards when reexamined in the welcome calm and less hurried pace of the
postwar world.

By 1943, oceanographers had taken their place on the navy's war-fighting
team. Although those who would never admit that science had a place on the
front line still harbored some doubts, the advantages offered by the many published aids, the performance of the BT and SBT in combat, and the devotion of
the scientists working directly with naval personnel had long since overcome
official skepticism.

The early frustration of the scientific community gradually faded. Ewing,
Iselin, Fleming, and others now felt sure that most of what they did actually
played a role in planning and saving American lives. In a letter to Admiral Furer
composed in 1943, Columbus Iselin sounded a note significantly different from
that frequently found in his correspondence three years earlier. While discussing BT matters, he observed: "On the pro submarine side the developments have been even more spectacular in that the usefulness of the ideas has
now been proven in combat. Thus there is no longer any feeling on my part that
the Navy is failing to take advantage of the physics and oceanography of underwater sound transmission."[9]

The importance of underwater sound alone produced an amazing array of
advances in human appreciation of the ocean. In addition to Hydro's surveys,
the diurnal warming of the upper layers of the ocean and the dynamics of the
relationship between the atmosphere and the ocean surface received much
greater attention. The submarine shared its operating environment with eco-
nomically important schools of fish and daily experienced currents, pressure,
density, waves, and deep-ocean circulation patterns. Wartime oceanography
either expanded knowledge in all of these areas or laid the foundation for
seminal postwar research. Sound signals and their naval importance suddenly

created a wealth of information on reverberation, refraction, and reflection underwater. Scientists at Woods Hole, the UCDWR, and the CUDWR–New London explored the influence of ocean layering and both bottom composition and geography on the behavior of sound beneath the waves. They fully realized that these same sound signals would, after the war, provide a way of studying the ocean and the layers of silt and crust at its bottom. Although the attention of the navy and its scientists usually focused on deep water, the infinitely more complex inshore problems also drew their attention. This proved particularly true during harbor studies, research on wakes, photography of wrecks and mines, and in studying the difficulties of the shallow-water reverberation problem presented to sonar listeners.

Nor did productive research offer results and future potential only in physical oceanography, physics, geology, and geophysics. In one of the most important discoveries of the war, marine biology played a vital role in explaining a phenomenon that repeatedly confounded navy sonar operators. On 24 June 1942, Scripps biologist Martin Johnson received a request from the commander in chief of the U.S. Fleet to determine the source of changing levels of ambient noise interference perplexing Pacific Fleet sonar operators, who regularly encountered a noise not unlike the crackling sound of dry burning twigs. At a wavelength of twenty-four kilocycles, navy listeners had to deal with this crackling at a level of thirty decibels at different depths, and varying in location and intensity during daylight and darkness.

Johnson's research produced remarkable and intriguing results. He discovered that two species of shrimp produced snapping sounds that turned into the cacophony of white noise that bothered submarine and surface sonar systems. The intense concentrations of this marine life in certain tropical and subtropical areas of the Pacific made them a genuine technical problem for listeners. As it turned out, *Crangon dentipes* and *Synalpheus lockingtoni* topped the list of a variety of marine noisemakers that plagued underwater sound monitors. These two groups of snapping shrimp tended to rise from the depths closer to the surface at night to feed, and their behavior also changed with temperature variation. Johnson usually found them over coral, shell, or rock-covered sea bottoms. He went on to uncover other noisy types of marine life, including fish like the Croaker, which produced croaking and drumming sounds.[10]

Johnson's research provided scientists with their first insight into what would become known as the deep-scattering layer, which interfered with underwater sound transmissions and varied according to geography, time of day, and season. While the shrimp discovery formed only part of the phenomenon, this insight awakened the navy and science to the role of life forms in sound transmission and drew more biologists into the navy's undersea warfare investigations.[11]

In addition to their work with Hydro surveys, underwater sound, published aids for operators, and marine biology, scientists at Scripps and the UCDWR also completed the first comprehensive textbook on the oceans. Entitled *The Oceans: Their Physics, Chemistry, and General Biology,* this book was the end product of a joint effort by Harald Sverdrup, Martin Johnson, and Richard Fleming begun in 1938. Published four years later, the navy certified its quality and the degree to which it depended on the latest research and scientific insights by limiting its wartime distribution and imposing a security classification. The authors could not have hoped for a better peer review.

Demise of a Dialogue?

Despite the success achieved by effective translators like Ewing, Revelle, and Vine and the extensive wartime use of oceanography, both the naval and scientific communities clearly recognized as artificial and possibly temporary the effective dialogue made possible and necessary by the national emergency. For the most part, naval officers and oceanographers still constituted two distinct communities of practitioners. Each possessed a common identity supported by a peculiar language and culture, using well-defined methodologies to pursue familiar, validated goals. Indeed, given its multidisciplinary nature, a number of scientific cultures existed concurrently within oceanography. A code of standards and behavior deeply imbedded in each culture rigorously governed behavior and the extent to which one community might find common ground with ideas and participants from another. Individuals spent their lives motivated, rewarded, and indeed identified by their professional framework.

As the war came to a close, most scientists and naval officers recognized that they had barely begun to remove the barriers. Only a select few navy veterans, particularly in the ASW and prosubmarine specialties, had thus far become true apostles of oceanography's combat and support applications. The circumstances of the interwar period still seemed very near. Would oceanography sustain a dialogue with the navy at war's end sufficient to permit scientists like Ewing to satisfy their postwar craving for knowledge with naval support and encouragement? In 1944 and 1945 even the best-informed scientists and officers could not tell.

These apprehensions dominated the last two years of the war for those involved in oceanographic work within the navy and the NDRC. The strategy for coping with this uncertainty took the form of hard work, education, and communication. Essential projects that had become well established since the outbreak of the war continued with an eye toward rapid but careful research as comprehensive as the pressures of war would permit. The various agencies involved then communicated to the navy results considered groundbreaking and

significant.[12] Iselin occasionally proposed the idea of appointing liaison officers to make sure that the needs of the operating forces reached laboratories like
those at Woods Hole, and that proposed solutions effectively made their way
back to the fleet. Unfortunately, this role more often fell to scientists in the
field, to freshly minted navy BT observers, or to officers from Hydro and Bu-
Ships influenced by Revelle and Mary Sears. Iselin at Woods Hole and Sverdrup at Scripps also worked to sustain the comprehensive nature of their contributions by arguing that similar work at both institutions did not waste
resources, but rather provided a more complete picture. In the case of underwater sound transmission research, Iselin persuaded the Navy Department that
sound conditions off the California coast near San Diego and La Jolla did not
provide the diversity offered by the northern and central Atlantic waters within
the reach of WHOI scientists. The study of waves and swells continued on both
coasts for the very same reason, although the methodologies widely differed.[13]

In 1944–45, the wartime oceanographic effort explored wakes, the revelations in BT data, the possibilities of underwater photography, and underwater
sound behavior with uninterrupted intensity. All the while, leading scientists
disclosed to each other in revealing personal correspondence their apprehensions about the effectiveness of wartime communication with the navy and
their concerns for the future.

At the beginning of March, 1944, Iselin diplomatically but directly conveyed his anxieties in a series of letters to Roger Revelle. Since Revelle operated
within BuShips Code 940, Iselin felt more comfortable bringing his concerns
to the navy less formally via this friend and fellow professional who had many
scientific and naval connections.

These epistles to Revelle and other confidants like Maurice Ewing revealed
the WHOI director's personal attitudes and his longing for both an end to the
war and a clear, reassuring picture of the future. In one instance, Iselin asked
Revelle to persuade his uniformed colleagues to husband their supply of young
trained technicians by exempting them entirely from the draft or by granting
them commissions and issuing orders that would keep them occupied with the
critical work at hand. Retaining the people who made the oceanographic program work remained a difficult proposition and a source of constant distress.

In July, 1944, the WHOI director expressed to Revelle his feelings of uncertainty about granting knowledge of the BT to Canada. If Iselin did not trust
the French Canadians—a reservation that arose from the French defeat in 1940
and the Vichy experience in Europe—neither did he feel comfortable restrict-
ing the use of something as valuable as Vine's version of the bathythermograph.
For Iselin, the world conflict burdened productive prewar scientific relationships with an unwelcome awkwardness and tension.

In yet another letter to Revelle he emphasized the necessity of keeping research vessels well equipped, available, and on a schedule flexible enough to cope with the unexpected opportunity that often presents the greatest bounty. Awkwardness and the unexpected seemed to rule the day.[14]

In bringing these concerns to Revelle, Iselin not only revealed the difficulty of doing wartime oceanographic research, but also his growing fatigue with the relentless pace and his anxiety for the postwar era. In the following months, regardless of the subject, the burden of an uncertain future continued to play a large role in Iselin's correspondence. On 9 November he wrote, "The future of this particular laboratory is at present much on my mind and will depend of course on how successful we will be in building up an able staff." If he and his colleagues proved less than effective in establishing a continuing dialogue with the navy during the war, the future appeared grim both for new projects and the retention of a seasoned faculty. How could ocean scientists properly respond to future naval needs, a national emergency, or the provocative and promising data collected during the war without an adequately trained and regularly paid team? Who would provide the funds? Could one optimistically rely on universities and research foundations to commit roughly twice the support they made available before the war? Given the inflated wages and costs of operating *Atlantis,* interwar budget levels would not begin to sustain postwar research. If Iselin could bring himself to feel that private sources would find the money, the situation would appear more positive and certain. But then what of the possible conflict precipitated by private or academically based oceanography called upon to respond again to government needs in the event of a future national crisis? He found the uncertainties and complications bewildering: "My own worry, which may be unwarranted, is that we have gotten too used to being subsidized by the government and that private funds will not be forthcoming. I find that [it] is most difficult to think through this problem. You can't wait to get the contracts and then go looking for the staff. The Navy will want the services of the best men you have. If you encourage them to work for the Navy, your output of pure science will suffer seriously and so on. I think that I will retire permanently to the Canadian woods."[15]

Iselin knew full well that for many scientists, both in and out of uniform, peace meant a return to individual pursuits. The WHOI director thus dreaded the unpredictable effect of the war's end on the personnel and projects of wartime oceanography. He understood the desire to resume interrupted prewar work and the difficulty many had in sustaining their link with the government. As before the war, some scientists remained uneasy about officially sponsored research. They were concerned about government control and direction of research, as well as the threat of weapons proliferation. As WHOI director, Iselin also wondered how universities and research institutions could accommodate the need for research excellence in the ocean sciences and the ever-present possibility of a future national emergency. Some new peacetime accommodation with the navy seemed logical for the oceanographic community, but what form would it take? Iselin's facetious comment about retiring to the wooded beauty of Canada, where he occasionally took vacations, no doubt referred to an inner need he felt for a place where the assault of current and future concerns would not feel like being caught in a cross fire. Having to wait, work, and worry without sufficient authority to affect future developments constantly plagued him.

Iselin had good company. Many in the field shared his anxiety and frustration. The often frantic pace of wartime research and the effort to communicate the value of their work to the operating forces at once fulfilled and frustrated oceanographers. Frequently feeling marooned in Miami's DuPont building at the headquarters of the Gulf Sea Frontier, WHOI oceanographer Gordon Riley occasionally permitted his anxiety about the endless procession of naval imperatives and the time and care required by quality research to spill over into his letters to Iselin. In his January monthly report written in the form of a very informal letter to the director, Riley sounded discouraged and aggravated, which resulted in an immediate response from Iselin expressing concern. Fearing that he had allowed a depressed moment to color his entire report, Riley promptly wrote back to place his comments in perspective. In so doing, he provided an excellent sense of the pressures affecting a wartime oceanographer in the effort to sustain a dialogue with the navy while trying to maintain the standards of a scientific professional. He had both good and bad news, which no doubt failed to surprise his correspondent: "I believe you have the impression, fostered perhaps by my January report, that things are pretty well bogged down here and that I would sort of like to close up and get out. But that isn't really the way I feel about it; any desire to get back to Woods Hole, at least during the next few months, is purely a personal one. [We] . . . and everybody else in jobs like these, have our ups and downs, and in spite of our best efforts to the contrary they color the letters we write. This undoubtedly makes it harder for you to figure out just how we are getting along and how useful it is."[16]

Riley then proceeded to provide a personal picture of the scientific effort, which few of his colleagues would have contested in 1944. His composition provided an image of success, frustration with the navy, and constant efforts to improve communication and to educate. He combined this with an expression of professional ethical concerns. Riley and many others feared that the pressing need for answers to combat problems had compromised the quality of their science. Yet despite all of this, they knew the war made this toil necessary. They thus kept working and reporting. In his 8 March 1944 letter to Iselin, Riley reviewed his situation at Gulf Sea Frontier, sounding upbeat but determined to draw an accurate picture for the WHOI director:

In the first place I'd say that during the year we have been here we've honestogod sold the program to SCTC [the Gulf Sea Frontier training command], the A/SW Dept., and Operations. They ask our advice and help, and they are all most cooperative whenever we want something. . . . The men of GSF ships know something about the BT, and a very few are really sold on it, but they don't use it as much as they should, and practically never do they use the message procedure. We could and should improve that situation a little, but I'm convinced that nothing short of renewed [U-boat] activity in this area would completely solve that problem. The number of [BT] slides is directly proportional to the scaredness of the skipper.[17]

What the oceanographers had to offer appealed to many officers, but it had not yet become part of naval doctrine. Only the latter would ensure oceanography's role in the development of strategy and tactics.

Next to his research, Riley viewed education as his primary task. He expressed dissatisfaction that his team had not yet completed its set of instructional ocean charts for 1944. Indeed, he felt it would take until June to complete the project, and he knew the navy needed the maps immediately. Along with the charts, the Miami group wanted to supply instructional materials on their use, as well as provide further information on the oceanography of the regions in question and engage in a discussion of the problems involved in determining effective submarine detection ranges. He felt self-conscious that the entire project moved far too fast, often with too little data. Even so, this well-considered letter to Iselin had an air of reality and optimism about it: "As to research—well, I'm not proud of it, but I really think we've made some progress in the problem of the effective range, and there's no reason to think it will get any worse as time goes on, and it may get better."[18]

Although he made it clear that he would welcome a replacement if Iselin had a qualified person available, Riley suggested that he stay and expand the services that his office might provide as the data it compiled increased and improved. He thought that Iselin might consider suggesting to the navy a branch office in Key West to support the Convoy Control Center and the U.S. Navy Sound School.

Working with disturbing rapidity, granted only a few shipboard hours, and faced with scores of officers to educate and convince, Riley occasionally wondered if his work even faintly resembled science or education. He also wondered when he missed deadlines and took extra months to complete charts and oceanographic aids if his team had done any good at all for science or the ships and personnel going into harm's way. Serving both was neither physically nor ethically easy.[19]

SOFAR and the Deep-Sound Channel

The early history of Maurice Ewing's discovery of the deep sound channel illustrates both the productive wartime dialogue with the navy and the reasons behind Iselin's concerns for the future. While doing refraction experiments sponsored by Woods Hole and the Geological Society of America aboard the *Atlantis* in 1937, Maurice Ewing made a startling observation. He had no listening equipment on board, just seismographs lying three miles down on the ocean bottom. Aboard ship, his staff assembled the bombs that provided the sound source for his refraction experiments, which were designed to probe the geological composition of the ocean floor beneath the sediment. Filled with curiosity as his experiments began, Ewing put his ear to the ship's rail to listen for the rate of echo from the explosion of ten-pound TNT blocks. In roughly eighteen seconds, the explosion echo repeatedly traversed the distance between the ship and the bottom, making the round trip seven times by Ewing's calculation. Allowing for surface and bottom reflection to reduce the intensity of the signal each time it made the trip, the sound from the explosion traveled twenty-one miles before dying out. The Lehigh University physicist estimated that each reflection had reduced the intensity of the sound by roughly one-tenth. If his approximation came close to the truth, a sound might retain its initial clarity and intensity in an ocean layer that would minimize reflection loss. Proper amplifiers and hydrophones could boost the signal another thousand times for the human ear. If research proved this hypothesis correct, the ocean layer that would permit minimal reflection could very well transmit sound over thousands of miles for the well-prepared listener. On board *Atlantis* that year Ewing postulated the existence of this natural condition, calling it the deep sound channel.

R. J. McCurdy mentioned the sound channel hypothesis in his report on the submarine temperature-depth recorder submitted on 21 November 1941. While admitting that he needed much more data, McCurdy suggested that the submarine seemed the perfect platform from which to listen and take advantage of the sound channel's possible transmission properties. If a submarine could use the ocean environment to improve detection range without betraying its own position with active sonar, American ASW and submarine forces would gain significant advantages. Naval officers would have to seek the same intimacy with the ocean as tradition and professionalism demanded they have with their vessels. To do less would place them at a critical disadvantage.[20]

Late in the summer of 1943, Admiral Furer visited Iselin to discuss Woods Hole's ongoing programs for the navy. In a memorandum for the record filed upon his return, the admiral commented with some enthusiasm on a method of underwater sound communication explained to him by the WHOI direc-

tor.[21] This method employed a layer in the ocean that offered amazing transmission possibilities for sound at a minimum speed with nearly unlimited
global range. Iselin brought this phenomenon to Furer's attention because
Ewing and company had turned their attention once again to the deep sound
channel.[22]

In 1942, Ewing resumed work on this oceanic eccentricity, which he had revisited only briefly in 1940 during BT research for the NDRC at Woods Hole.
According to Lamar Worzel, natural and manmade sound signals trapped in
the channel "never hit the surface, never hit the bottom. So the only losses you
had were the absorption losses in the ocean, and for low frequencies they are
very small. About 1943, he [Ewing] wrote a report to the Bureau of Ships that
this would be a system of signaling over along range that could be used, and
the original letter, as I remember it, he recommended that it be used on airplanes."[23]

Ewing combined his extensive experience at sea between the U.S. East
Coast and the Mid-Atlantic Ridge with new data arriving daily from the Hydrographic Office and SBT/BT information to determine the nature and possible utility of the sound channel. The Lehigh physicist certainly appreciated
the scientific significance of the channel, but he felt that utility alone would be
enough to persuade BuShips of the need for further research. In his report on
the channel filed with BuShips on 12 July 1943, Ewing suggested that the navy
could easily detect from a range of about two thousand miles the sound signal
from a small bomb detonated at or near the sound channel's deep horizontal
layer or axis. Due not to seasonal variation but rather to the excessively warm
waters of the Gulf Stream and the Sargasso Sea in the Atlantic, the channel axis
occurred at four thousand feet rather than the twenty-five hundred feet characteristic of the Pacific Ocean. In the summer of 1943, Ewing formally proposed
to the bureau that the navy could communicate in code at hitherto impossible
ranges by employing explosive detonations in the sound channel.[24] The advantage to his proposal lay in the fact that the navy and Woods Hole already possessed all of the experimental data, theoretical knowledge, and necessary
equipment to make the system work.

However, BuShips did not react in the way Ewing had hoped. More at
home with the vulnerability of radio communication, Capt. H. A. Ingram of
the bureau's Radio Division, Code 330, recommended against further testing
on the basis of vulnerability to interception and jamming. Even a recommendation from Columbus Iselin failed to change Ingram's views.[25]

Despite the navy's less than enthusiastic to Ewing's confirmation of the existence of the deep sound channel and his recommendation for operational
applications, exploring the potential of this phenomenon remained high on
Woods Hole's list of research priorities. This time, however, the landmark

discovery literally became submerged in a wide range of underwater sound work authorized by the NDRC and the navy as the war neared an end.

Early in the war, as part of the underwater sound program, both Ewing's team at Woods Hole and Richard Fleming's at the UCDWR had discovered and investigated echo skip distances to enhance the range at which sonar might prove effective. According to Fleming, Ewing had demonstrated off Key West as early as 1942 that echoes from bottom reflection as deep as nine hundred feet might find their way to the surface at regular intervals from the sound source. The predictability of this behavior formed the foundation of postwar convergence-zone sonar, which greatly extended the detection and listening range of sound gear used on ASW vessels and submarines.

Range had become all-important. Experience in the Battle of the Atlantic demonstrated that victory in the undersea war depended upon early detection of the enemy. By early 1944, existing JP, JK, QB, and QK sound systems had already achieved supersonic ranges of up to twelve thousand yards, with some sonic contacts at twenty thousand yards under ideal conditions. Skip distances promised much greater detection ranges and revealed predictable ocean characteristics that gave naval officers and scientists a greater sense of comfort and intimacy with the environment below the surface. These discoveries also guaranteed that the ASW and submarine communities stood on the threshold of amazing advances in undersea warfare strategies, tactics, and technology.

Next to BT work and the skip-distance project, sonar and subsurface detection specialists realized that Ewing's channel research held the greatest promise. Thus, despite the disinterest voiced by BuShips's Radio Division some months before, Capt. Rawson Bennett of the bureau's Underwater Sound Design Division, Code 940, and his oceanographic aide, Lt. Comdr. Roger Revelle, stepped into the breach and continued navy support for Ewing's sound channel research.[26]

In January, 1944, Bennett authorized new tests that Woods Hole and New London would jointly conduct off the East Coast. Hampered at times by foul weather, Gale White of the New London laboratory cooperated with Doc Ewing's WHOI team in two series of experiments during March and April. The USS *Saluda* (IX-87) served as the sound laboratory afloat for the scientific teams. Built in 1919 by the American Shipbuilding Corporation of Hog Island, Pennsylvania, as a wooden-hulled, yawl-rigged yacht, the *Saluda* entered naval service on 17 October 1942. The submarine chasers SC-665 and SC-1292 and destroyer escort DE-51 were assigned to support the *Saluda*. The companion vessels carried explosive charges used by the scientists as the sound sources for the deep-channel experiments. As the escort crews set off sixteen-ounce dynamite charges at eighty-foot depths and four pounds at four thousand feet, repeating the sequence at hundred-mile intervals, the scientific team aboard the *Saluda* recorded the return echoes. For additional data and to provide knowledge of intermediate layers, the submarine chasers detonated depth charges at three hundred feet and tossed grenades over the side for near-surface explosions.

Ewing built his detection system around two hydrophones deployed from the ship. A standard Brush Development Company C-21 hydrophone was used to listen for shallow-water echoes. He then chose the same type of device, only this time enclosed in a castor oil–filled pressure case armed with a preamplifier, to listen at greater depths. The crew lowered the shallow instrument to eighty feet, and then lowered the C-21 in the pressure case to the axis of the Atlantic Ocean sound channel at about thirty-five hundred feet.

In each case the results astounded both the scientists and informed naval officers. In the case of the DE 51, Ewing tracked the destroyer escort out to 900 miles—the distance limit imposed by its orders. According to the report on these tests the Lehigh physicist prepared with Lamar Worzel: "The extreme limit for audibility of signals was not reached [at 900 miles] as the signal was still well above the noise level of the deep hydrophone. Multiple reflections were received by both the deep and shallow hydrophones to distances of 200 miles. The sound channel sounds bore out the theory almost in detail. The end of the sound channel transmission was so sharp that it was not possible for even the most unskilled observer to miss it."[27]

Tests run to explore the potential of stationary hydrophones positioned on the ocean floor did not immediately achieve the same success as the *Saluda* devices, but only for an unanticipated reason. A WHOI team placed a geophone on the ocean floor at forty-two hundred feet connected to a listening station on Eleuthera. This tropical island, located off the coast of Florida at roughly twenty-five degrees north latitude and seventy-five degrees west longitude, at first presented a rather inhospitable home for a listening station as the sharp offshore coral heads kept damaging the cable connection to the geophone.[28]

Engineering and materials problems notwithstanding, the experiments eventually succeeded beyond expectation. In the report's concluding comments, Ewing tried both to demonstrate the scope of capabilities afforded by the deep-channel phenomenon and to describe possible naval applications. That scope, by any standard, proved amazing: "In recent work not included in this report a half pound bomb was heard with abundant signal strength at over 800 miles and four pound bombs have been heard over 2,300 miles. . . . An extrapolation of the data indicates that ranges up to 10,000 miles are possible with this size charge."[29]

As Lamar Worzel later remembered it, the 2,300-mile experiment was quite spectacular and the methodology often rather strange:

One of the early experiments, I was on Eleuthera Island when we were able to send a ship over to Dakar in Africa from Eleuthera, and we listened to the . . . signal all the way to Dakar, which was 2,300 miles.

This was still a four-pound bomb. We still had 20 dB of signal left when it got to Dakar, into the shallow water, and it had to climb over the mid-Atlantic ridge, which means . . . some of the paths were interfered with by the ridge, and some of them were steepened by the ridge and some were then flattened out again, so that the signal was a more complicated thing after it crossed the ridge.

But [it was] . . . still distinct and we could still time it and so forth.

One of the interesting things we had then was Lieutenant—he was a lieutenant then—Edgar L. Newhouse, who was the project officer at the time, took a bunch of bombs in an airplane to fly across the ocean, to drop it for us in one of the experiments, and he couldn't get permission to do this by any other means.

At the time, . . . [airplane] toilets just opened outdoors. So he took a suitcase full of bombs on a government flight—I don't re-member whether it was MATS [Military Air Transport Service] or what—but on a government flight, and every hour he would disap-pear into the toilet and drop one of these bombs down the toilet hole into the ocean. And we got our array of sound signals across the ocean that we wanted.[30]

The possibility of active or passive detection at ranges measured in thou-sands rather than scores of miles amounted to nothing less than a revolution in human appreciation of sound behavior in seawater and would eventually re-quire a complete redefinition of ASW and submarine warfare.

In their enthusiasm, Ewing and Worzel suggested using the sound channel as part of an air-sea rescue system to recover downed flyers that eventually became known as sound fixing and ranging (SOFAR). With this system, a downed pilot might drop a small explosive charge in the ocean from his raft or, indeed, the sinking aircraft might carry a charge to the channel axis. Upon ex-ploding, the extraordinary channel transmission range would make triangulat-ing on the detonation from three listening stations very simple. The navy had only to install floating or bottom-mounted hydrophones with cables attached to shore listening facilities. Ewing and Worzel also hoped that the ability to tri-angulate on a deep-ocean position would arouse the interest of submarine officers who were naturally curious about submerged navigation and passive, or listening, sonar.[31]

While this plan had applications in both war and peace, it did not appeal to the navy's leadership. As the war came to a close, only those projects quickly convertible to offensive systems caught the attention of high-ranking leaders like Admiral Lockwood, the commander of Pacific submarine forces. Thus, while FM sonar possessed very strong appeal, the SOFAR project did not. It did not matter that the discovery of the sound channel might increase the value and range of existing sonar by an order of magnitude. Admiral Lockwood resisted any suggestion of investing time and money in long-term development in favor of the quickly deployable. In June, 1944, he remarked to Capts. Alan McCann and Frank Watkins: "The discovery by Woods Hole that small underwater charges can be heard at tremendous distances is very interesting but I do not see in it an immediate application to submarine warfare. Therefore I would not spend too much time or money on it in this war. Maybe it will be useful in the next. The time required for development and production in almost every case is so great that I feel we must confine ourselves to consideration of those which can be produced in reasonable time and which have an important bearing on submarine warfare." The admiral also placed the Long-Range Aid to Navigation (LORAN) system on production hold because the SJ radar then on his Pacific submarines provided "sufficiently accurate fixes even when our improved accuracy in celestial navigation is not useable."[32]

These reactions reveal a great deal about the collaboration between ocean science and naval decision makers in 1944–45. Lockwood felt comfortable settling for SJ capability over LORAN and thought nothing of setting SOFAR aside in the same breath. In his own mind, the admiral could not separate the offensive use of the ocean's natural characteristics from the development phase usually necessary for new hardware. The latter could well take more time than the Germans and Japanese had left, but the implementation of the former would require only current technology and a program to educate officers and seamen in the new possibilities open to them by existing natural conditions.

In making their SOFAR proposal, Ewing and Worzel failed to suggest direct, offensive, combat applications for the sound channel. This proved a major mistake. In 1945, naval officers wanted weapons to defeat a nearly exhausted enemy. Deep-sea communication systems and air-sea rescue, while appealing, seemed passive and failed to promise a more rapid victory. Furthermore, the naval culture did not prepare its officers to recognize ocean characteristics as immediately available enhancements to combat systems already deployed. Neither were most officers inclined to seek the kind of oceanographic understanding that would lead to a quick appreciation of the sound channel's potential. The QLA sonar, on the other hand, seemed familiar, tangible, and rapidly deployable; characteristics that historically and culturally appealed to naval leaders facing a determined foe.

By 1944 it had become clear that the naval and ocean science communities would need more time in a high-pressure environment similar to that generated by the war before bridging the cultural gulf between them. While the war provided a foundation recognized by Iselin, Ewing, Vine, Revelle, BuShips's Rawson Bennett, and submariners like Lawson Ramage, there was still a long way to go.[33]

Indeed, in the last year of the global contest, funds for sound channel research actually began to dry up. On 15 March, 1945, Iselin, physicist George Woollard, McCurdy, and Worzel met with the navy at Woods Hole to discuss the future of the deep-sea sound program. The navy had already instructed Ewing to remove his people and equipment from Eleuthera and according to Lt. Comdr. Walter C. Sands of BuShips, budget plans for fiscal years 1946–47 did not include any money for sound channel work. Iselin suggested using the $50,000 remaining in the 1945 budget originally dedicated to the Eleuthera operation for further exploration of the channel and its properties. Ewing wanted to demonstrate the feasibility of SOFAR or some other channel-related work in Admiral Lockwood's backyard. Sands agreed to this as long as the navy had cable available in the Pacific area. He also made it clear that the navy would not provide a ship for laying cable. For that, Ewing and company would have to approach the army. These circumstances delayed for two years Woods Hole's completion of the first experimental Pacific SOFAR array.[34]

Although the budget for sound channel research did not disappear entirely, it took time for the navy to look beyond both its immediate needs and the cultural gap between the naval and scientific communities. The Bureau of Ships's Code 940, even after the postwar departure of both Bennett and Revelle, managed to find enough money to support some channel research. However, it would take nearly four years, and a rancorous Capitol Hill debate over naval roles and missions, coupled with the advent of the Cold War, to bring the navy's submarine operators and ASW officers to Doc Ewing's doorstep. When they did, few realized that the ace eventually played against the Soviet submarine fleet over the next forty years had been theirs for the taking as early as 1937.

10

Transition, 1945–46

[O]n my way to the train station . . . a small kid on the sidewalk
there in Woods Hole took his cap pistol and pointed it at me
and said, "Bang! Bang! You're dead!" and shot me with a
couple of caps going off. Well, this was just exactly the sort of
thing we had been looking for.

—J. LAMAR WORZEL Oral Commentary, 10 December 1980

IN DECEMBER, 1944, COMDR. ELI REICH FOUND HIMSELF COM-
pletely out of his element. Ordered into the cold winter of Washington, D.C.,
this Pearl Harbor–based submariner from Admiral Lockwood's ComSubPac
staff rushed to complete his official business so he could return to his work and
the warmth of Hawaii. Among the many issues that Reich addressed in his
trip report to the admiral upon his return, one disturbing note in particular
prompted Lockwood to fire off a three-page letter to the Office of the Coordinator of Research and Development. To his knowledge, the Japanese had not
yet surrendered. Until they did, Lockwood refused to give up his scientists.

Reich reported that the NDRC had begun the process of demobilization
and would soon shift the numerous tasks of Division 6 to naval bureaus, laboratories, and some private contractors. Lockwood earnestly reminded Admiral
Furer that American submarines still had a war to fight and that the most dangerous days might lie ahead. Lockwood's submarines now operated in the
shallower waters both off the Japanese coast and between the home islands
and Korea. Scientific tools and advice seemed more important than ever. He
argued, for example, that FM sonar installation and training had only just

begun. He needed the mine and torpedo detector under development at the CUDWR's New London laboratory and wanted the NDRC to complete the work on JT sonar and underwater telephones before closing up shop. Other pressing issues included harbor beacons, sonic decoys, and acoustic coatings for submarine conning towers as a defense against Japanese search radar. Without the NDRC laboratories and their technicians and advisers working with the fleet, the final months of the war against Japan could prove more costly than necessary.

Admiral Furer moved quickly to allay Lockwood's concerns. On 3 February 1945 he calmly outlined the planned disposition of naval projects and NDRC contracts for civilian scientific services, assuring the ComSubPac that he would not lose vital research and development projects or technical advice from field experts.[1]

In this exchange roughly seven months before V-J Day, both admirals demonstrated the great promise and fundamental challenges naval patronage would present to the oceanographic community in the postwar era. Although Lockwood could not help but worry given his wartime responsibilities, he clearly depended upon scientists working in undersea warfare fields and accepted them as a necessary and desirable component of the national defense. Although he occasionally had difficulty seeing beyond the current conflict, as in the case of SOFAR and the deep-sea sound channel, while the war continued he looked for every way possible to destroy his enemy and survive.

Lockwood reflected the thinking of his wartime naval colleagues who came to appreciate the value of oceanography through both combat experience and the vital efforts of translators and officers who came to view applied science as an asset. Many senior officers and younger frontline veterans vigorously advocated naval support for oceanography. Virtually no one suggested returning to the old common practice involving the Hydrographic Office.[2]

The postwar challenge to naval oceanography thus took the form of an unusual crisis of plenty rather than the far too familiar crisis of poverty. Although Iselin, Sverdrup, and their colleagues feared a return to prewar conditions complicated by much higher personnel and ship-operating costs, the navy's early commitment to postwar oceanographic research offered only limited comfort. While the navy seemed more approachable and its significant resources presented new and encouraging peacetime possibilities, the war precipitated fundamental changes in ways both promising and disturbing. Available funds now approached the billions of dollars, and the consequences of exponential growth driven by naval requirements and federal funding forced the ocean science community to address issues directly related to professional identity, scientific practice, and disciplinary culture.

From 1945 to 1950, while the navy struggled through the political blood-letting of the defense unification crisis and redefined its role in the national defense structure, the oceanographic community voiced its distress over the dynamics of postwar naval patronage. Most scientists went directly to the essential questions. To what extent could oceanographers determine their own scientific course and sustain professional standards in the face of overwhelming naval funding opportunities and demanding defense research priorities? In 1945, oceanography could boast only a short history as an independent scientific specialty. As before the war, this young, multidisciplinary science remained an awkward departure from the classic fields of scientific endeavor, with their ancient practices and expectations as well as cultural demands and rewards. Only a few marine scientists during and after World War II identified themselves as oceanographers, but many trained investigators studied the ocean. Thus, those strongly advocating oceanography as a naval asset at the same time found themselves debating the nature of this specialty with colleagues from many disciplines. Might the navy eventually play a greater part in defining that science than its trained practitioners? Should civilian oceanographic activities make a determined effort to solicit private sector funding in amounts that would give better balance to their research programs? Given the size of the naval commitment and the amazing expense of doing postwar oceanography at sea, would industry participate? The private sector might well recoil at the level of funding required in the ocean sciences. If the navy became the only major fiscal reservoir for oceanography, then science might have to live permanently with a situation analogous to the frantic pace of 1940 and the hasty science of the war years. In addition, how could oceanographers steer the navy away from its emphasis on applied research toward a balanced program? Applied research certainly led to the deployment of new weapons and systems more quickly, but the basic or pure variety expanded options and offered potential, constantly redefining the possible. Would the navy fund the latter as well as the former?

In 1945–46, the oceanographic community realized that it had only begun to engage in effective dialogue with the navy. In battle, officers now desired their presence, advice, and methods to enhance combat effectiveness. However, by 1945, the question had become less one of heterogeneously engineering a role for oceanography in naval warfare than of trying to influence the future course taken by the ocean sciences in the context of Cold War naval commitments. In the balance lay the professional identity of ocean scientists and the very nature and sustained quality of their work.

The NDRC's Conversion to Peacetime

By early 1945 most of the laboratories and activities engaged in undersea warfare for the NDRC had a clear vision of their immediate future. As chairman of the OSRD, Vannevar Bush worked with James Conant, his successor at the NDRC, and John Tate, the chief of NDRC Division Six, to smooth the transition from war to peace. Given the navy's intense interest in the work done by Division Six, integrating the various laboratories and projects into the navy's existing research and development activities presented only a minor problem.

The oceanographic community lost few wartime ventures in the process of transitioning to a peacetime environment. At the beginning of March, 1945, the CUDWR's New London laboratory passed under NRL authority despite a call from some submarine officers for its transformation into an independent naval submarine research laboratory. After the NRL director, Rear Adm. Alexander van Keuren, assured submariners of prompt and careful attention to their priorities, he spent the next three months combining New London's work with that imported from Frederick Hunt's slowly demobilizing Harvard Underwater Sound Laboratory (HUSL). Van Keuren used these components and missions to lay the foundation for New London's U.S. Navy Underwater Sound Laboratory (USNUSL), a vital component of postwar undersea warfare research activity.[3]

Harvard University did not sustain a laboratory devoted to sonar and undersea warfare after the war, and many scientists and students returned to prewar research rather than make the transition to the USNUSL.[4] Eric Walker, who worked with Hunt at HUSL, took some of the Harvard staff with him and created the applied physics laboratory that first served Pennsylvania State College, and eventually Pennsylvania State University. With Woods Hole absorbing its sonar analysis group as part of a contract with BuShips, CUDWR's last remaining operation in New London, the sonar-training group, remained at the New London Submarine Base operating under a contract between the Bureau of Personnel (BuPers) and Columbia University.[5]

On the West Coast, BuShips took over the UCDWR contract from the NDRC and continued the university's work at Point Loma until 30 June 1945.[6] The following year, BuShips's chief, Vice Adm. Edward L. Cochrane, negotiated a new arrangement with the University of California that created the Marine Physcial Laboratory (MPL) under the eminent physicist Carl Eckart.[7] In a 31 January 1946 letter to university president Robert Gordon Sproul suggesting the creation of the new laboratory, Cochrane clarified the navy's motives, the mounting importance of oceanography to critical naval operations, and the basic task of the infant MPL: "The Bureau of Ships is convinced that a continuation of this fundamental research forms an indispensable part of a farsighted naval defense program . . . the relative difficulty of detecting underwater objects by means of sound indicate that the submarine will continue to occupy an important position in naval warfare. It is believed that a detailed quantitative knowledge of the fundamental factors affecting the performance of underwater sound gear is a powerful guarantee that anti-submarine and pro-submarine techniques used by the U.S. Navy can be made most effective. . . . It is therefore the intention of the Bureau to promote basic research on the physics and acoustics of underwater sound on a long-range basis."[8]

Commander Roger Revelle, still detached from Hydro on temporary duty at BuShips's Code 940, played an essential liaison role for this contract by acting as one of Cochrane's representatives to the university during the yearlong period of negotiation and clarification of contract specifics. Revelle also nominated Eckart to serve as the MPL's first director and constantly pushed the university for the earliest possible decision.

As negotiations progressed with the full support of Harald Sverdrup and his staff at Scripps,[9] Cochrane agreed to shift contract responsibilities and $50,000 seed money for the MPL project to Admiral Bowen's Office of Research and Inventions (ORI). With the demobilization of OSRD/NDRC and the navy's expanding role as a prime mover in the ocean sciences, Secretary of the Navy James Forrestal created the ORI on 19 May 1945 to oversee postwar naval scientific research. The ORI assumed control of the Office of Patents and Inventions (OPI), the NRL, and Admiral Furer's wartime OCRD.[10] Since the University of California had already inaugurated negotiations with Admiral Bowen for a research contract, Sproul wanted the very flexible ORI contract provisions and accounting policies to govern all of the university's navy work. Funded by BuShips, the ORI, and the university, the MPL opened on 1 July 1946 with Eckart at the helm in quarters at Point Loma that belonged to the demobilizing UCDWR.[11]

Attached by the navy to the Point Loma radio and sound laboratory, several of the remaining UCDWR engineering groups formed the basis for the postwar Naval Electronics Laboratory (NEL), devoted to sonar training aids and submarine acoustic countermeasures. Frequency-modulated sonar, the most important of the UCDWR's wartime undertakings, did not remain in California for subsequent development. This all-important project migrated east to the USNUSL for further refinement.[12]

As the leading oceanographic laboratory, Woods Hole renegotiated its NDRC contracts with BuShips and emphasized underwater sound as its premier naval research field. Woods Hole and the USNUSL provided the East Coast complement to the SIO, the MPL, and the NEL. All other private commercial contractors followed a similar route, shifting their work from the NDRC to one of the naval bureaus or activities. The Navy Department also decided to establish formally an evaluation center to set and monitor standards for measurement, data collection, and instrument calibration. As the U.S. Naval Underwater Sound Reference Laboratory (USNUSRL), this activity remained at its Orlando, Florida, wartime location and continued work on the standards and measurements initiated before Pearl Harbor by the navy and the NDRC at both Orlando and Barcroft Reservoir, Virginia.

In addition, the navy also continued those field services valued so highly by Admiral Lockwood and provided by the New London laboratory, the UCDWR, select contractors, Scripps, and Woods Hole. In February, 1945, the coordinator's office planned three field teams derived from a stable of twenty-five technical experts. The first group worked for the remaining months of the war at ComSubPac headquarters in Pearl Harbor. Another was sent to the New London laboratory as it reorganized into the USNUSL, and the last advised the CNO in Washington.[13]

From a legal standpoint, most naval activities entered into new contracts of larger scope to provide both lawful shelter and funding sources for an array of wartime projects. At Woods Hole, for example, a fresh BuShips contract, NObs-2083, took the place of wartime contract OEMsr-31 and redefined responsibility for most of the work supported by the old NObs-17583 and Navy Project NS-140. Eight war tasks thus fell under the new bureau contract, including underwater sound, submarine diving and ballasting, ship-bottom fouling, development of instruments related to undersea warfare, waves, surf, meteorology, and consultation services for the USNUSL.

In a few unusual cases, a home seemed hard to find. The CUDWR's sonar analysis group—led during the war by Lyman Spitzer, Jr., from his offices in New York City's Empire State Building—traveled a bit before finding a final residence under contract to Brown University in Providence, Rhode Island. When CUDWR demobilized, BuShips took over the group, placing it first in Code 940 and then transferring it physically to Woods Hole in support of the underwater sound work at that laboratory. This arrangement lasted until 1947, when the National Research Council took over sonar analysis for a short time before diverting it to Brown in the autumn of 1948.[14]

The demonstrated value of oceanography to ASW and prosubmarine operations assured scientists and institutions of continued naval attention. The care with which the navy integrated NDRC veterans and projects into the postwar naval establishment confirmed the recognition within its ranks that applied oceanographic research merited a significant role in naval warfare. The difficulty thus was no longer simple recognition, role, or patronage levels. In the absence of the OSRD and NDRC, both the navy and civilian science had to develop a relationship that would form the foundation for policy decisions governing most oceanographic activity in the United States for the foreseeable future. But what form would that relationship take? Who would make policy? How much personal and professional freedom could oceanographers expect to exercise?

Converted and Curious

Demonstrating a large measure of appreciation for the ocean sciences born of wartime combat experience, BuShips's Admiral Cochrane initiated the flow of critical financing and technical support for the new Marine Physical Laboratory in San Diego with a letter to the University of California on 31 January 1946. In signing that letter and a series of others that emerged from his office that same day, Cochrane launched a policy as well as a laboratory. Given the bureau's interest in a ship's design, construction, and survivability, Cochrane's personal interest in oceanography and underwater acoustics was not surprising. While the MPL rejoiced at having BuShips's support and inaugurated its program in underwater sound and submarine geophysics, Woods Hole also received one of Cochrane's 31 January letters. The bureau viewed Iselin's institution as the best possible East Coast counterpart to the MPL's activity. The admiral pointed out that "representatives of the bureau have already approached the University of California informally with the request that the University carry out a long-range program of research in underwater acoustics. . . . It appears to the Bureau that the participation of the Woods Hole Oceanographic Institution in a parallel program along these lines would be highly desirable. Carrying out fundamental research at several different institutions would provide an important guarantee against the stagnation that might result if the work were concentrated in a single locality, and would also broaden the variety of scientific talent concerned with the underwater sound problem."[15]

Cochrane's timely actions thus assured the survival of the demobilized UCDWR acoustics programs with its transfer to the MPL. His actions also guaranteed the continuity of Woods Hole's underwater sound programs, with particular emphasis on further research into the deep-sea sound channel, an asset with great potential value understood by very few at the bureau or at ComSubPac in 1944–45.

Cochrane's policy of strong support for underwater acoustics continued BuShips's natural concern with submarine quieting and the development and deployment of effective sonar. The bureau's initial peacetime interest in SOFAR did not derive from a sudden appreciation of the diverse warfare applications of the deep sound channel. The SOFAR's potential and the strong Cold War motives that would drive the navy, science, and industry to fully exploit it remained in the near future. Rather, the bureau intended to experiment in the Pacific with Ewing's original SOFAR proposal by setting up a system for air-sea rescue that the Coast Guard would eventually inherit when the navy and its contractors proved the system was reliable.[16]

Not quite six months after Cochrane's letters to Woods Hole and the MPL, SOFAR had assumed the mantle of a well-financed official navy project. The Bureau of Ordnance (BuOrd) agreed to produce five thousand SOFAR bombs for $580,000 in cooperation with BuShips, with the former supplying the explosives and ordnance expertise as well as roughly 8 percent of the cost. The Bureau of Ships planned to use these bombs in an air-sea rescue zone between the Hawaiian Islands and the West Coast. Monterey and Point Arena in California along with Hilo and Kaneohe in the Hawaiian Islands would serve as listening-station sites to effect location by triangulation for crippled ships or downed flyers. This would give the Coast Guard and naval rescue personnel the best coordinates for a recovery.[17]

Woods Hole agreed to provide the scientific advice to facilitate the installation and the monitoring. Because Maurice Ewing had accepted a position at Columbia University in 1946 and had departed Woods Hole to move his base of operations from Lehigh University to New York City, it fell on Lamar Worzel to provide the SOFAR project officer, Lieutenant Commander Sands, with advice critical to the evaluation of proposed Pacific listening-station sites.[18] In September, Carl Hayes and Jack Stanley left Woods Hole for Mare Island Naval Shipyard near San Francisco. George Voigt and Thomas Condron followed them in October. Meredith Williams relieved Hayes at the end of the year.

Woods Hole personnel spent the end of 1946 and early 1947 participating in detailed hydrographic surveys of the proposed station sites at Monterey and Point Arena. They also conducted bathymetric surveys of the submerged locations tentatively selected for the SOFAR hydrophones. The survey team could not find a sharply defined sound-channel axis off Point Arena, but a course of minimum sound velocity did occur between fifteen hundred and twenty-five hundred feet, fixing the Pacific sound channel at a much shallower depth than the corresponding phenomenon in the Atlantic. The navy provided the destroyer escort USS *Fieberling* (DE-640) for the survey work, as well as helpful information gathered by USS *Sumner* the pervious year to advance the analysis of location suitability.

The USNUSL in New London developed the monitoring equipment and hydrophones and participated in the fundamental research with Woods Hole. Inexperienced with the deep sound channel, the NEL staff worked closely with the East Coast institutions, preparing to assume the role of SOFAR manager and maintenance activity. As the San Diego–based laboratory gradually acquired the necessary expertise, it also made sense for the bureau to transfer responsibility for Pacific SOFAR oceanographic support from Woods Hole to Scripps in La Jolla.

Worzel gave Sands the benefit of his experiences on Eleuthera, Ewing's wartime prototype SOFAR listening station. First he warned Sands about errors of position. Each hydrophone needed to rest on a known set of coordinates. The installation team thus required data from careful surveys and needed well-defined points of reference on shore, accurate radar sets, and proper optical instruments. The best and most complete gravitational and astronomical observations would ensure that errors of up to a mile, which had occurred at Eleuthera, would not reoccur. Worzel also insisted upon adequate BT measurements and water-bottle samples. Deepwater temperature and chemistry would provide information on the refractive quality of the water at the sound-channel axis.

On the engineering side, Ewing's long-time assistant suggested simultaneous installation of various types of cable and one of the new magnetostriction hydrophones. This would permit concurrent testing and future selection of the best possible materials. The Eleuthera episode also indicated that hydrophones suspended above the bottom offered a better field of coverage and more readily avoided the shadow zones caused by submerged geologic structures. In all of his advice to Sands, Worzel did not fail to praise the USNUSL's contribution to improving sound recording devices far beyond his early Eleuthera experience.[19]

Unfortunately, SOFAR bomb procurement caused the ultimate bottleneck in the entire process. From the time it first began working with explosives as a sound source in the late 1930s, the Ewing group had always devised its own systems for deploying the charges, and in some cases even designed its own detonators. Called upon in the early spring of 1946 to approve the bombs that would make the SOFAR search and rescue system work, BuOrd declined to certify the detonating systems for the Mark 15 depth charge sound source designed at Woods Hole. It also objected to the demolition-block pressure detonators designed by Worzel and his WHOI assistant, Harry Robinson.

The latter case employed an unusual method of activating the detonator that made BuOrd personnel nervous and demonstrated the style and habits of the Ewing group in its relentless pursuit of research goals. Worzel and Ewing selected a type of tubing that would collapse with fair regularity at the axis of the sound channel, roughly four thousand feet below the surface in the Atlantic Ocean. Worzel later recalled that they first lined the inside of the tube with the heads of matchsticks to precipitate a friction spark or fire when the tube collapsed. This would ignite a number twelve nonelectric detonator selected for the job. Because lining the inside of each tube was such tedious work, Ewing requested some of the flammable paste the Diamond Match Company used in manufacturing matches. With this in hand, they could simply paint the substance on the inside of the tubes. When the company objected because it could not supervise the use of the combustible mixture to ensure safety and reduce liability, Ewing and his staff looked elsewhere.

On his way to the old Woods Hole train station for a trip to the DuPont Company in Delaware to obtain something that might emulate the match paste, Worzel made a remarkable discovery that completed the WHOI detonator. Although the substance worked well, it aroused safety concerns at BuOrd:

> [O]n my way to the train station . . . a small kid on the sidewalk there in Woods Hole took his cap pistol and pointed it at me and said, "Bang! Bang! You're dead!" and shot me with a couple of caps going off. Well, this was just exactly the sort of thing we had been looking for; the cap pistol ammunition makes a spark and so forth. So I grabbed the kid and said, "Where did you get this?" Well, he was so astonished and so upset that for a while he couldn't talk, and finally he calmed down and I calmed down, and he told me what store he had bought the cap pistol ammunition in. So I went down to the railroad station and called Harry Robinson . . . and told him where to buy the cap pistol ammunition. . . . When I returned . . . Harry . . . had made tests by putting in a series of five caps into the firing end of the collapsing cylinder and every single one of them had been successful. . . . These we used and successfully fired many, many shots, and did the important experiments on Eleuthera.[20]

With Woods Hole, the USNUSL, and BuShips taking the lead, the navy installed its first SOFAR network during the twelve-month period beginning 27 June 1946. Along with the funds authorized for the Pacific air-sea rescue system, BuShips found sufficient money to continue the research on Eleuthera. For many months, Ewing's installation on the island, by then under naval control, remained the only land-based station in the Atlantic Ocean exploring the deep sound channel.[21]

Wartime Observations

ARELY THREE MONTHS AFTER WEST COAST RESEARCH PLAN-
ning began in 1941 and only two weeks after R. J. McCurdy's report on
the temperature-depth recorder, the Japanese attacked Pearl Harbor in
the Hawaiian Islands. In the seventeen months of preparation afforded the
oceanographic community before 7 December 1941, American oceanographers
and the navy had only begun to explore the advantages of establishing an effective dialogue that would take them beyond the interwar common practice.

While the latter integrated scientists and naval personnel to a modest extent, it merely demonstrated the possibilities of a professional dialogue to those
who chose to listen. For the navy to exploit its combat environment to the
fullest, the two communities of practitioners would have to begin sharing a
common professional language and a set of mutual standards and preferred
goals against which they would measure performance and the value of research.
Indeed, both ocean scientists and naval officers would find it difficult to define
their respective activities without reference to the other in such an environment. Unfortunately, a relationship of this nature did not develop between
officers and oceanographers during World War II.

Instead, the demands of mobilization and conflict combined with the entrepreneurship of key scientists to demonstrate clearly the importance of the
ocean sciences to successful naval operations. As a result, in a way both very
similar to the decisive action that brought the NDRC into being and reminiscent of the interwar activities of Thomas Wayland Vaughan and Admiral Gherardi, national need and personal initiative took control. Individual scientists
and officers helped oceanography slowly assume a place in the war effort even
though a considerable professional and cultural gulf between the two communities remained.

During this hectic period, Columbus Iselin, more than any other ocean scientist, fashioned the role oceanography would play in the opening months of

American involvement in the world conflict. His frequent trips to Washington even before the NDRC first met demonstrated that the young WHOI director saw in the international struggle both the pressing need for a more effective national defense and an important part in that defense for oceanography. Neither was he blind to the political and economic benefits that might accrue to Woods Hole from early involvement in defense research.

Iselin not only understood the kind of research that might prove valuable, but as one of the most effective cultural translators, presented his case successfully before a wide variety of people in a seemingly endless procession of meetings. All the while, he displayed a natural rapport and easy manner that enhanced his ability to communicate with subordinates, university colleagues, naval officers, and national policy makers. This talent, for example, enabled him to work effectively with the multifaceted Frank Jewett—neighbor, patron, WHOI trustee, and leader of science and industry.

Iselin also knew how to listen. He could meet with officers like Comdr. Robert Briscoe and Lt. Comdr. William Pryor, as well as naval scientists like Harvey Hayes, and come away with a remarkably clear picture of the circumstances within the navy and the task facing oceanography. In his work with Briscoe, Pryor, and other naval officers, Iselin discovered both cultural obstacles to effective collaboration with the navy and a deeper understanding of the professional community he hoped to penetrate.

He thus quickly began to appreciate the quandary of very able line officers often handicapped by their own traditions, practices, and standards. The traditions of naval warfare had led Pryor to order that the BT be used in sea exercises off Key West without employing the results to his advantage in a submarine search. In the heat of action or in training, war fighting did not yet include scientific instruments—even for one who had come to appreciate the latter. Naval culture could often prove harder to overcome than the forces of nature.

Working at Harvey Hayes's side at the NRL, the WHOI director practiced the art of dovetailing his own agenda—this time underwater sound research—with a long-standing navy program already in place. The relatively new discipline of oceanography instrumentation design was both a vital tool and a persuasive asset. Hayes realized this in 1923 when his sonic depth finder drew the navy's attention to oceanography's broad potential for the first time. In addition to furnishing important data, other simple and scientifically powerful instruments like the BT brought Iselin the attention and credibility he needed to begin penetrating the culture of traditional naval warfare. The established instrument makers had designed few tools with this kind of scientific and naval significance. If the frontline navy would regularly use the BT at sea, Iselin believed oceanography would find a permanent place in naval warfare.

His effort to heterogeneously engineer a role for both oceanography and his institution in the American preparation for war proved as educational for Iselin as it was for the officers and scientists inside the navy. While exploring the potential in WHOI–U.S. Navy cooperation, Iselin received an education in navy customs and conventions. Given his familiarity with the common practice created by Hydro and civilian science before 1939, as well as his own interwar work with the navy on the afternoon effect, Iselin understood the limits of the navy's interest in oceanography and some of the pitfalls that came with dependence upon the federal government and the naval service. However, merely achieving this degree of understanding did not prepare him for actually trying to practice applied oceanography with naval personnel aboard warships at sea. This required far more than his considerable scientific expertise, seamanship, and refined patrician social skills. He never found an informed audience like the staff at Hydro or the sympathetic style of Frank Jewett among those who had to perform dangerous convoy escort duty on board navy destroyers in the U-boat-infested North Atlantic.

Working at the Key West sound school with Lieutenant Commander Pryor gave the WHOI director actual experience with these problems and a key to possible solutions. Iselin soon realized that the war-fighting community required not only tangible evidence of oceanography's utility, it also demanded swift, culturally familiar answers immediately applicable to naval problems. Thus, while he appeared determined to wait "until they begged for more," his continued effort to cement the relationship demonstrated a decision to provide for the navy's needs. He knew that the submarine war would repeatedly bring naval officers to his door. When it did, he had no intention of playing hard to get. Iselin not only believed the national defense demanded WHOI's commitment, he also felt attracted to a powerful patron that could ensure his institution's present and future.

Fortunately, the WHOI director could build his entrepreneurial and translation strategies upon ingenuity, remarkable insight, and elegant solutions almost on demand. In luring the Lehigh University group to Woods Hole, Iselin both enhanced WHOI's scientific authority in an area critical to naval warfare and promoted the creativity, flexibility, and relentlessness that were Maurice Ewing's trademarks. He watched student-physicist Allyn Vine both study the physics of underwater sound and bridge the gap between a scientist and his data. Vine's technical and engineering magic exhibited to an unprecedented degree the value of the engineer and technician in fashioning the instruments and gear necessary for an oceanographer to achieve the goals of a scientific project. His variation on the Rossby-Spilhaus BT provided a foundation for the underwater sound research effort, offered a simple way of increasing vital oceanographic data, and offered naval personnel a tangible demonstration of oceanography's critical role in sea warfare.

Beyond Vine's BT and underwater sound research, the presence of the Ewing group also affected in other ways Iselin's decision to continue aggressively building a relationship with the navy. Their habits at sea helped provide Iselin with a useful perspective on the perennial issue of naval control over scientific inquiry. Always concerned about the long-term effect of official patronage, many ocean scientists wanted the navy's assistance and realized the need to participate in the national defense, but ethically questioned the long-term effect of applied research driven by tax dollars on their community and work. Of all those working at Woods Hole during the war, perhaps Maurice Ewing felt this concern the least. He had long since realized that he had to surrender very little of his independence, even to a powerful patron. In the tradition of the interwar common practice Ewing, Vine, Worzel and their team constantly employed piggybacking to satisfy both personal and national defense goals. Laboring around the clock at sea for the navy, the Ewing group would do research to satisfy its own scientific agenda in parallel with the work called for by their defense commitments. With the increasing demands of mobilization and the practical effect of Iselin's lobbying taking hold during 1940–41, the amount of personal scientific research they could accomplish on navy-sponsored voyages decreased. Even so, their private research never ceased, and working with the navy brought to their attention areas of inquiry they might not have addressed otherwise. As Doc Ewing exclaimed to Iselin at the end of 1943: "I see the most exciting projects I have ever dreamed of for doing my kind of work after the war is won. I just sit and pant whenever I think of it."[1]

Columbus Iselin recognized very early the opportunities for oceanography in the navy's preparation for war. While certainly not alone in urging ocean scientists to participate in national defense projects before the Pearl Harbor disaster, no other oceanographer played such an early and determining role in engineering naval appreciation of applied oceanography. Iselin initiated a wartime relationship between the navy and the oceanographic community that, while still relatively immature in 1945, certainly placed both his science and his institution in the heat of battle and helped set the stage for their very significant role in the postwar world.

The cultural translation employed by Iselin and his more effective wartime colleagues in persuading naval officers had little to do with rendering "navalese" intelligible to the scientific community. As anthropologist Clifford Geertz described it, this kind of cultural translation "is not a simple recasting of others' ways of putting things in terms of our own ways of putting them (that is the kind in which things get lost), but displaying the logic of their ways of putting them in the locution of ours; a conception which again brings it rather closer to what a critic does to illumine a poem than what an astronomer does to account for a star."

The navy and ocean science communities needed to appreciate an alternative view of the world. This went far beyond merely accounting for differences in vocabulary. Both scientific and naval professionals cast their lives in terms of their research or warfare at sea. They lived their respective professions as vocations and, whether practicing ocean science or waging naval warfare, it permeated every aspect of their lives. To quote Geertz once again, "It appears in their religion, their morality, their science, their commerce, their technology, their politics, their amusements, their law, and even in the way they organize their everyday practical existence."[2]

It fell to the translators to understand the depth and scope of cultural experiences and to provide essential lines of communication in a time of national emergency. Furthermore, as both translator and director of a struggling oceanographic institution, Iselin understood well the future scientific and material benefits of effective communication with the navy. He earnestly sought to practice good science, but he also wanted to secure survival and prosperity for his institution.

As Iselin and Ewing discovered, it took time to impress upon warship crews the value of applied oceanography. Throughout the conflict, civilian and naval apostles of the ocean sciences constantly exercised their entrepreneurial and translation skills, selling and reselling the applicability of those technical and natural assets developed or discovered by NDRC oceanographers. Fritz Fuglister of WHOI and Eugene LaFond of Scripps, managing the BT programs, found themselves frequently reminding the forces afloat to use the instrument and preserve the slides properly while preparing them for shipment to Woods Hole or La Jolla.[3]

Just as the forces afloat often proved very difficult students, set in their ways and hard to convert, the scientific community had much to learn about the realities of naval warfare. While both cultures drew their primary motivation for cooperation from the need to win the war, time and the desperate circumstances deepened their mutual understanding. Submarine warfare alone made a better appreciation of the ocean environment imperative. Effective submariners and ASW officers soon realized that applied oceanography improved a ship's chance of survival and increased the likelihood that crewmembers would again see their families after a difficult North Atlantic convoy crossing or a submarine war patrol near the Japanese home islands. For their part, many anxious scientists also earnestly hoped that the profitable wartime application of oceanography and the lives spared in combat would induce the navy to become the generous patron that Harvey Hayes had envisioned and that Theodore Roosevelt, Jr., championed after the Great War. If a return to the common practice proved their only alternative, oceanography's future looked bleak.

Driven by these multiple motives, many scientists found themselves doing more instruction and training than pure scientific research. To those used to doing personal research in the laboratory or at sea, these tasks usually proved boring and repetitive. For most professionals, educating officers and technicians, collecting data, composing reports and sailing directions, and doing everything on a tight and often frantic schedule did not feel like science. Gordon Riley spoke eloquently for those obliged to perform on demand, haunted by the feeling that their efforts might neither significantly increase knowledge nor help those facing combat.

Nonetheless, Riley, Martin Johnson, Dean Bumpus, Mary Sears, Allyn Vine, Lamar Worzel, Walter Munk, William Schevill, and scores of others remained on task buoyed by the results they did achieve. While the war raged in far away places, it doubtlessly seemed a bit unreal cooped up in office buildings and shore establishments performing fundamental, repetitive tasks in basic oceanography for naval personnel who occasionally questioned their value. To adapt a frequently used description of combat, wartime oceanography treated the majority of scientific personnel to long hours of routine and boredom only occasionally punctuated by very brief moments of excitement, discovery, and carefully executed science.

During those brief moments of discovery, and because scientists tolerated the boredom and continued to work, naval officers and their crews went to sea better prepared to face their enemy. The *Sailing Directions* provided vital environmental information for ship operators at the front. Educating BT observers made the collection of quality data for tactical application and postwar use increasingly easy and reliable. Marine biology, BT data, ocean-bottom geology, and marine chemistry further advanced the quality of the ocean intelligence published in navigational aids, charts, and sailing directions. These same efforts laid the theoretical groundwork for shipboard FXRs, sound sources that helped decoy enemy acoustic torpedoes; for research on the surface wake scars left by submerged submarines; for first generation sonic harbor defense systems; and for the FM sonar that so excited the imagination of Admiral Lockwood at ComSubPac headquarters.

The most stimulating and fruitful wartime oceanographic research stemmed from the insights and activity of the Lehigh University group. Driven by Doc Ewing's insatiable curiosity and tireless work habits, his group produced the cardinal discoveries and instruments that bore directly upon the major tasks of the naval war. Ewing, Vine, and Worzel became parallel pioneers with Francis Jenkins of the University of California in underwater photography by taking their prewar work and applying it to the navy's salvage and ordnance needs after Pearl Harbor. More significantly, however, wartime efforts to understand the behavior of sound underwater and the influence of ocean-bottom composition depended completely upon the Iselin-Ewing analysis completed in February, 1941. These insights proved essential in the development of secure echo-sounding equipment by Carl Eckart of Scripps. Harald Sverdrup's chapters on physical oceanography in *The Oceans* no doubt drew on this same knowledge.[4] The significance of ocean temperature stratification to the development of new sonars at the UCDWR and the CUDWR's New London laboratory—fields opened wide by Ewing's sound work and Vine's refinement of the BT—drew Walter Munk's attention to the ocean's temperature microstructure and gave proper emphasis to the bottom composition research by Francis Shepard and Kenneth O. Emery at the UCDWR. Sverdrup and Munk suspected that the structure of the ocean layers revealed by BT temperature readings might prove far more subtle and dynamic than indicated by an instrument designed to give an ASW officer a quick profile and possible assured range. Furthermore, once in the relatively shallow waters of the Sea of Japan, the materials lying on the bottom and the nature of the sediment layers beneath would promote or inhibit active sonar and passive listening by returning harsh and unintelligible echoes or absorbing sound waves. These proved critical factors both in the creation of FM sonar by the UCDWR late in the war and in the sound detection work conducted by New London and Frederick Hunt's group at the Harvard Underwater Sound Laboratory. Ewing even set the stage for the most important applied oceanography and geophysics of the postwar era by discovering the deep sound channel and postulating for submariners and ASW specialists the possibility of sound fixing and ranging (SOFAR) over thousands of miles of ocean.

Next to Iselin, Ewing, and Vine, Roger Revelle played the most significant wartime role in bridging the gap between oceanographers and naval officers. Like most of his colleagues, Revelle left La Jolla and Scripps to go to war at the USNRSL at Point Loma before his transfer to Hydro and BuShips. However, unlike most oceanographers, Revelle accepted an officer's commission and went on active duty. With the help of his Washington patron and future chief of naval research, Lt. Comdr. Rawson Bennett, Revelle moved into a central role in wartime oceanography. At Hydro, he helped coordinate the BT programs, introducing central administration to the data collection and a large measure of uniformity in both classroom instruction and shipboard practice. When he moved to BuShips on temporary duty from Hydro just before Christmas, 1942, Revelle found himself in Code 940, the bureau office concerned with advanced sonar design, and in an excellent position to further oceanography's cause among those who could best appreciate his message. His freedom to interact with both naval engineers and scientists and NDRC contractors working on naval problems provided very valuable professional and administrative experience. It also placed him in a position to act as a familiar point of contact for leaders in the field like Iselin and Sverdrup. As the war ended, the WHOI director wrote his series of epistles to Revelle in an effort to discern trends in naval policy with an eye toward alleviating his own anxiety about postwar federal oceanographic policy. The navy had become the richest and most powerful patron in the history of the ocean sciences, and for many essential players in 1944–45, Revelle increasingly became the personification of that benefactor.

When war seemed imminent Vannevar Bush, James Conant, Frank Jewett, and their colleagues created the NDRC in 1940 and seized wartime science from both the armed forces and the NAS and NRC. This maneuver thwarted the plans of other scientific groups, as well as the ambitions of Rear Adm. Harold Bowen of the NRL, who hoped to take control of naval scientific research in the event the United States again went to war. While certainly self-serving, these ambitions did not simply stem from a personal desire to advance in rank and exercise authority. Many within the naval community felt that the service should control scientific projects designed to give the United States an edge in naval warfare. The same naval culture that made translators necessary fostered a parochialism and insularity that rebelled against civilian authorities like the NDRC.

Faced with a fait accompli, the Navy Department at the very least wanted to maintain good relations with the Bush organization and to draw the attention of the civilian scientific community to high-priority problems facing naval forces at sea. Created in the early days of the war, the Office of the Coordinator of Research and Development—led at first by Jerome Hunsaker and later by Rear Adm. Julius A. Furer—played middleman. Stationed in the cultural gulf between the two communities, Furer and his staff emphasized essential naval priorities for the NDRC, looking out for the navy's interests in the largest and most comprehensive mobilization process the scientific community and the country had ever seen.

Unlike the authority of the OSRD/NDRC organization that did not cast its shadow beyond victory, the OCRD staff had an even greater effect on naval scientific inquiry as the war ended and the problem of postwar research organization arose. Furer gave his people wide latitude to promote and facilitate important professional relationships and establish naval research priorities. Intrigued by the problems often inhibiting cooperation between the naval and scientific communities, the admiral's crew—led by a pair of reservists, Lts. Bruce S. Old and Ralph A. Krause—formulated a scheme for the direction of postwar naval research during informal nocturnal brainstorming sessions in 1942–43. These discussions laid the conceptual foundation for oceanography's most powerful and flexible postwar patron, the Office of Naval Research (ONR).

First as the Office of Research and Inventions and then as the ONR after August, 1946, this naval agency institutionalized the common practice now raised to the tenth power in scope, assets, personnel, and potential by World War II. The major oceanographic institutions and leading scientists, awkwardly integrated with the navy by thirty years of curiosity, mission, economic deprivation, and national emergency, soon discovered the amazing potential of the informal network of relationships refined and enriched by the most destructive war in human history. The ONR would help the navy and civilian oceanographic community build on that discovery.

11

Crossroads, 1945–46

It is true of course that you could not expect Navy officers to
understand at the outset that in setting off an atomic bomb in
the water they would be carrying out a major oceanographic
experiment.

—COLUMBUS ISELIN to HENRY BIGELOW, 25 February 1946

Admiral [McCann] expressed a prompt willingness to cooperate to the fullest extent with any sound scientific program. He
was well up on SOFAR and wanted several technical points
made clear to him. We can be assured that he will back fundamental geophysics if we keep a fair balance of military pay dirt
coming in.

—ALLYN VINE to COLUMBUS ISELIN, 16 May 1946

IN JULY, 1946, A NAVAL-SCIENTIFIC TASK FORCE CONDUCTED
two nuclear tests at Bikini Atoll in the Marshall Islands under the code name
Operation Crossroads. A few months earlier, the first operational nuclear devices brought the Japanese war machine to a grinding halt and persuaded Em-
peror Hirohito to surrender. As the first significant postwar tests, the Crossroads project more precisely measured the awesome destructive power of these
weapons and demonstrated in a sobering manner their social and environmental significance. Military planners and civilian populations began to understand
better the implications of living with the "Bomb."

In the midst of these discoveries and revelations, Crossroads also provided
a suitable stage for yet another drama. As the OSRD/NDRC demobilized, the
navy and civilian scientists began to redefine their working relationship. Now
they had to build bridges without the intercession of a wartime agency. Five
years earlier, the NDRC effectively bypassed the maze of politically charged
professional and ethical concerns that repeatedly blocked effective cooperation.
That bypass no longer existed, yet scientific concerns remained—and wartime
experiences had turned the navy into a greedy suitor. At the same time, the mer-

curial nature of the immediate postwar era turned oceanography into a certain but nervous bride. Few doubted that a marriage between the navy and the ocean sciences would take place. It remained only to define the nature and character of the union.

Apprehension spread in the scientific community as the navy took the postwar lead in the ocean sciences, defined prime areas for research, and expended unprecedented financial and material resources. Hardly disinterested spectators, Iselin, Revelle, Sverdrup, and other leaders in oceanography craved a clearly defined advisory role for naval and civilian scientists. They wanted scientific inquiry to assume the same status as strategy, tactics, engineering, and weaponry. Iselin, in a February, 1946, letter to Bigelow, demonstrated both his uncertainty as an institution director facing the postwar world and his determination that oceanographers should play a strong advisory role: "I had hoped that when peace came the technical people who had played the game would be listened to and might even be given some small measure of authority. It is true of course that you could not expect Navy officers to understand at the outset that in setting off an atomic bomb in the water they would be carrying out a major oceanographic experiment."[1] Someone had to tell them. Iselin would work for the navy, but not simply as a technician for hire. He and most of his colleagues wanted to help navigate a course into the future. They wanted to participate in formulating policy.

While oceanographers voiced these concerns, Operation Crossroads began to take shape. Democratic senator Brian McMahon of Connecticut first suggested testing an atomic bomb against large naval vessels in the late summer of 1945. McMahon insisted that the two devices dropped on Japan with minimal opportunity for careful observation did not provide sufficient data on the actual power of the bomb and the extent of the radiation contamination that followed in its wake. Admiral Ernest J. King, at that point still commander in chief of the U.S. Fleet and CNO, refined the senator's idea further by recommending two detonations, one above water and one submerged. Major General Curtis E. LeMay of the Army Air Forces chaired the ad hoc committee created in November, 1945, to set the planning in motion. After King and LeMay laid the groundwork, the Joint Chiefs of Staff (JCS) appointed ordnance and missile specialist Vice Adm. William H. P. Blandy as the commander of Joint Task Force One (JTF-1), charged with carrying out the test.

Oceanographers and their supporters within the navy proceeded along two fronts. As the military prepared to evaluate the destructive authority of nuclear weapons, oceanographers, led by BuShips's Roger Revelle, loudly proclaimed that this series of tests would amount to a gigantic air and ocean experiment demanding careful scientific study. On 13 December 1945, an ocean sciences group presented the case for Operation Crossroads oceanography to representatives of Blandy's staff and a group from the Manhattan District nuclear program. Revelle, supported by Hydro's Lt. Mary Sears as well as BuShips's Norman Holter and Lyman Spitzer, successfully argued for a Bikini ocean science program focusing on ocean currents, waves, instrument development, and the effect of both radiation and the force of the blast on marine life.

Two significant memoranda emerged from the meeting. The first laid the foundation for contractual relations between BuShips and the major American oceanographic laboratories. A second memorandum sent to Hydro discussed the problem of radiation diffusion through ocean currents and setting the proposed biological survey in motion.

Carving out a significant advisory role for the oceanographic community presented considerable difficulty considering the navy's goals in this case. Even so, the 13 December meeting represented a first step that leaders in oceanography hoped to follow quickly with others. As the Crossroads project began to take shape, Columbus Iselin voiced concerns to Bigelow based upon recent wartime experience. His comments displayed a combination of frustration and professional determination: "The Army and Navy treat their own laboratories just about as badly as they treat us, so my complaint is not due to the fact that we are a civilian organization. . . . However, until some scientists become admirals or some admirals decide to give science a fair hearing, it will be just touch and go each time whether or not we are able to pull off a reasonably good job. A small part of the Navy asks you to do some work and the whole rest of it works night and day to make it difficult."[2]

Iselin's remarks emerged not from disappointment and despair, but rather from realism born of experience. He knew that working with the navy, in spite of good intentions, would prove difficult and awkward at best. Although an appealing vision of studying the ocean without the patronage and professional problems presented by the navy doubtless appeared in his dreams, the attraction of federal funding and the strength of naval research priorities constantly remained before his waking eyes. Along with Sverdrup, Revelle, and Carl Eckart on the West Coast, Iselin wanted to take a turn influencing those at the helm. He knew postwar oceanography needed navy money, but it also needed a large measure of freedom and flexibility. Serving both the navy and oceanography did not lie outside the realm of possibility if civilian scientists participated in shaping policy.

Soon after the 13 December meeting, Admiral Blandy appointed Roger Revelle director of ocean science planning for Crossroads. Revelle assumed responsibility for the skilled people and abundant resources assembled to conduct the predetonation survey of Bikini Atoll and the evaluation of conditions after the blast. His senior colleagues, especially Sverdrup and Iselin, looked favorably upon the appointment. Through a personality as familiar and approachable as Roger Revelle they could easily place their concerns before Blandy and perhaps influence naval activity and intent.

As the director of a civilian scientific institution, Iselin sought a postwar relationship with the navy that granted a significant role to oceanographic professionals. He thought that Crossroads offered an incomparable opportunity to place competent people in a position to influence national planning and policy in the ocean sciences. In March, 1946, he characterized the oceanographic component of Crossroads as "the biggest thing scientifically that will ever happen during our productive years." He thus found Revelle's position exciting and full of possibilities. The WHOI director also offered direction to others along these same lines. He advised his old friend and colleague Mary Sears to monitor and influence as best she could the formation of the new Oceanographic Division at Hydro authorized the previous month. Along with Sears, Iselin supported the UCDWR's Richard Fleming to head the new division. The WHOI director also did his best to convince Roger Revelle that he should apply for an important position at Admiral Bowen's newly established Office of Research and Inventions. In a letter to Sears in early March, Iselin commented that he would "be spending a day or two in Washington next week and I will have a talk with Roger. He sometimes listens to me. He should unquestionably take the ORI job."[3]

Doing so would place Revelle in a position to influence decisions governing the disbursal of naval research funding after his task at Bikini concluded. Iselin wanted experienced, savvy oceanographers to guide naval spending in a way that would benefit both the national defense and scientific priorities. In this way, scientists might have maximum freedom while working for the naval patron. He also put considerable pressure on Mary Sears to remain in uniform for a time so she could shape the new oceanographic staff at Hydro. Iselin expected both Revelle and Fleming to leave Washington very soon for the nuclear tests in the Marshalls. While they worked with BuShips and JTF-1, the WHOI director argued that Sears could play a significant role at home helping shape the composition of Hydro's Oceanographic Division. He suggested that she "stay on in the Navy until after Bikini and while in uniform, with everybody else in the Pacific, you should not hesitate to make the decisions. There is no reason why you should not shape the events. You are the one person who has stuck to his job and there can be no question of your trying to feather your own nest."[4]

Iselin shared his concerns and goals with most of his colleagues in the oceanographic community. Crossroads presented a unique opportunity to explore the significance and effects of nuclear weapons as well as a chance to begin the postwar era with the most ambitious oceanographic experiment ever undertaken. Scientists had to take the initiative.

While Iselin and like-minded colleagues worked on persuading the navy from without, Revelle did a very effective job from within. Working from his temporary position in BuShips, Revelle designed an oceanographic program for Crossroads broken into three parts: physical oceanography, marine biology, and geology. Bikini's physical oceanography included current surveys, oceanographic sections, special radiological diffusion experiments, and studies of waves, wind, and tides. Scientists would use current meters, dye markers, current poles, and drift bottles in the lagoon framed by this roughly circular atoll to measure current motion and water exchange with the open sea. As usual, oceanographic sections would demand the collection of water samples at many depth levels for chemical analysis, as well as BT readings for temperature variation. Wave experiments might prove particularly valuable given the most obvious and immediate effect of the Able and Baker detonations.

For their part, biologists would examine the atoll's animal and plant life, explore nearby islands in search of radiation effects, and compare their findings with data from not too distant, but largely unaffected Pacific islands. All of this work would take place before the explosions, after the tests ended, and periodically over the passing months to check long-term effects and the rate of repopulation. Subjects of the survey would include littoral and land animals, reef and lagoon fish, algae, seed plants, and both lagoon and deep-sea plankton. Revelle also incorporated a project to study the pelagic fishes like yellow-fin tuna and bonito in the deep water close to the lagoon. The Bureau of Ships approached scientists from Scripps, Woods Hole, the National Museum in Washington, and varied academic institutions to staff these projects.

The U.S. Geological Survey (USGS) led the geological effort. This aspect included examining bottom sediments, taking core samples, and photographically recording the ocean bottom, as well as studying corals and the general geology of the atoll.

Most of this research schedule began to emerge very early from Revelle's oceanography subsection of BuShips's Code 940. He worked there with his Scripps colleague and fellow reservist Lt. Comdr. Marston Sargent and BuShips senior physicist Norman Holter. In the weeks before the planned Able detonation, first scheduled for 15 May 1946, this oceanographic nucleus of three quickly grew into a team of eighty-one officers and civilians working for JTF-1 in Washington and at Bikini.

Plans to measure the considerable waves generated by the explosion first arose between September and November, 1945, when Revelle and Holter first reacted to Admiral King's Crossroads announcement. The explosions at Bikini would generate a shock wave in and out of the water that many technicians believed might overwhelm existing instrumentation such as wave and pressure recorders. New instruments emerged from discussions between Woods Hole's

Allyn Vine, Theodore R. Folsom of Scripps, and Waldo K. Lyon of the NEL. Scripps's John Isaacs helped by assembling a photographic system to provide still images and television to verify and amplify the data accumulated by the more conventional instrumentation.

These experiments, instruments, and the larger strategy in physical oceanography emerged from discussions between Code 940, Woods Hole, the British Admiralty Research Laboratory, the UCDWR, the NEL, and the University of California's Department of Engineering even before Revelle's assignment to JTF-1 on 5 February 1946. Code 940 gave direct responsibility for the physical oceanography program during Crossroads to both Hydro and WHOI. Hydro provided personnel, survey ships like the USS *Bowditch* (AGS-4), and general project coordination, assisted by Richard Fleming. Woods Hole contributed to the planning and called upon its experts in physical oceanography and instrumentation, including Allyn Vine, William Von Arx, William Ford, Hal Turner, Gordon Riley, and Thomas Austin to conduct experiments and survey work. In addition, Dean Michael P. O'Brien of the University of California engineering faculty arranged for the construction of a scale model of the Bikini lagoon at Berkeley to run simulation tests in support of Crossroads research.[5]

As the day for the first test approached, Revelle and his team found they suddenly had unwelcome time on their hands. On 22 March President Harry Truman pushed the test schedule back six weeks so members of Congress might finish their legislative activity in time to observe the tests. As Blandy rescheduled Able for 1 July, many of the scientists began to wonder if they would ever get home. Although a few naval personnel had arrived in the test area as early as January, the bulk of supplies and naval and civilian personnel did not reach the atoll until mid-March. By the time the president ordered Able rescheduled, many of the scientists had spent a week or more in the test area and had accomplished a great deal of work on a hurried schedule. Most expressed mixed feelings about working with the navy in this remote location for nearly two more months.

As Crossroads unfolded, the logistics and administrative details proved difficult, but the work progressed with greater speed and efficiency than many expected. Like most of his WHOI colleagues, Thomas Austin wrote letters to Columbus Iselin providing details of their travels and progress. Iselin assigned Austin to the WHOI physical oceanography crew at Bikini. This project operated under Gordon Riley as senior scientist and used both the *Bowditch* and the submarine tender USS *Fulton* (AS-11) as its base of operations. While Austin made it clear that gathering equipment often proved difficult, he assured his boss that from the beginning of his journey to the test site, Bikini had top priority:

> Gordon [Riley] and I took various invoices and scouted the [Mare
> Island Naval Ship] yard for odd and sundry items the men needed.
> Found nearly all the desired gear, however, it was the old story, 10
> copies, crossing 10 desks, each holding it up for a while, ultimately
> giving us the gear, as they had originally planned to do. Again, the
> shore-based personnel were of the greatest help. Comdr. King
> Couper and Commander Seymour, the COMSERVPAC supply
> officer, were our mainstays. If the various departments said that they
> did not want to deplete their supplies by giving us what we desired,
> we called either Couper or Seymour and they in turn released the
> item in the desired quantity. The word CROSSROADS is a magic
> word. We have exceeded the ship's allotment in many items by
> simply mentioning that word. It would be a handy thing to have
> around the lab after this is all over.[6]

Describing the trip out to Bikini and early scientific work, Austin criticized conditions aboard the *Bowditch* and strongly suggested that the ship's officers and crew retarded their scientific work. In his 23 March letter to Iselin, Austin contrasted this problem with the professional performance of the personnel on board other Hydro survey ships just recently arrived at Bikini. In contrast to his disgust with *Bowditch*'s complement, his positive appraisal of the other vessels and crews stood out sharply: "The officers and men are most cooperative, and the working conditions are the best."[7] Through Austin, Iselin realized that his group might well complete its work within the next three weeks. If this proved accurate, both Lt. Comdr. Clifford Barnes of the Coast Guard Reserve, in charge of the oceanographic survey section, and Revelle himself would find the six-week delay very uncomfortable. Barnes had already informed Austin that he expected the senior civilian staff to remain on site for both the delay and the tests. This clearly disturbed the WHOI scientist, who enjoyed his work but did not care to endure Bikini's remoteness any longer than absolutely necessary.

Eight days later, Gordon Riley felt that he had enough experience with research and living at Bikini to report to Iselin. He intended to make up for promised progress reports not yet submitted and to give the WHOI director some sense of the work environment at Crossroads. By 31 March the physical oceanographers had already completed roughly two-thirds of the Bikini Lagoon current measurements as well as much of the temperature work. The team had finished the planned BT observations as well as one cruise outside the atoll, but most of the dynamic calculations still remained to be done. Riley remarked that he had failed to write earlier because he had nothing interesting to report. Now he could provide engaging reading for Iselin on temperature and depth as well as a quick sketch of the fifty-five-degree isotherm on the west side of the test site. Their work had obviously progressed far enough to yield the first brush strokes of a comprehensive oceanographic portrait of the atoll. Riley felt comfortable with the data they had collected, but he offered the same concerns and criticisms that emerged earlier in Austin's note: "when Roger tries to persuade us to stay, he won't have exactly an unprejudiced audience. We want to do what is right, but all of us are bored and disgusted. Cooperation on the *Bowditch* has been practically nil, living conditions are poor, and working facilities worse. Worst of all we don't like to tell our wives we are going away for three months and then stay for six. Roger had better produce some very potent arguments."[8] It quickly became evident that Revelle would soon have to deal with restless, barely occupied men.

While the leaders of postwar civilian oceanography appreciated the difficulties of working in the Marshall Islands and the restlessness imposed by the delay, they also realized the importance of this event to the future of oceanography. On 20 March 1946, Iselin described to Revelle the WHOI-Hydro plans for physical oceanographic research and instrument development. If oceanographers had their way at Crossroads, they would become an asset in future tests and a partner in accurately predicting and coping with the consequences of nuclear explosions. Although he knew he was preaching to the converted, Iselin stressed oceanography's role in predicting the dreadful effects of nuclear weapons: "The justification for what may seem an overly elaborate oceanographic program is that only by exploiting this unique opportunity for testing out the circulation theorem can we hope to make reliable predictions as to the movement of contaminated water on some other occasion."[9]

In presenting his case to Revelle, Iselin hoped the message would reach Admiral Blandy and others in a position to compose naval oceanographic policy. If scientists became part of the policy process and continued doing their research with relative freedom and flexibility, Iselin felt that the navy would find them truly powerful and useful allies.

Meanwhile, Iselin tried to console and advise Austin, Riley, and other Woods Hole people at Bikini by insisting that they played an important role in shaping oceanography's future. In a 9 April letter to Riley, the WHOI director suggested that the momentary awkwardness and discomfort paled in view of their ultimate common goal:

> It is always very tough trying to do science from a Navy ship, as you very well know from previous experience. I would judge from your letter and from a sentence or two in Tommy [Austin]'s last one that the crew of the "Bowditch" will make things even more difficult as time goes on . . . An unhappy ship is not much fun for anybody . . . Certainly our experience with the "Mentor" this winter has been

perfectly terrible. They have not accomplished a day's work at sea
since before Christmas. The most recent shipyard bill was for
$70,000, and all because the next to the last motor-mac[hinist] did
not bother to oil the engine. . . . The argument for putting up with
such foolishness of course is that only through the government can
we hope to finance oceanography adequately.[10]

With people like Riley and Austin on site, Iselin had a hand in events and
some small measure of influence over naval activity. Doubtless, he and Sverdrup took some measure of comfort from the influential placement of their
people. As long as Riley, Revelle, Sears, Fleming, and others performed well
and lost no opportunity for persuasion, the future held promise. As Iselin
wrote to Riley: "Always remember that I will back you to the hilt in whatever
course you decide to follow. . . . There is one good piece of news which seem
official. Fleming has landed the job at Hydro. Moreover, but this is somewhat
less certain, I believe Roger [Revelle] has accepted the job at ORI. Mary [Sears]
is busy cultivating the new Hydrographer. Thus except for your troubles at
Bikini, oceanography seems to be a very healthy baby right now."[11]

Allyn Vine, the senior WHOI staff member participating in Crossroads,
used the delay of the Able shot to see if he could make the infant even health-
ier. During the six-week delay ordered by President Truman, Vine explored the
possibility of additional cooperative projects with the navy. His experience
with the BT, SBT, and underwater sound research during the war drew him
very close to the submarine community. Furthermore, his insatiable curiosity
and habit of relentless work in the tradition of his mentor, Maurice Ewing, led
him to accept the unfortunate six-week delay as an opportunity to recommend
new lines of joint scientific inquiry.

In late April, while he worked for Revelle as one of the senior scientists in
wave measurement and instrumentation, Vine discovered that the navy's
Pacific submarine command intended to operate with greater frequency in the
Bering Sea and on the Arctic margin. Along with his friend and WHOI-trained
BT technician Comdr. Butler King Couper, Vine approached ComSubPac and
proposed a research agenda for submarines scheduled to operate in the North
Pacific. He argued persuasively that the time at sea could serve both science and
the submarine community.[12]

In Vine's case, the habits of an experienced wartime translator were hard to
break. After speaking with senior officers at ComSubPac, the WHOI scientist
reported a very favorable response to Iselin. Indeed, the submariners even suggested that they could take a considerable amount of gear with them as well as
an occasional trained observer. Vine also induced ComSubPac to consider
sending submarine officers to Woods Hole in the spirit of the wartime BT train-

ing program. He reported to the WHOI director: "It seems to both King [Couper] and myself that this is a wonderful opportunity to collect data in remote parts of the world and what is I think even more important, for the Institution and the Navy to get squared away on a peacetime basis for research that we can both profit by. It may well be that ComSubLant [Commander Submarines Atlantic] will cooperate as well as ComSubPac but we have had a definite promise from here that work can be done."[13]

The program that Allyn Vine proposed to the Pacific submarine command included a very ambitious data collection effort. He had permission to advise submarine commanders on geographical areas in need of special study and to offer plans for programs that would fit within the time and mission assigned to each boat by the CNO. The research would include BT readings and water sample collection, gravity measurements for both geologic and position purposes, current measurements, animal and earthquake noise studies, sea ice investigations, and vertical current studies, as well as monitoring chemical and biological conditions and changes. While still at Crossroads, Vine proposed a light research regimen for the USS *Entemedor* (SS-360) on its impending patrol into the Bering Sea. Less than a month later he reported to Iselin that ComSubPac had sent the suggested research program for the *Entemedor* patrol to all Pacific Fleet submarines for evaluation and comment.[14]

Revelle, who arrived at the nuclear test site in mid-May, and Iselin, thousands of miles away in Woods Hole, realized early, along with Allyn Vine, the most important fundamental truth about the postwar relationship emerging at Crossroads. The navy formulated policy and possessed the assets civilian oceanography needed. On his visits to Pearl Harbor during the Crossroads delay, Vine met with Vice Adm. Alan McCann, who succeeded Charles Lockwood's as ComSubPac on 18 December 1945, and convinced him to request oceanographic assistance to arm his crews with a better knowledge of their operating environment. In peace, as in war, personal and frequently unofficial arrangements based upon trust and unspoken assumptions repeatedly ruled the day. Although the navy determined policy and possessed the assets oceanography needed to prosper, persuasion and advice offered by oceanographers clearly influenced key officers making the decisions that shaped naval involvement and patronage. If the navy accepted a proposition, as ComSubPac did from Vine, project scientists would have maximum flexibility and a large measure of crew cooperation as long as the ultimate naval purpose of the mission did not disappear in the civilians' eagerness to serve science and knowledge. As Vine put it to Iselin on 16 May 1946: "Admiral [McCann] expressed a prompt willingness to cooperate to the fullest extent with any sound scientific program. He was well up on SOFAR and wanted several technical points made clear to him. We can be assured that he will back fundamental geophysics if we keep a fair balance of military pay dirt coming in. This is the general opinion out here and I am sure that cooperation from the New London group will be nearly as good."[15]

Oceanographers had to recognize that the navy would insist on science that would serve the needs of naval warfare. The task ahead for Revelle, Vine, Iselin, Sverdrup, and others consisted of helping the navy to define its needs as broadly as possible. In that way, the relationship emerging from the war could serve both the national defense and civilian scientific goals.

Vine constantly sought new possibilities for research and suggested to Iselin that the navy had an interest in the Arctic intense enough to produce considerable funding for the future. As far as he was concerned, this commitment would move well beyond single WHOI-navy Arctic projects and might well require a department or center devoted to the oceanographic characteristics of the far north. He also urged the WHOI director to consider whether his institution should do Pacific oceanography, expanding beyond its traditional Atlantic base. Indeed, how should Scripps and Woods Hole handle the Arctic? Vine wondered if the top of the world was Atlantic or Pacific. As he split his time between riding ComSubPac submarines and preparing for Crossroads, the WHOI senior scientist told Iselin that the distinction seemed to blur at the top of the world and that "either WH [Woods Hole] or Scripps may have to annex the Arctic Ocean." With the same determination as Revelle but more energy, and with less anxiety than Iselin but an almost childlike anticipation and enthusiasm, Vine concerned himself less with naval or federal incursions into the scientific domain than the possibilities offered by accommodation and common effort. He also enjoyed the company of submariners, the intriguing problems they faced, and the interesting ways they employed his solutions. He communicated easily with them, once commenting to Iselin, perhaps metaphorically, that it took two days to get a typewriter through the Crossroads chain of command, but six minutes to obtain the same machine from ComSubPac. Vine never stopped looking for common ground and new bridges to build. The delay in Crossroads offered an opportunity to that end. At the end of April he wrote to Iselin: "We are all in fairly good shape. The Navy is treating us well. Work is slow but I have been busy all the time. I would also be busy if they delayed CROSSROADS two years."[16]

Members of the Scripps staff similarly kept Harald Sverdrup informed of events and conditions at Crossroads. Marine biologist Martin Johnson wrote letters filled less with local color and a sense of working conditions than with observations on doing scientific research and working with the navy. He suggested that the overcrowding aboard the *Bowditch* caused the unhappy conditions mentioned by Austin and Riley and in early April, but that moving some personnel to other ships reduced the tension among the staff and crew. Al-

though work occasionally slowed to a snail's pace because of weather and logistical difficulties, he had already made considerable progress on his sampling
routine at and near the test site. In the process, Johnson discovered sufficient
plankton and other sea life natural to the lagoon to use as a baseline for measuring the immediate and long-term effects of the tests. He also discovered un-
noticed coral formations and other physical attributes of the atoll that led to far
better charts of the islands than previously available.[17]

Tests Able and Baker were finally performed early on the mornings of 1 and
25 July respectively, providing important data on the military and environmental effects of first-generation nuclear weapons. The oceanographic work on the
islands, in the lagoon, and in the area surrounding the test site provided not
only information previously unavailable for Bikini Atoll, but also a basis for the
careful postdetonation examination of the area between and after the tests. After the first explosion, eight congressional witnesses and thirty-nine journalists
left the area and returned to the United States via Pearl Harbor. When the
Crossroads operation concluded, the staff of JTF-1 made preliminary plans to
survey the area in 1947 to monitor the long-term effects of the experiment.[18]

Crossroads provided the oceanographic community insights into the
navy's level of commitment to oceanography, and the possibility of a deeper,
more effective postwar dialogue. As Iselin pointed out, most oceanographers
in the early postwar years understood that they could not escape a certain degree of dependence upon the navy if they hoped to attain their research ambitions. Crossroads demonstrated that they could not only work with the navy in
peacetime but exercise considerable, informal, and frequently welcome influence on naval policy. While the scale of early postwar oceanographic activity
certainly rivaled wartime, the nature of the relationship emerging in 1946 recalled the informal style of the interwar common practice far more than the
wartime scientific research coordinated by the OSRD/NDRC.[19]

If the joint navy-civilian effort at Crossroads provided any measure, postwar cooperation depended upon many independent factors characteristic of
both interwar and wartime science. These included the productive activity of
translators, the practical experience of scientists now more familiar with naval
problems, and the influence of wartime line officers now senior in rank and
converted by combat experience to the importance and relevance of oceanography. The experience of Crossroads forced these factors to converge, laying part of the foundation for years of productive Cold War oceanographic
research.

12

Shaping the Postwar Dialogue, 1946–50

There is so much to be done that I cannot help but have a rather hopeless feeling. Poor Dick Fleming will have to bear the brunt of it. He was already good and tired when I saw him last, about ten days ago.

—COLUMBUS ISELIN to MARTIN POLLAK, 31 March 1948

WHEN THE WAR CONCLUDED, THE NAVY CONTROLLED VIR-tually all funding for oceanography in the United States, and its goals did not always appeal to civilian scientists eager to return to peacetime pursuits. At the same time, the leaders of the civilian oceanographic community realized that postwar science would require levels of funding for research and ship operation that far exceeded both the prewar norms and their postwar means. It quickly became obvious that the navy and civilian scientists needed the essential means of communication and cooperation—a dialogue—that had completely eluded them before the war. The NDRC no longer existed. Now they would have to build other bridges.

Oceanography at the Hydrographic Office

For the navy, any approach to joint planning had to begin at Hydro. In helping to lay the groundwork for Crossroads and exploring the future with ORI, the navy's hydrographer had a small but firm oceanographic foundation upon which to build. In 1943, Hydro established an oceanographic unit in the

Pilot Chart Section of its Division of Maritime Security. This creation emerged from a recommendation of the armed forces' Joint Meteorological Committee and a directive from the vice CNO that the navy conduct all oceanographic work for national defense. Mary Sears called the Pilot Chart Section home during her wartime tour of duty.

By late 1945, oceanography obviously consumed an increasing percentage of Hydro's time, whether in publications or surveys at sea. The discussions with ORI made this clearer than ever. Hydro had become a multidisciplinary activity integrating the work performed by Sears and her wartime colleagues with new special projects and programs, as well as priorities suggested by civilian oceanography leaders. Rear Admiral George Bryan's people combined intelligence on currents, temperature, sea swell, and ocean-bottom composition with extensive survey work, data collection by warships on patrol, and the results of unusual ventures like Operation Crossroads.

This realization prompted the Navy Department to approve planning for a Division of Oceanography within Hydro. The idea first emerged from Admiral Furer's office as early as January, 1945. Lyman Spitzer advised Columbus Iselin that Lt. Comdr. John T. Burwell, a member of Furer's staff, thought that a standing committee on oceanography within the navy "would not be a very useful one, since both Roger [Revelle] and Burwell felt that large committees without particular authority did not usually accomplish much. It was felt that it would be much more profitable to enlarge the Oceanographic Unit at the Hydrographic Office, and put at its head a naval officer who could perform most of the functions of the Standing Committee. The chief function of this officer would be to promote the various projects that the oceanographic unit might carry out; the technical details would be the primary concern of the civilian chief oceanographer."[1]

Writing to Bryan on 3 January 1946, a very interested Columbus Iselin appreciated the proposed civilian role and immediately suggested priorities for the enlarged oceanographic unit, commenting that Hydro charts might better and more clearly plot temperature and currents for the benefit of the ordinary seaman and interested layman.[2] Within the navy, ORI went beyond suggested reforms in policy and practice to propose a structure for the unit. Following Burwell's suggestion, Bowen recommended that it work under a senior civilian oceanographer in two major sections. The Compilation and Publications Section would prepare all oceanographic literature for the navy, including the vital pilot charts, while the head of the Research Section managed a Bibliographic Unit, Oceanographic Data Unit, and a Research Unit. Bowen also suggested creating an advisory committee drawn from the navy and other interested agencies to assist the hydrographer in planning the division's activities.[3]

It surprised no one that, after the secretary of the navy approved the new division, those who would shape ocean science policy naturally tried their hand at influencing the new activity. After all, Hydro's wartime experience and the oceanographic program it helped develop for Crossroads placed it in a position to play an important role in fashioning the future. In this context it is easier to understand Columbus Iselin's reaction, his letter to Bryan, his strong advice to Mary Sears, and his support of Richard Fleming of the UCDWR as the division's director. Placed in key positions, ocean scientists friendly to Woods Hole, Scripps, and other laboratories would help the civilian scientific community play an increasingly important, if indirect, role in oceanographic policy.

When the Navy Department established the Oceanographic Division of the U.S. Navy Hydrographic Office on 29 January 1946, Lt. Mary Sears stepped in as the first officer in charge pending the appointment of a civilian director. She had to wait for three months before Richard Fleming accepted the post in April, 1946. For many in the civilian community, it seemed the ideal situation. Despite his anxiety over the appointee's relative inexperience in administration, the WHOI director felt sure that Fleming would get expert assistance from Sears to ease the transition and bring him up to speed. Thus, Richard Fleming did not travel to Bikini with Revelle, but instead moved to Washington and Hydro. Two weeks before Fleming arrived to take up his new responsibilities, Iselin wrote Sears and asked her to play the role of midwife for Fleming and the new division. He even suggested that she could pick up a salary at Woods Hole for the trouble. Hydro's work and its sponsorship of civilian science in the future depended upon a good beginning: "I urge you not to slam the door too hard in Dick's face. If you could lick the problem of where to stay, could you not be part time in Washington next autumn? It will be important I believe during the first year at least for part of the Woods Hole-Hydro contract to be reserved for just the sort of assistance that Dick evidently has in mind. You can draw your pay here, but as may be needed give him a hand, as indeed we all should. The quality and quantity of the work turned out by Hydro during the first years will be critical in the whole plan."[4]

Fleming asked Sears for help in transition, but Iselin wanted more. For the WHOI director, as for Sverdrup at Scripps and Thomas Thompson at the University of Washington, the "whole plan" also included asking Hydro to assume responsibility for the increasing civilian demand for knowledge about the oceans. This would leave the major research centers free to dwell on their science. In all of this, the civilian oceanographic community sought as much freedom and flexibility as possible while it worked to influence the nature and direction of the navy's oceanographic plans. For Iselin's sake and to provide Fleming with a good start, Mary Sears agreed to stay, but only until June, when she planned to leave for Denmark and a period of biological work far away from the navy. By then she felt certain that Fleming would feel comfortable in his new post.[5]

Unfortunately, the new director would not have time to feel comfortable. His tenure at Hydro's Oceanographic Division presented an astounding variety of challenges and crises. More than once, Fleming had to fight for the life of oceanography in the Hydrographic Office against congressional budget cuts, navy war-fighting priorities, and the preferences of hidebound officers.[6]

Soon after its creation, the division participated in the scientific aspect of the navy's Operation Highjump in Antarctica. Like Crossroads, which concluded earlier in 1946, Fleming suddenly found himself tasked with quickly composing an oceanographic program, complete with support staff, field scientists, and very scarce instruments. In doing this, he depended heavily on the advice and cooperation of Scripps, Woods Hole, and the University of Washington, as well as resources within the navy.

He quickly came to understand why postwar institution directors were so concerned over the expense and difficulty of doing oceanographic research. Fleming knew all too well the small numbers of truly skilled scientists available for polar oceanography. The numbers of students at schools like the University of California, Scripps's parent institution, had just begun to grow with the conclusion of the war. Furthermore, Hydro had instruments available only in limited numbers, so Fleming needed to borrow many critical devices. He also contracted regularly with private institutions to provide instrumentation expertise and survey support staff to assist the few people Hydro could send to sea.

In addition, he labored under a shortage of properly equipped survey ships. Only recently had the Navy Department assigned two former attack cargo ships (AKAs) to hydrographic work. Built in 1945 by the Walsh-Kaiser Shipyard of Providence, Rhode Island, the USS *Pamina* (AKA-34) and USS *Renate* (AKA-36) assumed new identities as 6,000-ton hydrographic survey ships under the names USS *Tanner* (AGS-15) and USS *Maury* (AGS-16). Portsmouth Naval Shipyard effected AKA-34's conversion to *Maury,* and the Norfolk Navy Yard worked on *Tanner.* These vessels provided the navy with a deep-sea survey capability and the logistical endurance to remain deployed for protracted periods. With the conversion of *Maury* and *Tanner,* the navy retired two older ships, the 5,400-ton *Bowditch* (AGS-4) and the 3,100-ton *Sumner* (AGS-5). The latter had served as the USS *Bushnell* (AS-2), a submarine tender assigned double duty as a hydrographic survey vessel in the 1930s. Both of the older vessels worked with the oceanographic community at the Bikini nuclear tests, with the crew of the *Bowditch* evoking more criticism than praise for its work with embarked civilian scientists. In spite of the promised ships, Fleming faced Highjump without modern survey vessels because neither emerged from the conversion process in time. *Maury* and *Tanner* did not begin their twenty-three years of service to naval oceanography until early 1947.

Beyond the uncertainties of staffing, instruments, and ships, Fleming also knew the risk of trying to do oceanographic research as part of a naval operation. Crossroads, as well as the preparation for Project Highjump in Antarctica, recalled the wartime awkwardness of performing effective scientific study in a situation driven by strategic and tactical priorities. The oceanographers working with the navy did have combat experience to draw upon and wartime performance to add persuasive force to their counsel, but these often counted for little in the face of a general naval culture less inclined to view the peacetime oceanographer as a professional partner. In a letter to Fleming, Columbus Iselin advised only a minimal investment in the Highjump project because of the uncertain scientific return: "To try to do good oceanography in the course of a naval operation, no matter what the part of the world, is very apt to be a losing game. As I see the situation, except for the bottom samples, the main value of the data will come from the two traverses of the southern hemisphere rather than from the observations made in Antarctic waters. I can think of plenty of useful and interesting things to do in the Antarctic, but they nearly all involve rather careful and prolonged research, and the undivided attention of a suitable vessel. The exploratory phase in Antarctic oceanography has been pretty well dealt with."[7]

Clearly, the problem did not stem merely from the navy's poor appreciation of the role oceanography might play in strategic planning and the execution of naval warfare. Reflecting postwar civilian scientific preferences, Iselin wanted to avoid the quick, "short fuse" research characteristic of the world war. Although the emerging antagonism between the Soviet Union and the United States would soon precipitate conditions uncomfortably similar to 1940–41, civilian scientists did not yet feel compelled by the Cold War to reconsider returning to deliberate work driven by prudent application of the scientific method. Such research took time and meticulous preparation. It also took money, research ships, trained scientists, and proper instruments in adequate numbers. The navy possessed these resources, but operations at sea rarely permitted carefully conducted long-term research. The quick response to navy needs expected from Fleming at Hydro did not fit the picture of postwar oceanography envisioned by its hopeful veteran practitioners. While they wanted more deliberate science, the navy often seemed bent on just the opposite.

Sympathetic to the need for carefully considered work but very much part of the naval service, Fleming and his Oceanographic Division tried to act as facilitator. He quickly reacted to naval needs driven by immediately perceived strategic and tactical considerations with oceanographic survey and research programs that would offer some opportunity for careful research.[8]

Fleming embarked upon this effort with limited funds and personnel. His early reports to the hydrographer provide a clear indication that he hit the ground running in 1946 with a program designed to bring the division a reputation for sound science. In November alone, roughly seven months after assuming his position, Fleming's office completed five hydrographic publications, with another in press, and had two Joint Army-Navy Intelligence Surveys (JANIS) in preparation. His people answered a dozen information requests, satisfied the needs of eight visitors or groups, and fulfilled Operation Highjump requirements for everything from personnel and instruments to funding and ships.

Shortly after arriving at the division, Fleming organized his activity by breaking it into four sections: Marine Geography, Technical Service, Program, and Oceanic Development. Two veterans of Revelle's Crossroads team soon joined the new director and his deputy, John Lyman, as section chiefs. Thomas Austin took over at Technical Service, and Butler King Couper directed Oceanic Development. Fleming made a habit of asking his section leaders to rotate positions to keep them familiar with the work of the entire division.

While enhancing the customary cooperation between Hydro and the civilian community characteristic of the old common practice, Fleming also sought to emulate the successful oceanographic expeditions of the civilian institutions. Along with Columbus Iselin, Fleming asked WHOI oceanographer Martin Pollak to fly back from the Mediterranean in April, 1948, after working for eight months as senior scientist on Woods Hole's very profitable *Atlantis 151* cruise. Fleming wanted Pollak's advice both on fitting out the converted AVPs, Hydro's newest survey ships, and on planning oceanographic research voyages.

Hydro's civilian customers also demanded Fleming's attention. As postwar founder and first director of the Lamont Geological Observatory at Columbia University, Maurice Ewing proved to be the most demanding. He not only wanted the Mediterranean ocean-bottom photos taken by David Owen's new Columbia University underwater camera, but also sounding data gathered when the R/V *Atlantis* crossed the Mid-Atlantic Ridge en route to Europe. He had asked for similar data taken during another *Atlantis* expedition a year earlier. Ewing, who had not yet acquired the R/V *Vema* for Lamont, remained one of the navy's regular and most exacting customers. Fleming needed to design surveys and research expeditions to serve the fleet and scientists like Ewing if only to demonstrate clearly the practical utility of the navy's investment in oceanography.

Fleming's years at Hydro were much like a storm at sea. Between outfitting vessels, training crews, providing information to naval and civilian customers, finding adequate instrumentation, planning cruises, and persuading naval officers of oceanography's utility, he had little time for anything else. Given these circumstances, Columbus Iselin responded to criticisms by Martin Pollak and critical inquiries about Fleming by John Coleman, executive secretary of the NRC's Committee on Undersea Warfare, by asking for time, patience, and consideration. Iselin appreciated the pressure on Fleming and the rapidity and efficiency with which an often-unreasonable navy and civilian scientific community expected Hydro's new oceanographic division to perform.[9]

As the Cold War intensified international anxiety, Fleming's monthly reports demonstrated the increased demands upon his division and its personnel. Writing on 31 March 1948 to Martin Pollak, who felt sick at the lack of professionalism among the hastily assembled AVP survey ship crews, Columbus Iselin commented: "There is so much to be done that I cannot help but have a rather hopeless feeling. Poor Dick Fleming will have to bear the brunt of it. He was already good and tired when I saw him last, about ten days ago."[10]

Fully aware that the daunting challenge could overwhelm the new division chief, Iselin wrote to John Coleman on 14 December to allay his fears about the division's program and director: "I have the greatest confidence in Dick Fleming and the last thing in the world I want to do is to increase his present difficulties. Thus, until the present lack of system has been proven unworkable, I think that we should keep quiet."[11]

Fleming had a task without an obvious or easy solution. Between demands for information and techniques applicable to strategic and tactical problems, he had a personal and professional mission to conduct oceanographic research. As it grew, the Oceanographic Division initiated research and survey cruises, acquired additional vessels, strengthened its ties with civilian laboratories, coordinated seagoing projects of interest to the ONR, BuShips, and the Office of Naval Intelligence (ONI), and provided work for a great many aspiring oceanographers.[12]

In many ways, the ill-starred Project AMOS proved to be the most ambitious project Fleming undertook while serving as a division chief at Hydro. As the recently victorious U.S. armed forces struggled to define the primary postwar threat, the security of the transatlantic convoy lanes remained paramount. Beginning in 1948, the Hydrographic Office worked with the USNUSL in New London to perform comprehensive surveys that would thoroughly integrate different types of environmental data for use in convoy escort duty. The CNO's office derived the project's acronym from "acoustic, meteorological, and oceanographic surveys." Fleming's division at Hydro took responsibility for coastal and transatlantic research emphasizing the meteorological and oceanographic aspect of the project, while the USNUSL took the acoustical approach. The project began in 1948 and lasted into the early 1950s, with Fleming actually leaving his post at Hydro for a time on loan to the ONR. He returned to the oceanographic division in May, 1950, to resume his duties.

Despite Hydro's effective and comprehensive survey program, AMOS never amounted to much. As the navy redefined the primary national maritime threat in terms of the young Soviet submarine force, AMOS came into conflict with the oceanographic and acoustic effort initiated to support the installation of the acoustic ocean sound surveillance system (SOSUS). The preoccupation with convoy duty surrendered to direct detection of a submerged adversary at previously unimaginable ranges.[13]

Regardless of the emphasis or the particular project, the increasingly hot Cold War at sea depended upon the accuracy of the environmental surveys conducted by Hydro. By the time a thoroughly exhausted Richard Fleming left the Oceanographic Division for the University of Washington to succeed Thomas Thompson in 1953, the navy had a more comprehensive oceanographic program, with an increasing staff, multiocean capability, and an amazing level of production.

Many within the navy very early viewed the creation of Fleming's new division as an opportunity. In response to his original ORI recommendation, Admiral Bowen also received permission from the secretary of the navy to organize and chair an advisory committee to support Hydro and the ORI. This group officially emerged from the same legislation that converted ORI into the Office of Naval Research in August, 1946, but Bowen began his deliberation on committee membership and goals with a select group of advisers on 15 May. When the president signed the ONR legislation, Bowen's planned advisory group became the Naval Research Advisory Committee (NRAC), and included Warren Weaver of the Rockefeller Foundation as chairman and high-caliber scientists like Detlev Bronk of the University of Pennsylvania, Arthur Compton of Washington University in St. Louis, Karl Compton and Philip Morse of MIT, and Lewis L. Strauss of the Atomic Energy Commission (AEC) as members. For Bowen, this committee and the new Hydro division emphasized the early postwar authority of the ORI and its influence within the navy and the civilian scientific community. For both the ocean science community and the Pentagon, the expansion of Hydro with ORI support also confirmed the navy's unique role as oceanographer to the armed forces.[14]

The admiral's desire to emphasize a function peculiar to the navy had political as well as naval and scientific significance. The intense and acrimonious roles and missions debates in the Pentagon and on Capitol Hill that would soon produce a unified National Military Establishment (NME), the ancestor of the Department of Defense (DOD), made it imperative that the navy project a service image both unique and cooperative. Essential tactical oceanographic services seemed one way the navy could easily shine in a future unified DOD. Indeed, barely two months after the National Security Act became law on 26 July 1947, the NME had its own scientific Research and Development

Board (RDB), chaired by OSRD veteran Vannevar Bush.[15] Based upon the
Joint Research and Development Board established on 6 June 1947 to coordinate army and navy research programs, this ambitious successor consisted of
two senior officers each from the army, the navy, and the newly established air
force, and established eleven advisory committees in preferred specialty areas.
While the Cold War was young, the NRAC occasionally played advocate before
the RDB, defending the competence and utility of naval research activities as
well as its effort both to hire qualified personnel and to foster productive relations with scientific institutions like the NAS.[16]

The Advent of the ONR

Next door to the BuShips Code 940 office that housed Roger Revelle and
marine biologist Marston Sargent from Scripps, Lt. Gordon Lill worked with
American and British engineers on advanced ASDIC and sonar for the last two
years of the war. Trained while an ensign in the first WHOI BT class by Maurice Ewing, Fritz Fuglister, and Allyn Vine, Lill spent much of the war working
on submarine BT installations for the commander of the Atlantic Fleet's service force out of Norfolk, Virginia. After V-J Day he returned to Kansas State
University for a master's degree in geology, then entered the doctoral program
at the University of California in Berkeley. With his money about to run out,
Lill planned to accept a well-timed job offer from Thomas Hendricks at the
USGS paying what was then the considerable sum of $3,000 per year. Before
he had time to take the job, events and Roger Revelle intervened. According to
Lill: "I went by to see Roger who was still in Washington and he said 'Hell
don't go to work for Hendricks, come to work for me.' I said well I had promised Hendricks I was coming. Then he said, 'Aw, I'll take care of Tom.'"[17]

Although Gordon Lill never finished his doctorate at Berkeley, he rose to
succeed Revelle as chief of the ONR Geophysics Branch, and became the
undisputed master of federal oceanography funding for the next decade.

The ONR emerged as ORI's successor on 3 August 1946. A logical extension of the planning by Lts. Bruce S. Old and Ralph A. Krause in Admiral
Furer's wartime office, Admiral Bowen wanted to convert the ORI into the
new agency by absorbing into the ONR both the NRL and the navy's nuclear
program, along with the extensive network of independent and university-
based contract research that supported it. Foiled in that effort by BuShips, Vice
Adm. Earle W. Mills, and Capt. Hyman Rickover, Bowen retired the following
year and left the ONR without a primary mission. Thus, by necessity, the research agenda funded by the embryonic contract program became the ONR's
full-time occupation.

In the early postwar years, the contract research program in oceanography rarely encountered the strong concern about federal influence and control that proved contentious in negotiations between Admiral Bowen's staff and scientists in other disciplines. Many within oceanography or observing from the sidelines, like Prof. Leonard Loeb of the Physics Department at the University of California's Berkeley campus, characterized naval oceanography as a unique situation. The navy and ocean science truly needed one another. In a letter to Sverdrup, Loeb commented that oceanographic "work requires support and facilities which only the Navy can furnish and they in return require the services of your scientific staff. Finally the work is of such a nature that it would be published without any secrecy or restriction. In view of the character of the radio and sound contracts not only for your work but for the advancement of the science of oceanography I am heartily in support of that program and believe it should be pushed to the limit. I think you will appreciate that there is a vast difference between that and the Navy's subsidizing cryogenics research or research of any discharge through gases."[18]

Harald Sverdrup appreciated the important distinctions made by Loeb and stood on the brink of changing his own opinion about potentially classified naval oceanographic research. While he preferred unrestricted grants, by the end of the Pacific war the Scripps director decided that scientists needed to know about the possible naval applications of their work. If this posed a moral problem, the scientist could withdraw. Until then, navy-sponsored ocean research should continue. This process and policy would encourage fruitful research and new possibilities for inquiry while leaving both the navy and its contractors free to choose the extent of their commitments. Sverdrup thus found himself in harmony with influential colleagues like Iselin, Ewing, and Revelle. The latter, on his way to the ONR in February, 1947, was replaced for a time at Buships by Allyn Vine.[19]

When Gordon Lill went to work for Roger Revelle in late 1947, both viewed a generous budget and the absence of established procedure as opportunity rather than uncertainty, and they proceeded to define ONR procedures, standards, and practices. Bowen and his ORI staff had already suggested the scope of naval interest and pioneered the contracting style that would become an ONR trademark. In the late summer of 1946, for example, the ORI contracted with Woods Hole for a wide variety of tasks, including some marine biology and underwater sound work. Iselin gladly signed an agreement that gave his institution wide latitude and $100,000. As director, he received discretion from the ONR to distribute the funds as he saw fit. Revelle and Lill built on this foundation and took advantage of its flexibility. When later asked to describe the expectations governing his position under Revelle in the ONR Geophysics Branch in 1947, Lill replied: "None. I invented it. Roger didn't tell me how to do anything. Neither did anybody else. I was making contracts and I wrote the descriptions for contracts in oceanography and meteorology and the solid earth sciences. They probably still use them, I don't know. I set up these contracts; I invented the stuff that went into the contracts. I called it creative writing."[20]

In the early years, Revelle, Lill, and their colleagues did not have to justify their grants or contracts to anyone. As one of the ONR's principal agents, Lill invested in people with reputation and experience and relied on their judgment to help determine the distribution of ONR funds.[21]

In the immediate postwar years, experience and good judgment resided by definition in those scientists who fashioned oceanography's relationship with the navy during World War II. A list of ONR's earliest major beneficiaries thus reads like a roll call of the most important wartime cultural translators, their colleagues, and students: Iselin, Ewing, Vine, Revelle, Riley, Munk, Eckart, and others. Gordon Lill knew their reputations, witnessed the quality of their work, and, very much like Admiral Gherardi before the war, viewed their research as both pioneering and of the greatest possible use to the national defense.

In its first decade, the ONR disbursed its funding as widely as possible, looking for long-range programs that would pay naval dividends over a long period. As Lill recalled:

> We developed certain things ourselves in the geophysics branch of ONR, and it was true in other branches. We could see things that we thought were important to go ahead and develop. We also accepted unsolicited proposals, some of which were not as unsolicited as you might think. . . . After a few years, we had more research proposals than we could possibly finance; we just couldn't get enough money. So it worked both ways. The Bikini Scientific Resurvey [of 1947] was Roger Revelle's idea as far as I know. We also conducted operations, Skijump in the Arctic, and we did exploratory work in seismology—listening to big explosions wherever they might occur around the world, including atomic bombs. And, we were responsible for starting the oceanographic buoy system that is in operation today . . . of course, we were [also] on board and committees and interagency meetings.[22]

Arthur Maxwell, who came to the ONR from Scripps while in the process of writing his dissertation, became Lill's deputy in 1955. Revelle had returned to La Jolla and Scripps approximately seven years before, and Lill moved into his place as geophysics chief. Maxwell's memory of early ONR policies, procedures, and award criteria confirms that particular members of the oceanographic community, largely the veteran translators of World War II, played a pivotal role in attracting Lill's attention and patronage as well as determining the distribution of funds to particular projects:

> The way it worked is that the director [of a research institution] would collect a bunch of programs, put all these together and he would submit it to ONR, and we would review all of these and decide essentially the ones that we thought were best. But when it came to funding it, we'd give him the money so that he had the final authority, even though we might pick and choose among the ones we thought best. The director really had the last say-so of how he wanted to spend the money. So he had a lot more freedom then than he has now. As long as the institutions were producing good results, we tended to renew. It was an era when budgets were growing. We encouraged at that time new oceanographic institutions to start. . . . I think that the overriding criteria we used was the capability of the individual involved. We didn't worry as much about what he was doing as whether he was good and capable of doing the job. . . . Right now, if you put in a proposal, you almost have to have solved the problem to get it funded. It's got to be that good. You're competing on the problem. And it wasn't that way. We looked around at individuals and what they were doing, and funded individuals . . . we funded them through the directors, but really we funded what the individuals were doing.[23]

Hardening of the arteries—as Gordon Lill called the combination of creeping bureaucracy, costly naval operations, and congressional surveillance—did not begin at the ONR in earnest until the end of its first decade. However, as congressional oversight increased and the operational navy jealously coveted the ONR's generous budget, a more formal and regular method of submitting proposals and judging results gradually became necessary. Even then, the scientists who ran the ONR's divisions wanted to minimize the paperwork and leave as much time as possible for research. To avoid excessive reporting and an unnecessary bureaucratic burden, Lill and Roger Revelle, then serving as associate director of Scripps under Carl Eckart, developed a new type of report to permit the proper justification of naval oceanographic expenditure without an endless stream of official correspondence. Revelle and Lill suggested that the chief scientist or principal investigator distribute the responsibility for composing the standard project report among the scientists on board the research vessel during the return voyage. By the time the ship tied up at the dock, its staff would assemble the finished parts into a completed report. Lill accepted this "ship report" as an adequate account of how scientists spent ONR money. The Geophysics Branch and most other ONR divisions also accepted peer-reviewed scientific publications as the equivalent of a final report. These techniques kept the scientists on task and minimized the administrative and accounting burden.

Thus, while the ONR earned a well-deserved reputation for extraordinary flexibility and enlightened investment in long-range oceanographic research, the grant process did have a well-defined set of controls. The ship reports provided one method of evaluation and accounting. Regular participation in projects or simple visits to laboratories and institutions offered another. The personal relationships Revelle, Lill, and Maxwell had with Iselin, Ewing, Sverdrup, Eckart, and other institution directors presented yet another. The latter technique reduced the ONR's investment risk considerably because the navy at large knew the men accepting and managing the money. These people offered their seasoned judgment and the reputations of their institutions as a guarantee that a naval investment would produce dividends.[24]

Building on this policy, Revelle and then Lill sought to fashion a ring of oceanographic programs around America's coastline directed by people they could trust. The ONR already funded existing programs at Woods Hole, Scripps, the University of Washington, and Lamont Geological Observatory. Maurice Ewing established the latter after his move from Lehigh and Woods Hole to Columbia in 1946. To expand this assortment of education and research centers in oceanography, funds controlled by Revelle and Lill helped establish programs at the Universities of Miami, Oregon, and Rhode Island, and Texas A&M University.

One distinguished American oceanographer virtually started his career as a young administrator in this ocean science program and participated in Lill's management system at two different times and in two ways. John Knauss, who would later create one of the best oceanography schools in the country at the University of Rhode Island, spent a very short period in the wartime navy training for an undergraduate meteorology degree at MIT. After completing his master's program in physics at the University of Michigan in 1949, he spent a summer in Europe only to discover on his return that he had lost a promised job at Los Alamos in New Mexico. Saved from unemployment by his naval reserve status, Knauss went back into uniform and journeyed to Washington, D.C., to take a course on wave forecasting. There he encountered Roger Revelle and Walter Munk, both visiting at Hydro from Scripps.

Knauss had served briefly with Munk and Waldo Lyon on Operation Blue Nose, an early postwar bid to explore the limits of ASW and submarine activity in the Arctic. The scientific party deployed on the Fulton Class submarine tender USS *Nereus* (AS-17) with the ComSubPac, Admiral McCann, on board. The *Nereus* worked with the submarines *Boarfish* (SS-327), *Caiman* (SS-323), and *Cabezon* (SS-334) near and partially under the polar ice pack. Knauss took water samples and temperature measurements, while Munk and Lyon worked on sonar, underwater sound, and submarine operating potential under the ice.

Upon hearing that Knauss had lost a job opportunity, Revelle informed him of a chance to manage the ONR Geophysics Branch for a six-month period while Gordon Lill traveled to Africa. Knauss inquired about the post and was quickly awarded the assignment. When Lill returned he decided to keep Knauss on for a longer period, and the latter had a chance to explore firsthand the evolving oceanography program at the ONR. Knauss later recalled that the ONR's formative process very much influenced his own: "That was when I really got excited about oceanography. I enjoyed the people and I enjoyed the work, so I stayed on until September of [19]51. I decided if I was going to be an oceanographer I had to get a Ph.D."[25]

Even though Knauss left in September, 1951, to study with Roger Revelle at Scripps and began building a formidable reputation in equatorial ocean-current studies, the ONR's long arm easily reached La Jolla. After he initiated his doctoral program and participated in the oceanographic aspects of Operation Castle's thermonuclear Ivy detonation in the Marshall Islands in 1952, the navy ordered Knauss back to active duty. His relatively short time in the navy during World War II, his presence on the reserve list, and the continuing war in Korea led to his recall. Knauss did not mind very much. The navy had paid for his undergraduate education and, besides, Gordon Lill had Knauss assigned to the ONR to take advantage of his previous experience.

During this two-year tour his jobs ran the gamut of ONR responsibilities. Since regular employees occupied all of the supervisory positions, Knauss drew a random sampling of assorted jobs related to ONR monitoring of contract work at naval laboratories and civilian institutions. In one case, Knauss traveled to the Chesapeake Bay on a cruise with a number of ONR contractors—including Robert Urick, the leading specialist in underwater sound at the USNUSL in New London—to explore the possibilities of detecting mines in shallow water in response to the navy's needs in the western Pacific. Knauss performed many of the water temperature measurements while Urick and others wrestled with the problem of using one-hundred-kilocycle echo sounders to locate mines in a coastal environment littered with debris from local communities, characterized by abundant marine life, and plagued by sufficient turbulence to produce false return echoes. During this same period, Knauss also went to sea with Wood Hole's Allyn Vine on an underwater sound project funded in part by the ONR. They worked in the Caribbean, operating off Bermuda, Guantanamo Bay, Cuba, and Jamaica.[26]

On more than one occasion, Knauss served as an ONR visiting inspector. Both Lill and Maxwell performed many of the same duties. Other ONR operatives, like Lt. Comdr. Dominic Paolucci, made extended trips, visiting a number of locations in a particular region each time. During the course of a trip along the "northeast corridor" in July, 1955, Paolucci visited Prof. Charles Stark Draper at MIT's instrumentation laboratory to evaluate the current work on weapons guidance systems. He then returned via the Lamont Geological Observatory, where he visited Ewing's second in command, J. Lamar Worzel, regarding ONR-funded research on submarine thermal wake detection. These visits also drew ONR personnel to meetings and symposia to keep tabs on the latest works in progress, especially on submarine design, performance, and silencing.[27]

Trust, flexibility, and communication guided the funding process while personal ties, informal visitation, and actual project participation confirmed the quality and direction of the work. Between 1946 and 1950, Revelle and Lill turned the loss of the ONR's primary mission into a remarkable long-range asset.

From the beginning, the ONR had a difficult time protecting its funding and preserving its mission. As the ink dried on the ONR legislation and the Planning Division tried to define the level and rate of commitment to various research fields, budget cuts threatened the ultimate productivity of the naval investment. Roughly four months after the creation of the ONR, the new agency faced austerity measures that reduced its $50 million in resources by approximately 50 percent. While planning for fiscal year 1948, the chief of naval research, Rear Adm. Paul Lee, tried to resist the reduction. He argued before the assistant secretary of the navy that the director of the budget might cripple the navy's contract research program in its infancy. After pleading for $28.5 million to meet the needs of current contracts and future planning, the ONR's projected 1948 research budget amounted to only $10.5 million. During the 1947 fiscal year, the ONR had $21 million available, twice the amount the administration's budget officers expected it to work with in the new fiscal year. After having derived such great profit from applied science and engineering during World War II, Lee could not understand the diversion of money away from programs and scientists with a proven record and a commitment to the national defense. He pointed out that the ONR had already assumed the role of central catalyst for defense research in the United States as well as educator and patron for those pursuing careers in science. Lee feared having to cancel contracts and obligations in 208 out of 396 current cases, releasing roughly 400 scientists and denying promised support to 1,000 graduate students. The duration of ONR contracts would fall from two or three years to eighteen months, in addition to a reluctant general reduction of commitments in all scientific fields.[28] This fiscal policy would make the long-range commitments initiated by the ORI/ONR impossible, damage the general credibility of the navy as a patron, and endanger its hard-won, productive relationship with civilian science. Karl Compton of MIT put it best in a letter to Warren Weaver, chairman of the NRAC, when he said, "If, just as this program has gotten well underway, confidence established, staffs recruited, and future plans laid, the whole program is thrown into a tail spin by drastic reductions, then I think the situation will be worse than if the plan had never been undertaken at all."[29]

Compton went on to note that only the ONR sponsored and funded scientific training when no other national program existed to boost oceanographic education. He pointed to President Truman's proposal for a $1 billion national training program to foster economically important job skills, while the ONR withered on the vine attempting to fund research and education. For Compton, "the proposed cut in the budget of ONR seems to be absurdly shortsighted."[30]

Compton reflected the view of the scientific community at large, which naturally saw investment in scientific research as the best possible contribution to the nation's security. Oceanographic institutions around the country proceeded to ally with the RDB and its panel on oceanography and the NAS's geophysics advisory committee to the ONR to bring authority and true influence to oceanography's cause. Independent navy studies done during these years, such as the 1947 Steelman Report, a research and development management study, called for a balance between defense and civilian research in all scientific and technical fields. In the midst of early postwar fiscal austerity, recommendations made by these respected committees and contract analysts helped validate both large-scale defense research and Vannevar Bush's July, 1945, call in *Science: The Endless Frontier* for a civilian science foundation.

Validation did not come easily, even with the considerable support the ONR quickly amassed within the civilian scientific community. Its total contract program budget languished considerably under $50 million as the budget reductions and political upheaval of the defense unification controversy, often called the "Revolt of the Admirals," descended on Capitol Hill in 1949. The CNO, Adm. Forrest Sherman, appointed Medal of Honor recipient Vice Adm. Oscar C. Badger to review the ONR's contract budget as a potential source of waste. With the Cold War becoming more intense and the fleet in need of additional funds, the ONR's very existence seemed in doubt.

Badger's conclusions, submitted to Sherman on 13 February 1950, suggested that although the ONR's programs should reflect more closely the requirements of modern naval combat, they seemed sound and well administered. Sherman, who clearly had hoped for a different conclusion, immediately appointed another review board to examine Badger's findings. Before the board could meet under the direction of Rear Adm. H. S. Kendall, North Korea invaded its southern neighbor in June, 1950. These events quickly resurrected congressional and naval memories of World War II science and its significant contribution to victory. With the Pentagon focused on the Asian conflict, the ONR's budget for fiscal years 1950 and 1951 doubled. This in turn ensured that the programs Admiral Lee, the NRAC, and many contractors wanted to sustain finally received funding that promised a much greater return. Even though the ONR had survived the review process and prospered thereafter, its administrators never forgot the navy's peacetime inclination to look more closely for an obvious, direct link between its mission at sea and the research the ONR chose to sponsor.[31]

Within the Academy: The Committee on Undersea Warfare

The lasting and mutually rewarding partnerships formed with the ONR did not represent the only ways of promoting oceanography or influencing naval policy on the ocean sciences. Some scientists interested in research that might serve the national defense also sought to sustain after V-J Day the relationship between the navy and the scientific community characteristic of the wartime NDRC Division 6.

Hitler's U-boats nearly brought Great Britain to its knees, and the American Pacific submarine force destroyed both the Japanese merchant marine and many essential Imperial Navy warships. Furthermore, the advanced submarine technology developed in Germany in the last months of the war placed in question the capability of any surface ship to hunt and kill a submarine. The German Type 21 U-boats employed extraordinary battery power and advanced hydrodynamics to permit sustained seventeen-knot bursts of speed for limited periods below the surface. No destroyer, destroyer escort, or submarine chaser could successfully track and attack a submarine leaving its sinking victim behind at that rate of speed. When this technology was revealed by the U.S. Naval Technical Mission in Europe during the final year of the war, it was seized by both the Soviets and Americans and ASW strategy and tactics became obsolete virtually overnight. The navy needed a greater knowledge of the ocean down to depths approaching a thousand feet to provide an environmental context for a complete reevaluation of submarine and antisubmarine warfare.

In a 1 November 1945 letter to ORI chief Admiral Bowen, the director of the UCDWR, Gaylord P. Harnwell, called for a committee to "maintain Naval liaison, determine membership, organize and conduct symposia, issue bulletins and summaries of proceedings." Like the demobilized Division 6, he wanted to facilitate communication within the field of undersea warfare and provide advice to the navy on the most promising avenues of scientific and technical investigation. Harnwell had already accepted an invitation from NRC chairman Detlev Bronk to direct a committee tasked with holding an undersea warfare symposium in cooperation with the Navy Department. The UCDWR director felt very strongly that the effort should go beyond a one-time meeting of undersea warfare specialists. Harnwell wanted to create an undersea warfare committee within the NRC. His idea struck a resonant chord, for a few weeks later Harnwell received independent support from John Coleman of the still extant CUDWR Sonar Analysis Group, who saw in the NRC a possible permanent home for his agency.[32]

On 16 September 1946, representatives from the navy and the scientific community met to give form to the undersea warfare symposium and Harnwell's proposal. According to James H. Probus, a participant in these formative events, those advocating the new committee sought "to keep alive the close contact and mutual understanding that had been established between the navy and the civilian scientists of Division 6. Their objective was to retain the experience and interest of these scientists, and to interest new scientists in the problems of modern undersea warfare."[33]

By this time, the ONR had taken over the ORI's functions and sent its representatives to the meeting as did the various navy bureaus, the CNO's office, and the Naval Ordnance Laboratory. The NRC and civilian science in general were represented by Detlev Bronk (then at Cornell Medical College), Harnwell (University of Pennsylvania), John Tate (University of Minnesota), Eugene F. DuBois (Cornell Medical College), Walter S. Hunter (Brown University), Philip Morse (MIT), and Columbus Iselin (WHOI).

This group approached Admiral Bowen on 25 September with a proposal for a permanent undersea warfare committee that would operate under the auspices of the National Academy of Science's National Research Council and seek to fulfill the mission described by Gaylord Harnwell the previous November. Bronk sent his letter via Capt. Lawrence R. Daspit of the ONR, and Bowen's positive response of 23 October 1946 led the NRC to establish the Committee on Undersea Warfare (CUW).[34] Eight days later the ONR offered a broad formal definition of the CUW's role and its most profitable fields of investigation in cooperation with the navy.

From the beginning, the CUW proved exceptional and influential. It reported directly to the NRC executive board, did not operate within any of the established divisions, and consulted directly with the navy bureaus and the ONR. Its domain included the most critical scientific and technical aspects of ASW and prosubmarine activity. Members of the committee sought interaction between the navy and civilian science on undersea warfare issues in a manner that might foster understanding, as well as cooperative and effective investigation, evaluation, and analysis. Very few major undersea warfare concerns fell outside the CUW's purview. For example, detection and tracking appeared at the top of a list of tasks the ONR wrote into the CUW's agreement with the navy. Then came attack measures, submerged communication and navigation, identification of friend or foe (IFF), human factors, and countermeasures. The ONR-composed plan for CUW activity also included close cooperation with government and private institutions involved in studying the ocean. The total initial budget approved by the navy came to $54,000.[35]

A 21 November letter from the ONR also offered the CUW assurances that it would receive full naval cooperation in its effort to become a rich resource for the submarine and ASW communities. As Capt. Norman Lucker of the ONR put it: "It is most gratifying to hear of the sincere interest of the members of the Committee in the Undersea Warfare problem. Representatives of CNO have assured us that they are now in a position to present the chief problems in the field and to give their views on priorities. This Office wishes to assure you that the Committee will have access to all of the information that we have which might be useful to them. Also that they will be most welcome at all of the submarine and anti-submarine conferences and meetings, and that they will be able to visit any naval laboratories or facilities which they deem advisable to visit in carrying out their functions. It is our view that the Committee should have a free hand in its methods of attacking the problems."[36]

The ONR also inaugurated a series of symposia on undersea warfare in September, 1946, to keep vital information flowing to the CUW. After the initial gathering, the committee took over these annual meetings and continued the process, holding the second at the Department of Commerce Building in Washington on 5 and 6 May 1947. The symposia revealed problems, expressed priorities, offered information about research in progress, and helped strengthen the committee's relationship with the navy in the process of trying to overcome the challenges of war beneath the ocean's surface.[37]

The ONR and the NAS signed a contract on 20 June 1947 formally defining the committee's broad responsibilities. John Tate, former head of NDRC Division 6, became the first chairman, with Harnwell serving as vice chairman. Other members included Iselin, DuBois, Morse, Hunter, W. V. Houston (Rice Institute), Frederick Hunt (Harvard), and T. E. Shea (Electrical Research Products, Incorporated, and a veteran of the CUDWR). John Coleman (Penn State) served as executive secretary.

Initially conceived only as a consulting body, the CUW actually went as far as participating in the early stages of instrument and vehicle prototype development. The committee's duties as defined by the contract with the navy included:

(1) Reviewing, analyzing, and evaluating the Navy Department's general program of research and development.

(2) Proposing additional fields of investigation within these general fields and additional projects for research and development therein.

(3) Collecting, collating, and disseminating scientific and technical information in these fields.

(4) As determined by mutual agreement, assisting in research and development, including participation in the design and construction of prototype and/or breadboard key components where necessary to predict performance and technical feasibility.[38]

Changes in the nature of undersea warfare and advances in submarine design made it easy to define the major areas of research but no less difficult to master the major problems. What kinds of materials would enable submarines to reach ever-greater ocean depths? How would the designs and hull structures have to change? What sort of sound transmission and detection conditions would American and Soviet submarines regularly discover in a deep, subsurface world accustomed mostly to natural noise? The postwar revolution in undersea warfare generated many questions that the navy, industry, and science expected to answer.

13

The Forest and the Trees, 1946–50

There is no immediate danger that the government's interests
in oceanography will diminish or that substantial amounts of
government subsidy will cease to be available to the private
laboratories. However, there is a grave danger that insufficient
attention will be given to developing a balanced oceanographic
program and that unless steps are taken, the research will
break up into units too small to be effective.

—COLUMBUS ISELIN, 1 September 1948

THE IDENTITY AND NEEDS OF OCEANOGRAPHY AS AN INFANT
discipline nearly disappeared in the flurry of ocean science programs and
activities emerging after the war. Many scientists studied the ocean, but
who practiced oceanography as opposed to marine biology or chemistry? Did
oceanography include geology and the emerging field of geophysics? What did
an oceanographer do as opposed to a physicist? Without this type of definition,
the young multidisciplinary science would never draw students, could never
develop the proper perspective and professionalism, and would never achieve
recognition from the scientific community at large. Lacking these assets,
oceanographers could not effectively respond to the civilian or military needs
of the future. If oceanography hoped to influence the navy and participate in
policy formulation, it had to have its own house in order.

While most oceanographers pondered these issues after the war, Columbus
Iselin's remarkably complete surviving correspondence provides a revealing
chronicle on the issue. He returned frequently in both professional and casual
letters to the need for careful disciplinary definition and the construction of a

culture resting upon university-based educational programs directed by experienced oceanographers.[1]

Many scientists did not see in oceanography's postwar circumstances the kind of future they wanted for the infant discipline. The war and the advent of the Soviet threat after the defeat of Germany and Japan ensured that the ocean sciences would never again lose the navy's attention. Unfortunately, ocean scientists in and out of the navy often gave the appearance of so many discrete contractors performing on demand for the federal government. Armed with considerable assets and offering lucrative research contracts, BuShips, Hydro, the ONR, and some of the naval warfare specialties seemed to hold the upper hand in determining oceanography's direction. The modest fellowships and grants that sent faculty and graduate students to sea on board vessels like *Atlantis* before the war could no longer support this activity. Research ship operations and the cost of instruments alone had already begun to exclude the small independent researcher in favor of federal and naval priorities. Piggybacking could compensate only in small measure, and the university professor or independent investigator remained at a disadvantage. In a note to Maurice Ewing about costs and the effect of a major naval presence in oceanography, a clearly worried Iselin remarked:

> I do not blame you for being fed up with the Navy. I feel almost as strongly as you do on this subject. The difference is that you can wash your hands of the situation and I cannot. The trustees signed a contract with the Navy and it is my job to see that we live up to it as best we can. Moreover, we are almost in the situation of no contract, no institution. In other words unless I continue to work towards some reasonable system of continuing government support, in a year or two we shall be busily exploring Buzzards Bay in the "Asterias." The Institution's own endowment funds have become a drop in the bucket. The interest rate has gone down to almost half and costs have doubled . . . my success in getting clean money in the future depends somewhat upon being able to please Revelle and company during the life of the existing [BuShips] contract.[2]

Even his navy friends could not always allay Iselin's anxiety about oceanography's dependence upon the naval patron. When the WHOI director sent him figures on the spending levels in the two active BuShips contracts, Lt. Comdr. Butler King Couper, still at the bureau in the months before his tour in Fleming's new division at Hydro, could only ask Iselin to wait. As a junior officer, Couper could not act on budget matters without Revelle's approval, underscoring the awkward position of oceanographers all over the country. Operation Crossroads preoccupied Commander Revelle in the spring and summer of 1946, and the spirited Couper characteristically expressed his frustration to Iselin in verse:

> *The aggravatin'est words of mice or men—*
> *"On Roger's return", are binding me again!*[3]

These circumstances worried a number of leading ocean scientists. The oceanographic community was distracted by the diversity and appeal of federally funded tasks while the discipline itself still lacked clear definition, broad university recognition, and properly trained practitioners in adequate numbers. When Crossroads concluded and many ocean scientists returned to their institutions, the federal coffers opened wider than ever. Hydro and BuShips continued their patronage, but now the ONR entered the picture—with Revelle managing the Geophysics Branch and Gordon Lill on the way. Congress also provided enormous sums of money for the USGS to work on the continental shelves. Iselin wondered how Woods Hole and smaller-scale institutions would cope "in a world in which research is bought in large quantities as if it were lumber or pig iron."[4] What of the very important private and more restricted lines of research? While many scientists approached private corporations for research funding, the best chance for support, especially given the expense of going to sea, clearly remained with government agencies. This remained true even with the severe federal budget reductions inflicted on the navy and science in the two years before the Korean War.

Iselin's correspondence during the last half of 1946 suggested that although welcome and plentiful federal funding certainly created more opportunities than interwar poverty, disciplinary confusion also reigned. Who made the fundamental research choices: civilian science, the navy, or both? How did all of the postwar work fit together? Was ocean science building a firm foundation of knowledge or simply filing a random series of reports? Did oceanographers want the various parts of their discipline treated separately as physics, biology, chemistry, and geology, or did greater integration make more sense? In this case, Iselin regretted the influence of older scientists like Princeton's Richard Field. The WHOI director thought that Field and many of the older hands seemed to find the postwar prospect of integration difficult after a career in a particular discipline and its scientific culture.

The issue of personnel raised even more questions. Since by 1946 most of the shipboard research fell to younger men, the shortage of competent, experienced people became painfully obvious. Only Scripps and the University of Washington had legitimate training programs. Did the government and the academic community plan to address this problem? Could these two schools alone train the next generation of oceanographers? Did current practitioners expect Scripps and Washington to create a culture for oceanography comparable with that of the classic scientific disciplines? What role would the professional scientific societies play? Did oceanography need its own professional society? If so, what would become of the American Geophysical Union? Attending the ONR's first undersea warfare symposium held in Washington in 1946, Iselin witnessed the confusion this kind of uncertainty caused and he found it disturbingly familiar. He wrote to Mary Sears: "You should have been at the Subsurface Warfare meetings last week in Washington. It was 1940 all over again, but with 90 civilians and 300 officers assigned to working out new problems. Pure science will have to wait a little longer I am afraid."[5]

Speaking at that symposium, Iselin brought these issues to the attention of the ONR and other participating government agencies and universities. He suggested that neither the present practitioners nor the universities realized the possible future demand for ocean scientists. From seafood to mining to national defense, opportunities were considerable, but adequate means for attracting and educating students did not presently exist. According to a representative of the NRC who brief notes quickly recorded in sentence fragments as Iselin spoke, the WHOI director lamented: "Little encouragement in Universities. There is no center at which the different phases of oceanography are taught. Oceanography does not get its share of able students from the universities. La Jolla (U. of C.) gives courses leading to advanced degrees; 30 students in the past three years. All have found jobs. But physics laboratories are more attractive."[6]

Iselin also noted that the distribution of funds for research seemed terribly skewed. He calculated that the government had thus far in 1947 invested roughly $600,000 in physical oceanography and the state of California lavished $400,000 on studies of the sardine harvest while many other research possibilities withered on the vine. Again, the ONR symposium note taker did not fail to grasp the point: "These expenditures are not too wholesome. Temporary. Don't train young men. Oceanography not being developed in a balanced manner. Biological and chemical problems suffer in relation to physics and geology of the sea. No support to speak of for general marine biology."[7]

While many oceanographers had a great deal of freedom, more work than ever, and liberal government funding, no government agency, civilian laboratory, or academic institution had assumed responsibility for the healthy and balanced development of the discipline. To approach this task, Scripps, Woods Hole, and the University of Washington would have to double their roughly twenty teaching positions and ask the academic community to recognize these professionals as peers. With an improving job market and university recognition, more students would hopefully investigate oceanography, and other universities might finally perceive a reason to add this field of study to their curriculum and create additional faculty positions. In September, 1947, the NRC observer heard Iselin repeatedly beat the drum for expansion, training, and carefully designed education programs: "Must have a few centers of oceanographic research where the subject as a whole can be developed in an orderly manner and where applications are not the main thing. Must attract first rate minds to the subject."[8]

Working for the navy and other federal agencies distracted the oceanographic community from the pressing professional business of creating a strong foundation for the discipline. Too often, immediate practical application seemed to rule the day while oceanography itself needed attention and nurturing. In parallel with their work for the navy, the nation's oceanographers and other interested scientists now began paying closer attention to a professional culture for the discipline that would build on the practice and expertise of the existing laboratory and university programs. Without this culture, the leaders of postwar oceanography in the United States could not promise a productive and competent future.

Iselin insisted that only the scientific community itself could begin creating a professional culture for the ocean sciences. Oceanographers could not rely upon the navy or any other agency to design curricula, administer laboratories, train students, or set and teach the standards for measuring excellence in oceanographic practice.

The creation of university programs and a disciplinary culture for oceanography persisted as a major issue for the WHOI director. His experience and success as a wartime scientific entrepreneur prompted him to address the issue as soon as peacetime offered an opportunity beyond the occasional speech or professional paper.

Iselin's initiatives began soon after the war, driven by his intense concern for oceanography's future combined in equal proportion with his ambitions and fears for Woods Hole. These motives clarify his vigorous participation in the early postwar debate over Revelle, Fleming, Hydro's new Oceanography Division, and leadership opportunities at the ONR. He believed that the scientists chosen to advance oceanography at Hydro and supervise the ONR's divisions should receive appointments roughly equivalent to university professors, with complete academic recognition. He also wanted the right person matched with the proper job, and he suspected—just as Mary Sears feared—that the navy's choice would, in the long run, serve neither Woods Hole nor oceanography. Therefore, while Iselin respected the talents and abilities of Roger Revelle, he viewed the BuShips officer as being too self-serving and too partisan toward Scripps to manage the Oceanographic Division at Hydro.

Iselin, not without his own parochial motives, perceived Revelle as a challenge both to Woods Hole and his vision of building oceanography as an independent, university-based discipline.

The WHOI director thus lobbied strongly in support of Richard Fleming's candidacy for the Hydro position in cooperation with Mary Sears, a Hydro and WHOI insider with a similar vision. Together they envisioned Richard Fleming working closely with Henry Bryant Bigelow and Harald Sverdrup, the only oceanographers in the National Academy of Sciences, and effectively influencing the navy's hydrographer on matters critical to a productive partnership with civilian oceanography. Their vision also involved a strong oceanographic discipline, validated by the university community, growing in parallel with the navy's continued postwar oceanographic activity. Both Iselin and Sears saw Revelle as a more freewheeling personality, better suited to the ORI and emerging ONR. In February, 1946, Iselin wrote to Sears:

> The second thing which seems to me might well be done is to suggest to ORI that Roger is the boy they have been looking for all along. . . . If it will make it any easier, tell [ORI] this is my suggestion. So far as I know, they have nobody with any knowledge of underwater acoustics or the geophysical sciences. If they want to make ORI stronger, Roger is the man. I am convinced that if they offer him a fancy job he will stay. It is just the kind of thing he likes to do and is good at. He will meet all the famous scientists in the world, yet he will not have to compete against them professionally. In ORI he will even be able to maintain more or less indirect control on oceanography which is another thing he wants. If you can pull these two things off your next job is President of the United States.[9]

Revelle had extensive scientific and management experience, but he did not impress Iselin as an academic pioneer or cultural cornerstone. Furthermore, Revelle constantly exasperated Sears with his vacillation between the ORI and Hydro positions. She wrote to Iselin perceptively suggesting that Revelle really wanted a promotion to associate professor at the University of California, but he feared that the university administration would turn him down. Sears wanted Revelle to accept the ORI post before he went to Bikini and Operation Crossroads so that Fleming could immediately get to work at Hydro. Despite her respect for Revelle, Sears agreed with Iselin that these jobs had nothing to do with personal promotion, but rather with a foundation for oceanography. She and Iselin found themselves caught between friendship, personality, the ambitions of all concerned, and the serious fashioning of postwar oceanography. As Sears commented to Iselin: "When all is said and done, however, I certainly think RR [Roger Revelle] has brains and admire the way he can think. He really has done a big job even if he has been exasperating and I think it is to be appreciated. On the other had he just won't mend his ways. The final decision [between the positions at ORI and Hydro] should be for the good of the project and not just to reward personal vanities. I think that this is about all RR wants the job for really and now that he thinks there is some opposition he is just like a small boy and will fight for what he thinks he wants."[10]

Careful evaluation of Iselin's and Sears's observations on Roger Revelle reveal less an assessment of individual ambition than a measure of the strong desire for disciplinary foundation and unity of purpose. In the early postwar years, confusion and uncertainty reigned among strong personalities on how best to achieve those goals.[11]

Working with Frank Jewett, his patron of many years, as well as Detlev Bronk of Johns Hopkins University, Albert E. Parr of Yale's Bingham Laboratory, Rear Adm. Edward H. "Iceberg" Smith of the Coast Guard, and NAS president Alfred N. Richards, Iselin hosted a conference at Woods Hole on 11 August 1948. Precipitated by an 11 December 1947 letter to Jewett and the latter's energetic response and support, this informal meeting provided an opportunity to awaken the Academy to the need for professional and cultural development in oceanography and to discuss strategies to achieve those goals. In his notes from the meeting, Richards made it very clear that the discussion quickly focused on a specific goal and placed the responsibility for action with the NAS: "What the talk is getting down to is the selection of a committee vitally interested in the broad extension of oceanography not only as a field of international interest but of cultural importance; that a survey of existing institutions and the state of knowledge as compared with 20 years ago be made by one or two younger men who shall submit their findings to the main committee."[12]

Clearly, this group of scientific leaders took as their paradigm the work of the 1927 Committee on Oceanography chaired by Frank Lillie. Lillie's group found the money and the reasons to create WHOI, worked with the navy, encouraged university-based research, and provided an early central focus for the development of oceanography in the United States.

In a very characteristic but less conspicuous way, Iselin followed the paradigm by playing a role similar to that assumed by his mentor Henry Bryant Bigelow nearly twenty years earlier. The WHOI director accepted the task of turning the main points of the 11 August meeting into a memorandum that Richards and Bronk could take before the NAS and NRC. He passionately argued for a balanced and coherent national program in oceanography that would set standards, create programs, attract students, and then educate them to address the naval, military, and civilian aspects of the challenge posed by the ocean. He wanted for oceanography that same cultural coherence enjoyed by the other major scientific disciplines. Doubtless, everyone who attended the Woods Hole meeting knew of the many singular opportunities in oceanography offered by the ONR, BuShips, and Hydro, as well as civilian agencies and some university programs. By their very existence, these prospects demonstrated that many "trees" flourished but no person, group, or agency had yet properly defined, described, or helped fashion the forest. Oceanography needed a culture complete with a defined specialty, language, accepted practice, training, and complete professional system of demands and rewards. Dated 1 September 1948, Iselin's memorandum ended on an urgent note: "There is no immediate danger that the government's interests in oceanography will diminish or that substantial amounts of government subsidy will cease to be available to the private laboratories. However, there is a grave danger that insufficient attention will be given to developing a balanced oceanographic program and that unless steps are taken, the research will break up into units too small to be effective. We need something approaching an oceanographic university and it seems practical to develop this from the existing laboratories. Money is needed for fellowships, for senior positions and for the operation of vessels, especially in the field of marine biology. If a strong central core of pure science does not exist, applied oceanography will collapse for lack of new ideas.[13]

In November, Richards presented Iselin's distillation of the Woods Hole conference to the NAS.[14] He received permission both to assemble a short-term committee to define problems and issues and, after determining a source of financial support, to sponsor a more permanent panel. The latter would probe oceanography's needs and make recommendations for the future, very much in the spirit of the Lillie Committee. The NAS Committee on Geography and Geophysics appointed the panel on oceanography, which in turn named a working group on 15 December 1948 to prepare a draft "Integrated Plan for Research and Development in Oceanography." The group consisted of Iselin, Parr, Richard Fleming, Princeton's Harry Hess, who retained a naval reserve commission after leaving the service at war's end, and Carl Eckart of the infant Marine Physical Laboratory at Point Loma—all serving under Chairman Athelstan Spilhaus, dean of the University of Minnesota's Institute of Technology.

With an eye toward affecting the fiscal year 1951 budget debates on Capitol Hill, the Spilhaus group worked with six other teams appointed by the panel on oceanography. The teams examined needs and programs in a wide array of oceanographic specialties, from internal motion, waves, and tides, to sea ice and the optical properties of water and underwater photography. The group tasked with preparing the integrated plan employed the teams' findings to provide the NAS with a foundation upon which to build. Their proposals, formulated as a general description of the best avenues for future research, provided the basis for discussion at a conference the Academy would sponsor—a defining event for ocean research and the oceanographic community.

Shortly after the Spilhaus working group submitted its report to the NAS in late January, 1949, Richards wrote to Henry Bryant Bigelow with a proposal: "It is to this effect: That you and Iselin select a group of interested experts and arrange for a meeting the outcome of which will be clearly formulated questions or topics, the adequate exploration and development of which would constitute a long-range national program for the development of oceanography."[15]

The NAS president wanted to formulate a series of proposals that might persuade the Rockefeller Foundation to support new and significant ocean surveys under the auspices of the Academy and its oceanography panel. His letter to Bigelow concluded with the observation that Iselin's 1 September memorandum and Frank Jewett's initiative and enthusiasm offered a good starting point.

Bigelow's reply, based upon a conversation between he and Iselin, supported the initiative and nominated people who would make a significant contribution to Richards's proposed meeting. He suggested prewar leaders, veteran wartime translators, and some scientists just emerging into prominence. Among others, the slate included Ewing, Iselin, Thompson, Canadian A. G. Huntsman, Bronk, Admiral Smith from the Coast Guard, Merle Tuve from the Carnegie Institution's Department of Terrestrial Magnetism, Fleming from Hydro, Carl Eckart from the MPL, and WHOI's Alfred Redfield, who chaired the Joint Research and Development Board's panel on oceanography. Woods Hole's first director did not nominate himself. Bigelow thought that his age, teaching obligations at Harvard, and his role in the activities of the Lillie Committee twenty years earlier precluded his participation. He suggested that any scheme emerging from current planning should reflect the needs of the 1950s and beyond, not the priorities of one involved with devising strategies for the 1930s.[16]

Based upon Bigelow's nominations, Iselin recommended to Richards that the meeting take place in Washington on the heels of the ONR-CUW Symposium on Undersea Warfare scheduled for 16–17 May 1949.[17] Many of those scientists suggested by Bigelow lived in the Washington area and most of the others would come into town for the symposium.

Richards agreed and wasted no time sending out invitations. In his letters to the prospective participants, the NAS president described the central purpose of the gathering and its critical place in the process of defining the future of oceanography: "The proposed meeting is designed to be an intermediate step on the way to an appeal to an appropriate foundation for funds with which to finance a detailed and authoritative survey of present status and future aims."[18]

In responding to questions from Richards, Columbus Iselin commented that reading Bigelow's 1931 report for the academy entitled "Oceanography" would sufficiently prepare participants for the conference, which would be held at the Wardman Park Hotel in Washington.[19]

While Bigelow's effort two decades earlier sought recognition for oceanography within the scientific community, Wardman Park went much farther. The group did not simply revisit Bigelow's analysis or try to imitate the activities of the Lillie Committee. At the Wardman Park conference, major figures in the earth and ocean sciences concluded that oceanography needed coordination, formalization, and its own culture to take advantage of current opportunities and to promote others. The continuity of federal funding, only occasionally uncertain, did not emerge as a primary consideration. Oceanography urgently needed university recognition and joint planning among current programs that would provide experts with respected faculty positions and draw students to the study of the ocean. Iselin did a synopsis of the discussion and the NAS produced a 350-page verbatim transcript of the conference. In the course of his synopsis, Iselin shifted the emphasis away from the views of his long-time friend Maurice Ewing. The Lamont director insisted that agencies and institutions should simply channel the programs and funding toward the most able men in the field and give them the freedom to do the best possible research. In the concluding paragraphs of his conference summary, Iselin related a concern voiced by A. G. Huntsman to drive home his central plea and, indeed, the central concern and issue of the conference: "Dr. Huntsman expressed some disappointment that we had not arrived at anything new. In answer to the question, 'Where do we go from here' we seem to have concluded that we should simply go on. Mr. Iselin pointed out that perhaps the reason that so little time had been devoted to discussing new programs of oceanographic research was that the directors of the various laboratories are reluctant to see expansion occur before the professional standards in the existing lines of research have been raised. Oceanography is still clearly in its infancy. The first requirement seems to be to keep the infant healthy, rather than worry about how he will look when fully grown."[20]

The members of the civilian oceanographic community found themselves faced with a wide variety of opportunities while still unsure of their own identity. Who was an acoustician? Who worked in chemistry? Who preferred biology? Were any or all of these people oceanographers? Before they could make the most of contracts with BuShips, Hydro, the ONR, and other agencies, they needed a firm hold on their own identity. The 1931 Bigelow study merely drew attention to the existence of oceanography and achieved a measure of recognition for the burgeoning field from the traditional disciplines. With demand moving faster than they could respond, oceanographers needed a set of clearly defined standards and practices as well as the first threads of a cultural fabric that would hold the community together and facilitate communication and familiarity. Other sciences already possessed these assets. For oceanography, this level of disciplinary maturity remained a goal that most participants of the Wardman Park conference wanted to pursue more vigorously. Thus, as the navy's oceanographic awareness and interests expanded and the Cold War took form, ocean scientists also struggled, but in a far more organized and coherent way, to fashion a professional culture.

After he returned to the MPL, Carl Eckart prepared a report on the Wardman Park discussions for the NAS.[21] Emphasizing the wide array of neglected issues and phenomena waiting for serious study, Eckart argued for intellectual freedom, university and laboratory training programs, well-balanced research agendas, and adequate funding carefully distributed between naval, military, and civilian oceanography. The Academy published his report in 1952 as *Oceanography 1951: Report of the Committee on Oceanography of the National Academy of Sciences.* Eckart and Revelle prepared the final text of the report, and Mary Sears supplied her essential editorial touch after additions by Iselin and Alfred Redfield. Thanks to Jewett, Iselin, Richards, Eckart, and others, oceanography took another step toward disciplinary maturity.[22]

As these developments unfolded, the navy more precisely defined its Cold War mission with its "Study of Undersea Warfare," directed in the spring of 1950 by Rear Adm. Francis Low of the wartime Tenth Fleet. This study defined the potential of Soviet power combined with captured German Type 21 U-boat technology as the navy's primary postwar adversary at sea. The prospect of a formidable Soviet submarine fleet posing a future menace to the Atlantic sea lanes roused the political commitment, congressional funding, and naval resolution necessary to conduct critical new research on the deep ocean as a combat environment.[23]

The Low Report revelations fueled both oceanographic programs and the anxiety of those concerned for the future of the discipline. Massive naval assets and generous spending in carefully defined fields neither integrated the various disciplines within oceanography nor spawned the disciplinary culture so desperately needed. If the discussions at Wardman Park and the tone of the report that followed often demonstrated a measure of unity and obvious common goals, Maurice Ewing's attitude suggested that some leading ocean scientists still differed on how to attain those goals.

Neither did harmony reign within the armed forces. It soon became clear that preservation of the essential rights and prerogatives of each armed service emerged as one major result of the defense unification crisis that created the NME in 1947. Challenged by the infant air force in an acrimonious roles and missions debate, the navy emerged even more determined to preserve its surface, air, and submarine roles in the national defense. In spite of the need, unity, cooperation, and integration did not rule the day in either the armed services or the oceanographic community.

This situation recalled Vannevar Bush's impatience with the endless debate on the question of naval-scientific cooperation and integration in 1940. Oceanography's cultural translators now resorted to personally engineered alternatives based upon trust, individual relationships, and the force of mutual expectation. Personal relationships and professional confidence helped leading oceanographers chart a course around formidable institutional barriers.

This is not to imply that efforts to integrate better the work of bureaus, laboratories, and universities ceased. The CUW continued working with navy laboratories like the NEL on projects related to applied oceanography. The Joint Research and Development Board also sought ways to integrate Defense Department research activities. These plans usually suffered from an institution's desire to protect its own prerogatives and limit commitments to other agencies. Gaylord Harnwell, the CUW chairman, discovered this in his early postwar interaction with BuShips. He voiced his concern just a few weeks before scientists and naval officers gathered on 12 October 1951 at a timely conference organized by the CUW at the NEL. The participants discussed ways of improving cooperation between the civilian scientific community and the military. In a letter to NAS president Detlev Bronk, Harnwell remarked that BuShips seemed the most backward of the navy's research activities, unaware of its cultural and communication problems, and sharing this fault with a number of other Defense Department research sponsors.[24]

In reality the NAS/NRC exhibited some of the very same flaws. Harnwell, CUW secretary John Coleman, and Detlev Bronk wanted to cast the NAS in the role of arbiter between university-based science and the navy. The Academy's links to the university community might help free scientists for extended periods to labor on naval projects. Would universities agree to this possibility? Those at the ONR thought they had already addressed that question; what of the existing contract arrangements between the ONR and major academic institutions all over the country? Was the NAS trying to reinvent the wheel? Was the naval initiative not acceptable? This was the pre-NDRC interwar cooperation dilemma revisited. What price cooperation and collaboration and on whose terms? Indeed, were these even the most pressing questions? What about the issues raised at Wardman Park? These efforts and later initiatives designed to create a central research establishment for the Department of Defense under the sponsorship of the Joint Research and Development Board's Basic Research Group foundered on the gulf—cultural and political—between institutions struggling to cast postwar oceanography in their own image and likeness.[25]

During the first postwar decade, reminiscent of the NDRC's birth in 1940, individuals shaped postwar solutions to oceanography's major questions while their institutions struggled. In an environment dominated by universities, scientific associations, professional and disciplinary cultural expectations, and the often strict and parochial views of naval commands, scientists and officers discovered that pressing questions proved much less formidable on the personal level. Institutional reluctance opened the way for a postwar world in which key individuals and the opportunities created by personal and professional affinity shaped the development of oceanography. The dynamic that sustained postwar naval oceanography frequently and significantly rendered institutions and naval commands nearly incidental. Emerging from the war informed, experienced, and close-knit, civilian cultural translators set oceanographic priorities, managed funding, and created an eccentric culture for oceanography in response to the pressing needs related in *Oceanography 1951*.

The convocation marking the late-June, 1954, opening of Woods Hole's new navy laboratory—a facility since renamed for the institution's third director, Rear Adm. Edward Smith—demonstrated well the early dynamics of growth within American naval oceanography after 1945. Outwardly, this was an institutional celebration. The NAS sponsored the convocation to christen a new WHOI laboratory financed by the navy and dedicated to oceanographic research. Support from Iselin's colleagues and fellow wartime scientists on the CUW like Frederick Hunt and Harvey Brooks brought recognition from the NAS/NRC and guaranteed a stellar roster of convocation speakers for the event. The convocation certainly suggested a coherent identity for oceanography, further identified its scope and expectation, and made a statement about ocean science practice and standards of excellence. The composition of the convocation and the language used in its announcement, program, and literature implied that the NAS had successfully assumed the role of facilitator, bringing the navy and the oceanographic community together in a constructive relationship.

In reality, nothing of the sort had happened. This role, sought by the Academy and NRC before the war, remained elusive. The sponsoring institution, whether the navy, the NAS, Woods Hole, or any other, merely provided a stage for a group of mostly civilian oceanographers, translators all, who fashioned the culture of oceanography as well as its policies and priorities. Their institutions either facilitated their efforts or erected additional obstacles to overcome. In the case of the Smith Laboratory at Woods Hole, the determination necessary to define the need and the funds to make the project a reality depended completely on the relationship between Columbus Iselin, his successor Edward Smith, and the ONR's Gordon Lill.

Fourteen years earlier, only months before the United States declared war on Japan and Germany, Lill passed through the BT officer training program at Woods Hole under the guidance of Iselin, Ewing, Vine, Worzel, and others. As a science administrator he concluded early that accomplishments of significance emerged from personal collaboration between informed, talented, and determined people. Placed in a position to sponsor research by the ONR's unique postwar charter, Lill sought out these people and fostered this kind of collaboration all over the country. Work by the navy's sponsored scientists only confirmed for him that ONR investments in Woods Hole would provide dividends for the navy over both the short and long term.

Lill made the expansion and enhancement of oceanographic research and educational facilities a priority of his ONR Geophysics Division. The WHOI laboratory emerged as one of the first tangible results. He made similar investments in Scripps, and in later years established a policy of providing seed funding for new oceanographic education programs. Then, in their turn, Texas A&M University and the Universities of Miami and Rhode Island became beneficiaries.

Personal connections helped scientists and officers fashion a collective disciplinary culture in the absence of true central direction, established university programs, or a dedicated professional association. In spite of Gaylord Harwell's comment about the communication problems with BuShips during the first postwar decade, Roger Revelle's old haunt generously sponsored oceanographic research both independently and in conjunction with the ONR. If Harnwell found the bureau occasionally unreceptive, the chair of the CUW should have looked closer. Scripps, Woods Hole, and other labs worked closely at BuShips with a veteran of both WHOI's wartime BT program and Richard Fleming's office at Hydro, Butler King Couper. Iselin had a long history of corresponding with Couper that extended back to the early 1940s, when the latter's colorblindness nearly denied him a wartime naval commission. Couper brought Iselin, Vine, and other scientists from across the country closer to BuShips and fashioned a productive relationship even as the bureau's available money for oceanography declined relative to the ONR. As far as the working scientists were concerned, King Couper was BuShips oceanography.

The 1954 WHOI laboratory convocation serves well as a metaphor for the very personal dynamics that shaped postwar naval oceanography. In reality, the event staged by the ONR, Woods Hole, and the NAS celebrated the creation of a laboratory facility conceived by Iselin and Lill. Gordon Lill funded the facility on his own authority with the support of a CUW advised on oceanographic needs by Iselin and Hunt.[26]

With institutions trying to find ways to cooperate, struggling all the while to maintain their own authority and identity, individuals found a way to cross barriers both real and imagined. While the NAS, numerous research institutions, universities, and the U.S. Navy certainly provided assets and the necessary administrative support, individual vision made real by relationships with wartime roots fashioned postwar oceanography and first addressed the goals formulated at Wardman Park.

Carl Eckart, immediate postwar Director of the Scripps
Institution of Oceanography and the Marine Physical Labora-
tory, University of California. Courtesy Glasheen Collection,
Mandeville Special Collections Library, UCSD.

William G. Metcalf became one of WHOI's specialists in Arctic oceanography soon after the conclusion of World War II. Courtesy Woods Hole Oceanographic Institution.

Martin Pollack, chief scientist on WHOI's "Med" Cruise of 1947–48 (Atlantis 151) working with a cable-mounted Nansen bottle designed for taking water samples at various depths. Courtesy Woods Hole Oceanographic Institution.

Gordon Lill, while he was head of the
Geophysics Branch of the Office of Naval
Research, 1956. Courtesy Woods Hole
Oceanographic Institution.

Roger Revelle (center), associate director of Scripps, and John Isaacs (profile) on the
MIDPAC Expedition in 1950. Courtesy Barr Papers, Scripps Institution of Oceanography
Archive/UCSD.

This group photograph of the Nobska Summer Study participants was taken on the
Whitman estate in Woods Hole, summer, 1956. Columbus Iselin appears in the dark

jacket in the top row, center. The second man to his left is the CNO,
Admiral Arleigh Burke. Courtesy Woods Hole Oceanographic Institution.

Harold Edgerton, professor of physics at MIT, preparing one of his underwater illumination devices for deployment, mid-1960s. Courtesy Woods Hole Oceanographic Institution.

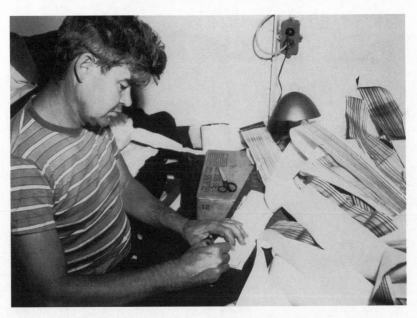

The relentless Maurice Ewing, founder of Lamont Geological Observatory, evaluating data at sea, probably on board R/V *Vema,* c. 1965. Courtesy Ewing (Maurice) Papers, The Center for American History, University of Texas at Austin.

The WHOI geophysicist and physical oceanographer John Brackett Hersey at sea performing acoustics experiments, probably on board R/V *Chain,* c. 1960. Courtesy Woods Hole Oceanographic Institution.

On 23 January 1960, Lt. Don Walsh and Jacques Piccard took the Navy's deep submersible *Trieste* seven miles down into the Challenger Deep in the Marianas Trench, setting an unbreakable deep dive record. The submersible is pictured here in the foreground with the support ship USS *Lewis* (DE-535) in the distance. Courtesy U.S. Navy.

Pioneering WHOI marine biologist Mary Sears (right) is pictured here in her office with an assistant, c. 1970. Sears, a retired Navy reserve Commander and World War II WAVE, became the founding editor of the authoritative journal *Deep Sea Research*. Courtesy Woods Hole Oceanographic Institution.

14

Back to Sea with a Flourish, 1946–55

The "Caryn" results at Bermuda were spectacular. This will not
mean anything to you, but it will to Roger [Revelle]. He can find
a copy of a preliminary report at NEL. It seems to me that submarine geology has been given the biggest kind of boost and
that the Bureau of Ships is back in oceanography in a big way.

—COLUMBUS ISELIN to RAYMOND MONTGOMERY, 1947

ITH THE CRITICAL QUESTION OF DISCIPLINARY IDENTITY
and the peacetime challenges and rewards of federal patronage
looming before them, the American oceanographic community
proclaimed its value by dramatically returning to the deep oceans of the postwar world for scientific purposes.

From the Ridge to the Med

Ewing and Iselin staged the first of these major expeditions in the summer
of 1947. With submarines diving ever deeper and the nature of the ocean betraying amazing sound transmission properties, ocean-bottom topography
took on great significance. With wartime U-boats descending to nearly a thousand feet and the deep sound channel ensuring long-range low-frequency transmission, the submarine community had compelling reasons to investigate the
ocean depths, the sound-absorption characteristics of bottom sediment, and
the array of geologic forms rising from the ocean floor in the Atlantic and
Pacific.

Doc Ewing had a passion for the single most critical aspect of Atlantic Ocean bottom topography. From 300 to 600 miles wide and nearly 10,000 miles long, the Mid-Atlantic Ridge runs from Iceland to the Antarctic Circle, dividing the Atlantic into two major basins. The ridge forms only one section of a global submarine mountain range that winds its way around the earth like the seams on a baseball. Its nature and origins captivated this relentless scientist, newly arrived at Columbia and only a few months away from establishing Lamont Geological Observatory.

In mid-summer 1947, Ewing led a joint expedition to the ridge hoping to unlock some of its secrets. The National Geographic Society (NGS), Columbia University, and Woods Hole funded the project, providing a scientific staff, the proper instruments, and *Atlantis*. The ridge cruise, christened *Atlantis 150* by Columbus Iselin, yielded information that opened up new worlds for geologists and laid the foundation for increasingly accurate maps of the ocean floor. Ewing took sediment cores, collected Fathometer tracings, and studied sound transmission behavior. His team explored an unfamiliar world characterized by mountains easily rising fifteen thousand feet from the bottom, surpassing surface phenomena like California's Mount Whitney, the highest point in the continental United States outside Alaska.

Although sponsored by the National Geographic Society and very open to public scrutiny, *Atlantis 150*'s work over the Mid-Atlantic Ridge silently supported both submariners and ASW specialists. In a 25 February 1948 report to BuShips, Ewing described the submarine optical experiments and camera refinements accomplished during the summer of 1947. He also drew attention to the reflection and refraction research that made strong statements about sound behavior and explosives at great depths during the course of the ridge cruise program.[1]

Ewing contributed to human understanding in ways needed more than ever by the undersea warfare community. As the tense relationship with the Soviet Union began to sour the early peacetime euphoria, a dialogue between the oceanographic community and the naval operating forces, especially those involved with ASW and submarine warfare, became critical. Hydro's Richard Fleming visited the Submarine Officers' Conference on 25 February 1948 and commented at length on the importance of oceanography to submariners. Rather than boring them by droning on about the wonders of modern science, Fleming came right to the point. He compared the dependence of aviators on meteorology with that of undersea warfare practitioners on oceanography and ended on a sober but promising note: "In comparison with meteorology and aircraft, I feel that oceanography and submarines have a long way to go. It is the job of the Hydrographic Office to help you in design problems, in testing, in providing information needed for strategic and tactical planning and in all ways

aid you in making most effective use of your boats, weapons, and their equipment."[2]

Measurements continued in the Atlantic during August, 1947, when
Woods Hole and the NRL cooperated on the calibration of low-frequency
sound sources for research. While Ewing went to sea to study bottom reflectivity for more than two months, Woods Hole went its own way and evaluated
new sonobouys in deep water. The USNUSL in New London explored submarine listening by placing a line of hydrophones on USS *Quillback* (SS-424)
and proceeding out to the deep waters of the North Atlantic. At the same time,
Richard Fleming expected the arrival of his two newly converted survey vessels
and intended to assign them the task of collecting important data on bottom
topography and temperature distribution in broad areas of the world ocean.

Continuing this burgeoning tradition of ambitious postwar expeditions,
on 10 December 1947 *Atlantis* once again left Woods Hole, only this time
bound for the Mediterranean Sea. Later christened *Atlantis 151* or simply the
Med Cruise, Woods Hole and the Hydrographic Office collaborated on a dual-
purpose project recommended by the ONR's Lt. Comdr. Charles "Monk"
Hendricks, commissioned by the navy, and funded by Hydro with transferred
ONR funds. Fleming wanted to satisfy his naval and civilian customers while
learning from an institution experienced in long-term deep-sea oceanographic
expeditions.[3]

The new chief of oceanography at Hydro also wanted to demonstrate that
wisely invested naval funds could provide official customers with knowledge
directly applicable to naval operations. In this case, the results proved particularly important. Chief scientist Martin Pollak and WHOI colleague Dean Bumpus probed the sound channel phenomenon and the transmission characteristics in the Mediterranean and Aegean Seas during the months when the
Marshall Plan and Truman Doctrine guided the American effort to contain the
Soviet Union in eastern and southern Europe.[4] The navy wanted as complete a
picture as possible of ocean-bottom topography and underwater sound conditions in the Mediterranean for passive sonar listening and active propagation,
anticipating a Soviet naval threat from the Black Sea.

Most of the *Atlantis* crew did not realize the importance of the cruise to the
navy. Bottom mapping had a variety of civilian applications, and other members of the scientific staff did chemical and biological research, studying water
temperature, salinity, nutrients, currents, and marine life. Some Med cruise scientists, led by Columbia's David Owen, also tested a new and innovative underwater camera designed by Maurice Ewing.[5] It was not until years later that
many members of the scientific crew—including Robert Abel, a young chemist
on his first major cruise—realized the primary purpose of spending nearly eight
months in a careful survey of the Mediterranean: "That's the "Med" cruise . . .

[its primary purpose was] ocean bottom mapping, ocean surroundings mapping, and underwater sound. It was strictly Navy granted, Navy oriented, and while the chief sponsor was the Bureau of Ships (sic), the rest of us guys did not know . . . ?!"[6]

Completed the following summer after six months at sea, the cruise crew and staff recalled years later that *Atlantis 151* was one of two major early postwar research voyages. It marked the institution's first solo effort after four years of war, and *Atlantis's* first journey back to Europe since its construction in Copenhagen and the opening of the Woods Hole Oceanographic Institution in 1930.

Iselin and Fleming asked senior scientist Martin Pollak to fly home early with data they thought would excite navy ASW officers and submariners. The *Atlantis* Fathometer ran continuously for more than twenty thousand miles and the staff did unclassified studies of the bottom sediment, the water chemistry, significant currents and waves, and even provided a selection of photographic images of the bottom from Ewing's deep-sea camera. In all, Pollak and Bumpus supervised forty-six deep hydrographic stations. After its return on 18 June 1948, *Atlantis 151* eventually became part of the institution's culture. Participating in the Med Cruise or merely recalling its significance marked a member of the Woods Hole staff as a postwar pioneer.[7]

In April, 1948, Ewing's Lamont group and geophysicists at Woods Hole planned another significant summer voyage that would employ explosives from the navy and geophones or advanced seismic instrumentation borrowed from corporations such as Humble Oil. While *Atlantis* concluded its Med Cruise, Ewing turned to *Saluda,* which BuShips had placed at Iselin's disposal for a few months. The Columbia geophysicist observed: "I understand the Navy has more need for deepwater, mid-ocean work than for the out-today-and-back-tomorrow kind. We like the deep water kind and are more likely to do it than anyone I know." While the royal "we" doubtlessly included the WHOI team, many of whom Ewing had trained, now it referred primarily to his on-board scientific staff, which came mostly from Columbia University and included Frank Press, Gordon Hamilton, and Ivan Tolstoy.[8]

While WHOI certainly participated in Ewing's sound work, it also relied on its own staff to do vital investigations for the naval undersea warfare community. As Ewing prepared to employ *Saluda* in the spring of 1948, John Brackett Hersey took Woods Hole's R/V *Caryn* to the Bermuda area for refraction and reflection experiments as well as exploration of the deep sound channel's capabilities. The navy's perception of a slowly emerging Soviet naval threat after 1947 served both Ewing and Hersey well. Rising tensions and the possibility that Joseph Stalin's navy might eventually supplement its modest fleet of short-range submarines with clones of captured German Type 21 boats kept sound transmission work and ocean-bottom geology very high on the navy's list of priorities. As with the Med Cruise, the navy had no interest in making undersea warfare research publicly known. It thus came as no surprise when Iselin sent a message through his executive assistant, Frank Ryder, to Hersey aboard *Caryn* asking him to act with his usual circumspection and to make sure his crew understood the need to stay away from the Bermuda press.[9]

For submariners, submerged endurance in deep water and sprinting ability became critical. With the rising possibility of a Soviet threat, BuShips initiated a program to convert wartime American submarines into vessels capable of greater submerged speed and sustained underwater operation. Much of the increased speed resulted from a process of severe streamlining combined with more powerful battery propulsion. As for submerged endurance, the advent of an American schnorchel device gave the modified submarines much more time below the surface. Submariners referred to the conversions as "Guppys" because they had *g*reater *u*nderwater *p*ropulsive *p*ower, emulating the most advanced German U-boats. Naval authorities assumed that the Soviets would understand the revolution in undersea warfare precipitated by the new submarine designs, undertake similar conversions, and initiate a new construction program. The Bureau of Ships began to build more potent submarines from the keel up during this same period beginning with the *Tang* class. While these developments presented unparalleled opportunities to increase speed, endurance, and depth of operation, it also alerted them to possibly dangerous Russian undersea potential. Innovative German naval architecture and novel propulsion technologies nullified the wartime American ASW advantage. This sudden sense of vulnerability focused the navy's attention on submarines, detection, and underwater sound more effectively than had anything since the darkest days of World War II.

American submariners and ASW crews, facing the possibility of Soviet Type 21 clones, needed all the advantages they could find. The kind of work Brackett Hersey conducted aboard the *Caryn* proved as vital as the deep scattering layer, the sound channel, and advanced sonar equipment. Hersey's research in the Bermuda area concerned the discovery he made with Allyn Vine that shallow low-frequency sound returned to the surface at predictable intervals. Known as convergence zones, these acoustic repetitions made extraordinary transmission and detection ranges possible much closer to the surface.[10]

In reporting to the NAS's Committee on Undersea Warfare in the early summer of 1948, Iselin celebrated Hersey's success aboard the *Caryn*. The WHOI scientist successfully detected submarines and surface vessels at the first two thirty-five-mile convergence zones relative to his position in the Atlantic. Although Hersey performed the tests under ideal conditions, he now had firm evidence confirming his hypothesis. He and Vine also realized the clear naval applications of their discovery. While noting that BuShips had already begun to adjust its research programs and sponsorship to explore low-frequency sound as a detection tool, Iselin quietly described the astounding promise of the discovery:

> The tests carried out to date in long range listening have been essentially practical tests and presumably limited to favorable conditions. On two occasions, a fully quieted, submerged submarine has been able to detect loud, surface vessels screw noises at extended ranges and in a more recent test, a carefully quieted hydrophone suspended from the ketch CARYN was able to detect a ten knot snorkeling submarine intermittently out to a range of 35 miles and also at the second intensity peak at approximately 70 miles. . . . While it was first considered that the important applications of this type of listening were mainly pro-submarine, the most recent observations demonstrate that in some circumstances the sea is much less opaque acoustically from the anti-submarine standpoint than has generally been believed.[11]

In a letter to Raymond Montgomery, a marine biologist affiliated with Woods Hole and Harvard who was spending some research time at Scripps, Iselin included praise for the Hersey group aboard the *Caryn*. While expressing his sincere excitement about the successful ocean-bottom investigations, seamount discoveries, and convergence-zone research, it seemed just as important that Roger Revelle should know, and Montgomery was in a position to report the WHOI accomplishment: "The "Caryn" results at Bermuda were spectacular. This will not mean anything to you, but it will to Roger. He can find a copy of a preliminary report at NEL. It seems to me that submarine geology has been given the biggest kind of boost and that the Bureau of Ships is back in oceanography in a big way."[12]

Iselin took solace from the *Caryn*'s achievement and the BuShips commitment. In the midst of the significant congressional threat to the ONR budget in 1947, the WHOI director noted that navy activities, particularly BuShips, kept informed of the advances made in low-frequency underwater sound and shifted their budget priorities to support the effort. Allyn Vine and Bill Schevill spent time at Pearl Harbor demonstrating to the ComSubPac staff the long-range advantages of low-frequency sound.[13] Indeed, the navy had plans to develop a series of new sonars—both active and passive, with variations deployable by aircraft—to exploit knowledge emerging from sound transmission studies.

Work by Ewing, Pollak, Bumpus, Hersey, and Vine persuaded BuShips to spend more time and money exploring the ocean floor and the effect of bottom terrain on undersea warfare and passive detection. The bureau also approached Iselin and offered to play the role of silent partner for the next two or three years, paying the annual $1 million cost of keeping *Atlantis* at sea. Iselin welcomed the arrangement, and for its part, the bureau requested only advice, research results, and anonymity. The public attention would focus on Woods Hole, Lamont, the National Geographic Society, and any other public sponsor. The most immediate illustration came with the second summer season of Ewing's research on the Mid-Atlantic Ridge with the National Geographic Society in 1948. As the cruise began, Iselin commented that the "Navy does not feel it necessary to have it become generally known that it is supporting such work."[14]

Navy projects remained the backbone of civilian oceanographic research. While this dependency recalled the problems debated at Wardman Park, it also created multiple opportunities that the civilian oceanographic community needed. Iselin and his colleagues across the country regularly attended both open and classified conferences hosted by BuShips or the ONR on the navy's problems and priorities. They wanted to influence naval policy on the oceans while responding to the navy's environmental requirements at sea. An informed partnership would serve everyone's agenda.

Despite the intense activity in 1947–48, oceanographers realized the immediate effect of budget debates on naval and oceanographic programs. With the ONR and other agencies under the gun, the discussion at the NRC's Committee on Undersea Warfare always returned to a proven solution to federal budget contractions. The CUW refashioned the interwar common practice on a grand scale. Closer cooperation and careful planning between BuShips, the ONR, Hydro, Woods Hole, the University of Washington, Lamont, Scripps, and other concerned activities remained the only effective antidote to enforced fiscal restraint.[15]

Cabot and the Gulf Stream

Operation Cabot offered one of the best examples of this postwar collaboration. For physical oceanographers, the Gulf Stream represents one of the most significant phenomena in the North Atlantic. Flowing north and east from the Caribbean, roughly parallel to the U.S. East Coast before moving out into the Atlantic toward Great Britain, the Gulf Stream acts like a river in an ocean, providing extraordinarily warm water where one expects to find only the icy cold of northern latitudes. This characteristic of the ocean profoundly affects marine life, food productivity, sea temperature, and naturally attracts the attention of commercial fishermen as well as the defense establishment.

Proposing a study of the stream for 1950, the Hydrographic Office unexpectedly found itself confronted by the availability of six ships. Woods Hole offered the *Atlantis* and *Caryn;* Richard Fleming's office contributed the USS *San Pablo* (AGS-30), a small converted seaplane tender, and the *Rehobeth* (AGS-50); William L. Ford, Fleming's counterpart at Canada's Defense Research Board, sent the *New Liskeard;* and the U.S. Fish and Wildlife Service freed the *Albatross III* for a short time. Using these assets, Hydro planned to stage a careful and detailed survey of immense importance. Here was a chance to accomplish the kind of extensive Gulf Stream study that Norway's Bjorn Helland-Hansen had proposed before the war as a cooperative effort between Norway, Germany, and the United States.

The possibilities excited physical oceanographers all over the world. Six ships could look more carefully at the Gulf Stream and its various eddies and meanderings than any single vessel might on a cruise five times as long. The project kept these vessels moving between Cape Hatteras and the Grand Banks from 6 to 23 June 1950, composing a Gulf Stream synoptic plot describing comprehensively the rate of change in the stream's position as well as the surface velocities of the water. Scientists involved also wanted to assess the stream's effect on the lower atmosphere, to explore the nature of the "meanders" that frequently departed from the main current path, and to study the formation of eddies when meandering cells of warm water wandered off in a new direction.

The Cabot team designed a survey approach that would not restrict the chief scientists to the traditional cruise grid. The latter approach permitted the accumulation of great amounts of data, but denied staff members the freedom to pursue particular questions raised by the ongoing research. With the *San Pablo* acting as flagship and Richard Fleming as chief project scientist, the six ships positioned themselves along the course of the stream at 150-mile intervals. This permitted the staff to gather data within the stream and then to move in and out of its course in a zigzag manner to determine boundaries, meanders, and eddies. Every thirty minutes the ships took bathymetric, meteorological, and other readings, carefully determining their position according to the Long-Range Aid to Navigation (LORAN) system, while the WHOI staff took surface current motion measurements with William von Arx's new Geomagnetic Electrokinetograph (GEK).[16] A second phase of the voyage called for taking simultaneous sections of hydrographic stations along predetermined lines of longitude while a third studied a restricted area to provide the navy with essential information for ASW and submarine warfare.

Woods Hole was designated to receive and analyze the data at the conclusion of a voyage well staffed with experts. Walter Munk of Scripps and Eugene LaFond, who took a postwar position with the Naval Electronics Laboratory, demonstrated their interest by initially responding to appeals for advice on how best to use this extraordinary list of available assets. Both LaFond and Munk suggested a few options for the exercise under the code name Operation Dream before the code name Cabot was officially adopted to identify the project. Munk joined WHOI's Henry Stommel and L. Valentine Worthington on the *San Pablo,* Thomas Austin of Hydro went to sea with the *Rehobeth,* William Ford was on the *New Liskeard,* Dean Bumpus was aboard the *Caryn,* and Fritz Fuglister and Martin Pollak led the group from the *Atlantis.*

This joint effort represented the first truly synoptic study of the Gulf Stream. It pinpointed a branching in the northern reaches of the stream and provided firm evidence that the entire phenomenon periodically shifted as much as eleven miles in a roughly north–south direction. Cabot also confirmed the existence of cyclonic eddies south of the Gulf Stream, as suggested by Iselin in 1940 and Woods Hole's Fritz Fuglister in 1947. The 1950 venture also confirmed earlier hypotheses by Iselin and Fuglister that eddies occasionally broke off and moved independently away from the main course of the stream. While scientifically significant, Operation Cabot also proved interesting to ASW and submarine officers dependent upon water temperature, currents, and air-sea interaction to detect, track, and possibly destroy enemy submarines.[17]

MIDPAC

Shortly after Cabot concluded, Roger Revelle took Scripps on its first independent venture into the deep ocean. He returned to La Jolla in the summer of 1948 from his naval service and the ONR eager to take advantage of the public and private assets available for ambitious postwar oceanography. In February, 1949, interim Scripps director Carl Eckart appointed Revelle his deputy despite the objections of some senior institution faculty, and the new administrator immediately began to look seaward. As Ewing and Iselin had made the Atlantic the oceanographic property of Woods Hole and Columbia University, Revelle saw the Pacific as Scripps's domain.

Barely eight months after his appointment on 10 October 1949, Revelle applied to the trustees of the University of California for $60,000 to participate with the NEL in a very ambitious oceanographic survey of the mid-Pacific. Building on the recent work done at Crossroads and the ONR-sponsored resurvey of Bikini Atoll, the joint University of California–U.S. Navy Electronics Laboratory Mid-Pacific Expedition, mercifully christened MIDPAC, marked Scripps's debut as a deep ocean scientific research center.

The expedition covered more than twenty-five thousand miles from California to Hawaii to the Marshall Islands. The magnitude of the effort served Revelle's purpose, which was to pull Scripps out of the company of coastal investigators and place it in the ranks of world-class deep-ocean research institutions. In his proposal to the university, the young oceanographer commented:

"The University of California's Scripps Institution of Oceanography has become the largest oceanographic laboratory in the world, yet up until now its investigations have been primarily confined to a narrow strip of ocean along the West Coast of North America. This was necessary in the past because its single ocean-going research vessel was not suited for sustained operations on the high seas, far from shore. . . . There are no boundaries within the ocean; it is a unit and no part of it can be thoroughly understood without knowledge of the whole. This fact has long been recognized by the scientists of our sister laboratory, the Woods Hole Oceanographic Institution, who have taken the entire north Atlantic and the Mediterranean as their province."[18] Revelle made it clear to the university trustees that he wanted to claim the Pacific for the University of California.

MIDPAC's scientific and technical staff went to sea on board two ships. The R/V *Horizon*, the Scripps vessel, with James L. Faughn as master and nineteen scientists aboard, left San Diego on the first expedition leg on 27 July 1950. The NEL's PCE(R)-857, a 180-foot naval escort vessel commanded by Lt. Comdr. D. J. McMillan, carried thirteen others, including participants from UCLA and the University of Southern California.

Revelle proved an able and amiable leader armed with determination and a healthy appreciation of both science and the media. While he knew exactly what he wanted MIDPAC to accomplish, he also intended to demand public attention. Nick Carter, the radioman aboard *Horizon*, remained on the air for many hours each day reporting the many accomplishments of the expedition in just the way Revelle wanted them heard. Revelle explained to journalist Andrew Hamilton that MIDPAC had merely "scratched the Pacific's bottom." Revelle's messages and his interaction with journalists led to heroic and fascinating descriptions of MIDPAC's considerable achievements. In December, 1950, Hamilton wrote that the expedition scientists brought back amazing examples of "a tremendous sunken mountain range and a long, submarine escarpment—one as imposing as the Rockies, the other as sheer as the east face of the Sierra Nevada . . . underwater structures, built by living creatures, that dwarf the Pyramids of Egypt or the Empire State Building . . . the shot heard around the Pacific—a Navy SOFAR explosion heard over 3,500 miles, the longest propagation of underwater sound ever recorded."[19]

In all of this Roger Revelle stood at center stage. Journalist Hamilton described the oceanographer as a tall, impressive man obviously suited to lead an expedition returning with impressive results. Revelle's project, Hamilton noted, represented the first major exploration of the deep Pacific since the stations taken by the R/V *Carnegie* in 1929. Revelle even received credit in Hamilton's story on MIDPAC for directing the navy's oceanographic research during World War II. Rawson Bennett, Columbus Iselin, and John Tate, among others, no doubt found the assertion amusing. This was indeed a heroic venture led by a new Jason with a very able crew of oceanographic argonauts.

Regardless of Revelle's tendency to exaggerate to the media, MIDPAC succeeded beyond expectation. In the course of steaming from California to Hawaii to the Marshall Islands and back, the staff uncovered an irregular bottom through constant soundings, even in terrible weather conditions. Eighty percent of the ocean floor exhibited some sort of rugged bottom topography, the remaining twenty proved flat. Several new sea mounts emerged from the sounding records as well as similar structures with flat, plateau-like tops called "guyots," a name chosen by geophysicist Harry Hess who first discovered them from his wartime supply ship while using a navy Fathometer. Hess named them for Arnold Guyot the nineteenth century Swiss-American geologist. The MIDPAC cruise also discovered a northeast to southwest ridge of submerged mountains south of Hawaii that scientists tentatively called the mid-Pacific mountains. These peaks came within 450 fathoms of the surface and at some points rose up from depths as great as twenty-three hundred fathoms. Like Ewing, Woollard, and Hersey in the Atlantic, Revelle kept the echo sounder pointed at the bottom and running around the clock.

He also adopted Ewing's practice of regularly taking core samples from the ocean bottom. The MIDPAC team took seventy-five cores, some approaching twenty-four feet in length, from areas of the Pacific ranging up to thirty-two hundred fathoms deep. Russell Raitt of the Marine Physical Laboratory also determined by explosive sound-source refraction that the ocean-bottom sediment between the mainland and Hawaii varied in thickness from fifteen hundred to three thousand feet. Furthermore, the expedition did not miss the opportunity to explore the deep scattering layer, only to discover several strata of marine organisms living above and below one another, each with its own diurnal cycle.

The MIDPAC cruise put Scripps on the map as an independent deepwater research institution. It also produced sufficiently important data, observations, and discoveries in conjunction with the NEL to draw public attention to the La Jolla laboratory and Roger Revelle, its rising star.[20]

Gravity Revisited

Earth scientists also returned with renewed vigor and navy support to new and significant research traditions initiated before the world went mad in 1939. Shortly after the Great War, measuring the Earth's gravitational field became fundamental to the study of geophysics and oceanography. Scientists who spent the interwar years trying to create an American version of the interwar Dutch gravity program conducted by Vening-Meinesz were given access to the

best instruments and ships and support from civilian and federal agencies to
continue the process.

The most relentless effort along these lines characteristically took place under the supervision of J. Lamar Worzel and his mentor, Maurice Ewing. Now
firmly in place in New York City, Ewing received permission from Columbia
University in 1948 to move his geophysics program from the main campus Geology Department in Manhattan to the recently bequeathed Lamont estate on
the Palisades. Employing the new Lamont Geological Observatory as a base of
operations, Ewing and Worzel embarked on a multiyear program to measure
the Earth's curiously irregular gravity field.

Shortly after his arrival at Columbia in 1946, Ewing began working with
Woods Hole to restore its Vening-Meinesz pendulum apparatus, a mechanical veteran of the 1936 navy-AGU expedition. In a concurrent effort, he and
Richard Field of Princeton persuaded Bell Laboratories to donate to WHOI
the crystal chronometer designed a decade earlier for use with the gravity device. With both the chronometer and the Vening-Meinesz apparatus at Woods
Hole, Ewing had easy access to the perfect tools for an extensive gravity research program.

Both Woods Hole and Lamont resumed gravity work very shortly after the
defeat of Japan. Supported at first by Roger Revelle at the ONR's Geophysics
Branch and then by Gordon Lill, research progressed in the Atlantic and Pacific
Oceans as well as the Gulf of Mexico. Iselin's people took gravity measurements
in the gulf during the summer of 1947 from the USS *Meteor*, then commanded
by enthusiastic would-be oceanographer Comdr. Charles Hendricks. Iselin
also encouraged the ONR to continue sponsoring George Woollard as he took
his gravity work from WHOI to the University of Wisconsin. Woollard
planned a global gravity measurement project that would establish base values
for the Earth's gravitational field at critical points all over the world. Woods
Hole remained Woollard's intermediary with the ONR well into the early
1950s.[21]

While Woods Hole's gravity efforts frequently took second place to its absorbing effort in underwater sound, Lamont focused completely on earth science and geophysics. Between June, 1947, and April, 1955, Ewing and Worzel
worked closely with the submarine community and the ONR to complete regional gravity surveys conducted aboard the navy's submarines. Although it
often proved difficult to obtain time on board the dwindling number of postwar submarines available, the Lamont faculty and their Columbia graduate students rode more than twenty boats over the nine-year period on roughly two
dozen separate gravity cruises. Lamont gravity measurements onboard the *Sea
Dog* (SS-401) and *Bergall* (SS-320) played a major role in a presentation made
by Ewing and Worzel at the AGU meeting in the spring of 1951. Lamont people

also rode the *Archerfish* (SS-311), *Balao* (SS-285), *Conger* (SS-477), *Corsair* (SS-435), *Diablo* (SS-479), and *Toro* (SS-422). Their observations covered substantial parts of both oceans, and by the early 1950s the Lamont team had also arrived at a method to permit reduction of its data using new IBM computing machines.[22]

Familiarity and personal trust frequently provided the key to these critical submarine deployments. With American strategists emphasizing submarine warfare and ASW as the top priorities of the postwar navy, the smaller fleet of the early 1950s occupied virtually every vessel in critical work. If Ewing and Worzel intended to continue their gravity research, they had to approach the admirals who controlled the submarine force in both oceans. In one typical case, Worzel and student Lynn Sherbert visited Rear Adm. Elton Grenfell, wartime commander of the *Tunny* and strategic planning officer on the staff at ComSubPac in 1942–43. At the time, Grenfell held a post at BuPers and served as a special assistant to the CNO. This Pacific submariner remembered the critical roles played by Worzel, Schevill, Vine, and others all too well. Faced with a request for time submerged for gravity research in the Atlantic, Grenfell approached his colleagues on behalf of the Lamont gravity program. His letter to the commander of the Atlantic submarine force, Rear Adm. Frank Watkins, proved instrumental in arranging early 1955 cruises on the *Archerfish* and *Balao* out of Key West to Ascension Island. While institutions and naval activities certainly provided assets and helped facilitate research, valuable personal relationships and friendships regularly translated words into opportunity and action.[23]

Action seemed imperative during this period, for the gravity work did not simply serve the civilian scientific community and human knowledge. The ONR and the submarine community had a special interest that they considered as important as learning more about the shape of the Earth and the nature of its crust. Lamont documentation describing the general purpose of the research duly noted: "When a suitable network of gravity observations exists, the precise shape of the earth can be determined. It is essential to extend the coverage of gravity observations as much as possible. These data are necessary for the precise direction of guided missiles."[24]

This period witnessed the first experiments at sea with Regulus air-breathing missiles. The Regulus I and II models were awkward to carry and deploy at sea because of their volatile fuel and large size. These liabilities demonstrated the need for rapid launching and stealth, paving the way for the quick ascendancy of the Polaris submarine-launched ballistic missile (SLBM) system. Increasingly precise knowledge of the Earth's gravitational field and its variations proved critical to the innovative inertial guidance system developed for the Polaris SLBM by Charles Stark Draper's team at the MIT instrumentation laboratory.[25]

At the Poles

Not all of the energetic postwar expeditions explored the secrets of the Atlantic, Pacific, and Mediterranean, or the ocean bottom. Columbus Iselin, in his correspondence with Roger Revelle only a few months after the war ended, brought up the military appeal of the polar regions as a potential operating area for submarines and military aircraft. The WHOI director felt strongly, consistent with his postwar views on the development of oceanography as a discipline, that civilian science should take the lead. He feared the military would determine the course of research for lack of civilian initiative or qualified scientists to take on the task. On 21 May 1946 he wrote to Revelle, then still at BuShips, that "What is needed is for a small group of first class men to read up on the Arctic and make a few exploratory trips. Practically all existing Arctic men in this country lack a general understanding of science or have a screw loose. Unless a qualified civilian group gets started thinking about the Arctic, the military people will have the cart before the horse. The practical applications and the engineering should wait until science has made some sort of a beginning."[26]

Iselin was not far off the mark. Submariners and the ONR had already turned their eyes northward. If the endlessly aggressive Comdr. Charles Hendricks was any indication, submariners debated the possibility of operations under the ice with some regularity. As WHOI's contact in undersea warfare at the ONR, Hendricks's views suggested that the navy seriously considered funding research to make such submarine activity possible given the geopolitical situation and the relative geographic positions of the United States and the Soviet Union.[27]

Interest in the Arctic stemmed from the recognition of political reality, not from universal naval interest. The geographic maxim that the shortest distance between two points is a great circle route held in the nuclear world as it had in the conventional. Both primary Cold War antagonists realized their vulnerability over the Pole and sought to minimize the threat. The first forays toward the North Pole by navy ships represented strategic and practical reconnaissance efforts. The navy wanted to determine the nature of the region and its suitability as a naval combat environment. Naval aircraft could fly personnel over the Arctic, and sailors could work on its surface, whether earth or ice, but could the navy expect combat against the Soviets in this forbidding region? Air bases, weather stations, and radar outposts seemed possible, but what of surface task groups accompanied by icebreakers? What about submarines? Although submariners like Hendricks and those at Submarine Development Group (SubDevGru) 2 in New London exhibited enthusiasm and the ONR encouraged research, many naval planners and those expected to fight from the deck of a ship frequently dismissed the idea as fantasy.

For years the Arctic remained a stepchild of the operating navy. Research conducted there, while usually requiring naval assistance, often took place only under the patronage of an Arctic disciple—an officer who felt strongly about the importance of the poles in naval strategy and soberly respected the proximity of growing Soviet naval capability. While every naval officer recognized the latter, few readily admitted the possibility of surface action against the Soviets at the top of the world. Arctic oceanography sponsored by the U.S. Navy, more than any other aspect of the science, lived or died according to the presence, enthusiasm, and influence of its officer patrons.[28]

As the Cold War's parameters slowly took shape, so too did perspectives on the Arctic's increasing strategic significance. Woods Hole collaborated with Coast Guard ice-patrol veteran Rear Adm. Edward Smith on a paper for the annual Undersea Warfare Symposium sponsored by the CUW and ONR in 1948. Discussing the scientific aspects of warfare in the Arctic, Smith lamented the primitive state of American and Canadian knowledge of this strategically important region. Postwar political and ideological tensions demanded research and investigation.[29]

Iselin was convinced that the ONR and the scientific community should encourage serious investigation. Generous naval patronage would certainly draw university faculty and students to Arctic studies. Demanding climate conditions and the frequent difficulty of air and ground travel made a carefully considered approach critical. Fully aware of the problem, Iselin supported the creation of permanent research stations on the Arctic ice. The station staff could study underwater sound in the region, temperature, salinity, dissolved oxygen, and internal waves, as well as current, water, and sea ice movement.[30]

Naval authorities did not need Iselin's suggestion. Driven by growing antagonism with the Soviet Union, if not its own strategic preferences and assumptions, the navy began to explore the Arctic in earnest. Research vessels available to Woods Hole and the other civilian ocean science centers could never conduct the necessary investigations because of cold, storms, and ice. However, the picture changed considerably with the involvement of naval ships and aircraft. As early as the summer of 1946, the Navy Department initiated its reconnaissance of the Arctic region with a small task group attached to the Coast Guard icebreaker *Northwind*. By the summer of the following year and the conclusion of Operation Highjump in the Antarctic, the Hydrographic Office resumed its northern survey as Operation Nanook II, supported by USS *Edisto*, a Wind Class icebreaker.

After returning from Operation Highjump with a measure of expertise in polar work, William G. Metcalf, a veteran and postwar reservist, once again donned his uniform and went north with Operation Nanook II. Hydro requested that he embark on the *Edisto* as the project's BT observer while the

Nanook II task group explored Lancaster Sound, Jones Sound, Barrow Strait, and Viscount Melville Sound. As the survey proceeded, Metcalf took BT measurements at the convenience of the ship's commanding officer. Since the navy viewed this venture as an American-Canadian strategic survey of the region, Metcalf frequently lamented the absence of sufficient scientific opportunity. He observed in his personal cruise log that "Any oceanography was to be done on a 'not to interfere basis' which means in essence that unless the skipper of the ship cannot possibly think of anything else he would rather do than stop his ship and wait while the oceanographer dangled things over the side, some scientific observation might be made. It was rare that the skipper was unable to think of something else he would much prefer to do."[31] Obviously, many officers learned very little from wartime experiences, while others realized very well the importance of environmental knowledge.

Despite Metcalf's unfortunate experience, the navy would have to pay attention to Arctic oceanography. Operation Nanook II took place in support of the effort to establish a line of defense early warning stations, an Arctic Distant Early Warning (DEW) Line, to alert the United States of a Soviet attack over the North Pole. In addition, the Nanook II survey of Resolute Bay later resulted in the creation of a major Arctic weather station and a joint American-Canadian base at that location. With Metcalf on board, *Edisto* journeyed to the Arctic Circle to resupply Eureka Station far up on the western side of Ellesmere Island, a very remote site for those early postwar years. Operation Nanook II remained in the Arctic from 15 July to 15 September 1947 and returned to the Canadian Arctic with Woods Hole personnel and Metcalf in tow during the same period in 1948. The region clearly had become critically important to the navy only a few years after victory over Japan.

In parallel with these efforts at sea, the navy established a research center at Alaska's northernmost point. In a letter to bureau chiefs and the CNO on 12 August 1946, ONR chief Rear Adm. Harold Bowen proposed the creation of an Arctic laboratory on the site of Naval Petroleum Reserve Number 4 at Point Barrow. In receipt of uniformly positive responses, the admiral organized a conference for 30 January 1947 to set the plan in motion. Based upon this initiative, the ONR transferred funds to the Bureau of Yards and Docks in April. Shelters to house the new Arctic Research Laboratory (ARL) quickly arose on the site, providing space for a laboratory, staff, and equipment. Research at Point Barrow began that same summer, not six months after Bowen's conference. A team from Cornell University and Swarthmore College under the direction of Laurence Irving came to the new laboratory to conduct biological investigations.

By the time the ARL assumed the organizational structure of a permanent scientific facility in February, 1948, it had a rotating staff of thirty-one, a resi-

dent director in Laurence Irving, an advisory board, and a physical plant of
three surplus Quonset huts that provided ten thousand square feet of laboratory, library, and working museum space. Reflecting the ONR's goals and
generous sponsorship, the ARL provided facilities at Point Barrow "for fundamental research in appropriate scientific fields related to the Arctic environment." The mandate given to the advisory board, chaired by John C. Reed of
the USGS, ensured a mix of basic and applied research as well as tasks fundamental to the navy's mission: "The Board recognizes as its principal function
the stimulation and promotion of basic scientific research under the auspices of
the Arctic Research Laboratory at Point Barrow, Alaska, in the interest of national security, and the recommendation of programs to the Chief of Naval Research; furthermore the Board will study and advise the Chief of Naval Research on such scientific problems as he may present to it."[32]

All proposals to work at Point Barrow went to the ONR for review and
then to the ARL for comment before the chief of naval research dispatched the
official reply. The ARL's initial proposed budget of $154,250 proved too ambi-
tious, but the ONR agreed to fund the operation at $125,000 annually. By the
spring of 1948, the advisory board nominees, including Scripps associate director Roger Revelle, received the chief of naval research's approval.

Soon after its appointment, the board approved a litany of research projects and areas of inquiry that formed the core of the ARL program. The subjects under investigation fell into two categories, natural and physical sciences.
The natural sciences offered a series of ten possibilities, including animal me-
tabolism, sensory guidance of bird flight, and botanical studies. Physical sciences like acoustics and studies of the nature, formation, and behavior of sea ice
more directly related to the naval mission in the Arctic.[33]

If debates occasionally arose over practice, complete agreement reigned
over purpose. In ways peculiar to the difference between scientific practice and
the requirements of military and political institutions, the ARL advisory board
and Hydro disagreed over the duration of faculty visits. The board believed
that a truly productive stay required at least a year at Point Barrow. Hydro,
obliged to recognize the navy's habits and priorities, suggested a two-month
rotation for visiting scientists. Although the ARL had to compromise because
of Hydro's influence within the navy and the ONR's sponsorship, none of the
participants debated the central purpose of the Point Barrow venture. Indeed,
they agreed completely with the points made by Iselin to Revelle two years before: "it is pointed out that the plan in this case is really more concerned with
familiarizing the oceanographers with the problems of the Arctic and more particularly the problems of Arctic ice. At the present time there are very few
oceanographers in the United States who have experience in the Arctic."[34]

The long-term program at Point Barrow emerged from a two-day conference sponsored by Woods Hole and Hydro on 19–20 November 1948. The ARL oceanographers would focus on the formation of freshwater and salt-water ice, the study of sea-ice dynamics, and further development of Arctic oceanographic techniques to facilitate future surveys from ships and aircraft. The conference participants, including the hosts, Scripps, and the NEL, agreed to keep at least two oceanographers at the laboratory at all times. Each of the four interested activities would provide trained personnel to staff the ARL at particular times of the year, with Hydro coordinating East Coast activities and the NEL those on the West Coast.[35]

Toward the end of January, 1949, Iselin sent Metcalf and John F. Holmes, both from the WHOI staff, to Point Barrow to evaluate the site and facilities for long-term oceanographic research, to determine the suitability of various kinds of instrumentation for polar work, and to conduct some basic scientific observations if time and conditions proved favorable. While suitably impressed with the facilities, the WHOI team immediately discovered the limitations of a program conducted at Point Barrow. Without support from the navy or air force, surface-bound scientists could never work much more than roughly a mile from the ARL site, given the movement and instability of the local shore ice. Deep-water research thus would require more suitable aircraft. Metcalf reported that the WHOI team did not have access to military airplanes at Point Barrow, and the local bush pilots refused to venture out over the deepwater ice. For truly profitable oceanographic research, Metcalf and Holmes concluded that the ARL staff would require regular access to air force or naval aircraft. Further-more, helping William Dichtel and George Lundquist of the Naval Ordnance Laboratory with their ice studies at the ARL convinced Metcalf and Holmes that transport and equipment difficulties in this inhospitable environment required instruments with great durability and a high degree of portability.[36]

Arctic research on the West Coast complementary to the ARL program usually originated with Scripps or the NEL. The navy's earliest advocate of submarine operations under the ice, Waldo Lyon, initiated his postwar work at the NEL after the UCDWR demobilized. In 1947 and again in 1949, the NEL worked closely with the Royal Canadian Navy and Canada's Pacific Oceanographic and Defense Research Board to conduct oceanographic surveys in the Chukchi and Bering Seas, north and south of the Bering Strait between Siberia and Alaska. According to the 1949 report composed by the NEL's Eugene La-Fond, "The purpose of the cruise was to make such oceanographic measurements which were required for the evaluation of experimental sound transmission data and to conduct other studies useful to naval operations in the Arctic."[37]

The 1947 voyage gathered general information on the region, whereas the 1949 venture explored many particulars. Under Waldo Lyon's direction as chief scientist, the 1949 NEL oceanographic expedition fashioned a diverse scientific program. The staff explored underwater sound, marine biology—especially in the deep scattering layer—the geology of ocean-bottom sediments, underwater canyons, turbidity measurements and sediment transport, as well as general circulation studies focusing on currents and temperature measurements. Three vessels participated in the project, including the USS *Baya* (ESS-318), a submarine devoted to experimental work under the command of Comdr. John D. Mason; Lt. Comdr. D. J. McMillan's EPCE(R)-857; and the HMCS *Cedarwood*, commanded by Lt. Comdr. John E. Wolfenden of the Royal Canadian Naval Reserve. The *Baya*'s scientific staff conducted most of the sound program, while the physical oceanography depended upon the Canadians. The balance of the program was split between the three vessels. While staying carefully east of the Soviet border, the physical oceanography and sound measurements provided welcome detail on a woefully understudied region benefiting from its sudden strategic importance.

Lyon's interest in the Arctic went far beyond his immediate function as chief scientist on board West Coast polar survey ships. He wanted to investigate the possibility of submarine operations under the ice in the Arctic Sea and shortly after the war established the Arctic Submarine Laboratory (ASL) at Point Loma with resources and real estate from the NEL.

The history of the ASL illustrates the variable nature of the naval commitment to Arctic oceanography. Once again, the survival of an oceanographic program depended upon the navy's faith in individual scientists or laboratories as well as professional trust and personal friendships. Waldo Lyon's pioneering under-ice efforts survived and prospered only with the proper naval patronage. In this case, his unique Arctic Submarine Laboratory survived or foundered according to the interest of the submarine type commander in BuShips or the needs of the NEL as the host activity. Patronage support from these sources often fluctuated according to a flag officer's personal perception of the Arctic's strategic significance and its viability as an operating environment before the advent of nuclear submarines.[38]

The U.S. Navy did not adequately investigate the centuries of Russian polar activity with sufficient care to realize that naval operations had a long history at the top of the world. Waldo Lyon later recalled that he

> couldn't get sponsorship in the Navy at that time because I was always working with, and always have worked with, the type commander, with the submarine force commander, and primarily started with Pacific, SUBPAC, Submarine Pacific, and it depended on who the person was, who it was at that time, and whether he had interest or not. We had real interest there from 1947 up through '53, and then

there was a change. [Rear Adm. Alan] McCann, who was the first one . . . as a junior officer, was the liaison with the transfer of the O-12, which became the *Nautilus,* under Sverdrup and Wilkins, and he always kept in the back of his mind—I'll never forget him talking about it, "Hey, they're doing it the wrong way. . . . I wrote him a letter or called him up, and [he] said, "Oh, sure, I'll give you a submarine," so he gave us five submarines. But we fixed one submarine, the *Boarfish* [SS-327], with the sonars, and that's when we first went under [the ice]. In fact, . . . McCann went along, stuck himself along on the *Nereus,* and he got on board the *Boarfish* when we tried the first time to go underneath, and he said, "I take responsibility. Do it." If you know the Navy, why, that was needed. So the first one was McCann, but that's the reason why he was interested. Then the second COMSUBPAC after McCann kept it going. . . . By the time 1953 came around, [Rear Adm. Oswald] Colclough was the SUBPAC, and he was a lawyer. No way. No more. "The Arctic is fantasy." Nothing to do with it. So then there was no sponsorship. We built the laboratory strictly on the basis of the commanding officer of the laboratory [NEL] and his interest. We did it all with surplus material and whatever we could scrounge to build the laboratory . . . the Arctic was a side issue you didn't talk about. You didn't even mention the word.[39]

One officer who did not need convincing retired from the Coast Guard to replace Columbus Iselin, becoming Woods Hole's third director in 1950. Rear Admiral Edward Smith, a native of Martha's Vineyard, played an instrumental role in the activity of the International Ice Patrol from 1921, when he became the author of that activity's annual report. As a lieutenant commander in 1928, Smith led the *Marion* Expedition, which spent seventy-three days surveying nearly half a million square miles of ocean between Greenland and North America. Arctic oceanography had an influential and respected advocate in the new WHOI director. While he proved neither as gregarious as Iselin nor as prolific and expressive in his correspondence, Smith kept Woods Hole in Arctic research and encouraged his peers around the country to expand their efforts.[40]

Less than a year after taking up the reins at Woods Hole, Smith involved his staff in an innovative, ONR-sponsored, two-year exploratory program in the Arctic employing aircraft rather than ships. Christened Operation Skijump, this venture used two navy P-2V aircraft and one R-4D to survey for polar ice landing sites and to develop the tools and instrumentation to investigate the physical oceanography of the Arctic Ocean. In 1953, Woods Hole's Valentine Worthington reported in the AGU *Transactions* on both the preliminary find-

ings of Skijump 1 in 1951 and the more detailed survey of the Arctic Ocean conducted by Skijump 2 the following year. Using Point Barrow as a point of departure, the range and durability of the R-4D limited the ambitions of the 1951
phase. Project technicians immediately developed tests for ice integrity to assure the safety of the plane once it landed. Skijump 1 established three stations
on the ice during April and May and focused on temperature and salinity measurements as an indication of water movement under the ice. Once on the ice,
the environment turned even the most ordinary data collection procedures
into a frozen purgatory. The team quickly discovered the difficulty of using
chainsaws and chisels to cut through pack ice five feet thick to gain access to the
ocean underneath. Water samples still in their Nansen bottles and not yet tested
froze on board the aircraft, reducing the reliability of the data taken.

In its second season, Skijump employed the P-2Vs both independently and
to extend the range of the R-4D by refueling. Skijump 2 occupied three stations
on the Arctic ice between 11 and 13 March 1952 and proceeded to gather sufficient temperature and salinity information to suggest a rather complicated circulation pattern featuring an anticyclonic eddy north of Alaska. In this they
challenged the simpler cyclonic model suggested years before by Fridtjof
Nansen, leader of the *Fram* Expedition, who drifted across the Arctic with the
pack ice aboard ship from 1893 to 1896. Although the R-4D had to be aban-
doned on the ice when its landing gear broke, this alternative to transportation
by sea proved economical and viable in the polar environment.

The second season also brought innovations in ice penetration. The scientists needed efficient cutting tools to open the ice sufficiently for the passage of
instruments. In 1952 the Skijump staff employed a five-horsepower, gasoline-
driven posthole digger. Putting the chain saws and chisels aside, the digger cut
a clean 12.5-inch hole through Arctic ice five feet thick in roughly ten minutes.
This proved a significant improvement over the 2.5 hours devoted to the same
task a year earlier. The diameter provided just enough clearance for the Nansen
water collection bottles the scientists chose to use. Skijump participants also
learned to heat the aircraft while it stood on the ice runway to avoid having the
water samples freeze in the bottles.

Although scientists applied themselves to Arctic research and the ONR vigorously supported these ventures, leaders within the scientific community
wanted a more determined and clearly defined commitment. The U.S. Air Force
established a meteorological station on a drifting ice island christened T-3.[41]
Would the navy do more than just place scientists and technicians from Hydro
and civilian institutions at the ARL and on board survey and supply ships? Edward Smith of Woods Hole argued for continued development of light and
heavy aircraft for Arctic work of greater detail and depth. Although William V.
Kielhorn and Lt. John Knauss of the ONR's Geophysics Branch suggested to

Columbus Iselin that the submarine could well prove more practical, safe, and productive at the North Pole, their agency continued to entertain proposals for varied approaches to Arctic oceanography. Research conducted from institutions on both coasts and from the ARL in Alaska continued to employ ships as well as aircraft and submarines.

For the disciples of Arctic research, the problem of broad support and commitment with depth and consistency remained crucial. The future of Waldo Lyon's laboratory frequently depended upon willing patrons and NEL commanders until the advent of nuclear submarines and the post-Sputnik Cold War science race made his work more visible and attractive. Even then, he felt constantly obliged to lobby for patronage and adequate funding.[42] Edward Smith earnestly supported both light and heavy aircraft for Arctic science and constantly pushed the navy toward deeper and more regular commitment. The semi-retired Columbus Iselin, as impatient as ever with painfully protracted piecemeal approaches, knew that the ONR offered no obstacle. With the Arctic an uncertain wild card in the navy's global strategy that few serving officers cared to play, the level of interest and commitment remained uncertain. Iselin urged Smith to draw a verbal picture full of clear images and sharp contrasts for the navy. It was time to exhibit the same determination in the Arctic that the navy displayed with underwater sound and deep-ocean surveillance. As far as Iselin was concerned, you received only what you paid for in money, time, and effort. No aspect of oceanography performed without adequate resources would serve the navy's mission. Iselin, the experienced oceanographic entrepreneur and now a member of the WHOI board, continued to employ the firm and slightly alarmist tone that had served him well in his years of debate with the navy:

> I feel that the system of sending along a few observers on a supply mission or other type of mission has assisted in training personnel and pointing up what can be accomplished aboard ice breakers, but the limited accomplishments in the past show that progress is not being made at anywhere near the speed required. The Navy must face the facts in this matter. Oceanography can no longer be relegated to a very secondary consideration. If the Navy wants adequate Arctic Oceanography it must be prepared to support the thing properly.
>
> Icebreakers must be equipped with extensive scientific facilities and/or assigned to oceanographic cruises with no interference from other projects. This is a theme which we at WHOI have been stressing for many years. (Martin Pollak and I hammered on this thing in the spring of 1950 and were hushed up at Hydro. It was true then and is even more true and much more urgent today.) The choice is

strictly the Navy's. It can . . . do the job right, spend a good deal more time and money than it is at present, and *get it's money's worth*. . . . We realize that we will stand accused of failing to recognize the Navy's other vast commitments which make it impossible to foster arctic oceanography on the scale described here. That is a decision only the Navy can make.[43]

Iselin suspected what Waldo Lyon knew all too well. The navy still had not come to terms with the significance of the far north and the extent to which it expected to confront its adversaries under the Arctic ice. This situation made Arctic oceanography for many years an uneasy and uncertain business for scientists and naval officers.

Initiating the IGY

The lost ambitious effort of the first postwar decade dwarfed the navy's arctic projects and truly exceeded the capabilities of the United States, requiring global participation. The International Geophysical Year (IGY) drew scientists from nearly every country with access to ships and instruments and an interest in any aspect of ocean, atmospheric, and earth sciences. As a critical component of oceanography, geophysics interested the navy as well as all of the civilian scientific institutions working for BuShips, the ONR, Hydro, and other naval activities. Earlier global projects, emphasizing both national scientific interest and international cooperation, provided a model for the IGY. Two Polar Years, the first in 1882–83 and the next in 1932–33, opened the North Pole to intensive study to the wider benefit of the world scientific community. Furthermore, atmospheric insights gained during the course of the Polar Years translated into millions of dollars worth of basic radio communications knowledge.

The International Council of Scientific Unions (ICSU), acting on an idea born of an informal discussion between British physicist Sydney Chapman and the president of Associated Universities Incorporated, Lloyd Berkner, authorized the IGY in 1952. The ICSU selected 1957–58 as the earliest period best suited to comprehensive study, anticipating uncommon solar activity and excellent eclipse prospects. Each participant established a national IGY committee to formulate its program and to integrate its particular goals with the larger project. Along with these national committees, the ICSU, the United Nations Economic, Social, and Cultural Organization (UNESCO), and the World Meteorological Organization (WMO) formed the backbone of the effort. In the Cold War environment this approach held promise because the IGY at once served each participant and the world scientific community. Particular programs reflected national ambition and specific national aims, but as part of the

IGY participants shared their research results. As the National Science Foundation (NSF) described it in its budget submission to Congress: "In short, the pattern is primarily one of unilateral activity, but coordinated into multilateral activity because significant advantages are to be gained from both simultaneity of observations and exchange of information."[44]

For the United States, the National Research Council appointed an IGY committee to oversee the integration of American scientific activity into the global program. Chaired by Joseph Kaplan of UCLA, the national IGY committee included, among others, Lyman Briggs from the National Geographic Society, Merle Tuve of the Carnegie Institution, and the University of Minnesota's Athelstan Spilhaus. As of March, 1954, twenty-eight nations expressed their intention to participate, a number that would eventually rise to sixty-seven. At this preliminary stage, the collected programs of pledged participants represented an approximate research expenditure of $100 million, with the United States committing about $12.5 million or one-eighth of the total. Oceanography and glaciology would benefit from $1,697,500 of the American pledge.

The NRC knew that the U.S. program would require a manager and paymaster. Created in 1950 and under the direction of chief scientist and ONR veteran Alan Waterman, the National Science Foundation agreed to assume this role. Representatives from the U.S. national IGY committee, NRC, and NSF appeared before President Dwight Eisenhower's National Science Board (NSB) and received permission to submit the IGY program to the Bureau of the Budget. The funding request appeared as a unique, nonrecurring item in the proposed NSF budget. In spite of occasional conflicts with the Defense Department over budget appropriations, especially with regard to the observation satellite program, American plans for the IGY actually received sufficient funding from an occasionally parsimonious Congress.

The American oceanography program consisted of two basic parts, ice-island observatories and deep-current studies. The IGY ice-island observatories staffed by scientists from the United States would, in the language of the program text, "obtain an understanding of sea level changes of both short and long periods, and their relationship to other phenomena in the ocean and atmosphere, by establishing temporary island observation stations in the Atlantic and Pacific Oceans."[45]

These stations would collect data on the effects of wind on sea level and record actual sea level oscillations with periods of five to fifteen minutes. The results of these studies would speak volumes about air-sea interaction, weather formation and patterns in various parts of the world, and the origin and development of heavy surf and tsunamis. The long-term nature of the IGY would also permit conclusions derived from seasonal and regional comparisons.

Deep-current studies, the other component of the U.S. IGY program, focused on global water movement and circulation patterns. Participants examined the movement into the Atlantic and Pacific of very deep water originating in the Antarctic. Equatorial current systems ranked high among scientific priorities in this case, as did the areas in the north central Pacific and Atlantic where the deep, cold water from the Antarctic converges on the warmer, more saline waters in both oceans.

In these aspects of the IGY, the navy worked with some of the premier oceanographic programs in the country to accomplish the project's goals. In the Atlantic Ocean, Maurice Ewing's Lamont Geological Observatory emerged as the primary scientific institution for the U.S. program. On the West Coast, Scripps played the same role, with support from the Universities of Hawaii and Washington.[46]

One aspect of the IGY remained beyond the capability of even the most sophisticated university programs. As a source of cold, nutrient-rich water flowing north, the Antarctic loomed large in many international IGY scientific programs. Knowledge of this region would further explain its often-significant influence on weather, communication, aviation, fisheries, navigation, current patterns, and many other aspects of ocean and coastal life. In support of the U.S. program, the Department of Defense agreed to provide logistical support for Antarctic research. Secretary of Defense Charles E. Wilson chose the navy as the service best suited to provide this assistance, and Rear Adm. George J. Dufek arrived in Washington on 16 August 1954 to assume command of the newly established U.S. Naval Support Force Antarctica.

Admiral Dufek immediately initiated planning for a scientific logistics base necessary for research and survival. With the IGY just over two years away and the Antarctic summer months of January and February the only time for safe and sure access to the Ross Sea via icebreaker, Operation Deepfreeze I had only a few, brief windows of opportunity. The Bay of Whales, then the traditional and most familiar location for establishing such a base of operations lay twelve thousand miles from Boston via Wellington, New Zealand, on the western flank of the Ross Ice Shelf. Every "Little America," beginning with Admiral Byrd's first three in 1928, 1934, and 1940, through Project Highjump's Little America IV set up shop at this location.

When the base reconnaissance team on board the navy's Wind Class icebreaker USS *Atka* (AGB-3) arrived at the bottom of the world in January, 1955, after more than a month at sea, it soon discovered that the Bay of Whales had disappeared. A victim of ice breaking away from the Ross Ice Shelf, this historic anchorage no longer existed and the *Atka,* commanded by Comdr. Glen Jacobsen, quickly had to find alternatives as the Antarctic summer came to a rapid close. The personnel of Deepfreeze I found a possible alternative at Kainan Bay and two others on the far side of the Palmer Peninsula in the area marked by the Weddell Sea.

The reconnaissance concluded just in time. The climate became truly severe at the end of the Antarctic summer, prompting Jacobsen to set sail quickly for Boston. The next phase, Operation Deepfreeze II, would have to wait for the following summer.[47]

As nature changed quickly and often without warning, so too the pace of political and naval developments tied operations and oceanographic research ever more closely together. As the *Atka* arrived off the broken remainder of the Bay of Whales in Antarctica, the Electric Boat Division of the General Dynamics Corporation in Groton, Connecticut, launched the USS *Nautilus* (SSN-571), the world's first nuclear-powered submarine. In August, 1955, Adm. Arleigh Burke became CNO, promoted over a sizable group of senior flag officers. He brought technical sensitivity to the office, a willingness to take risks, and an appreciation that the world was changing at a breathtaking rate. Submariners and oceanographers speaking of a revolution in undersea warfare and the ocean as an asset as potent as a torpedo or missile found that the navy was listening closely to what they had to say.

15

A Closer Relationship, 1950–58

Project NOBSKA . . . was of great assistance to the Navy not
only because of the direct and immediate impact it had on
many of our subsequent research and development plans and
programs, but because of the longer term benefits of the liaison that was effected with members and groups of the scientific community not previously or normally concerned with our
problems and activities.

—ROBERT W. MORSE, assistant secretary of the navy for
Research and Development, to FREDERICK SEITZ, NAS president,
3 January 1966

EVEN WITH THEIR COMMON DEPENDENCE ON THE OCEAN, A
shared victory over the Axis, and frequent project partnerships, a mutually satisfactory postwar dialogue between civilian oceanographers and
the navy remained elusive. Only between 1950 and 1956 did advances in communication and understanding, largely made possible by ONR personnel and
practice, finally suggest a maturing partnership with long-term, collaborative
promise. While matters of culture and jurisdiction still frequently prevented
effective communication, teamwork characteristic of that achieved during
World War II flourished on the individual level. Just as the latter was once the
key to wartime productivity, so too was it the key to the future.

The bridge that began to materialize across the cultural gulf between civilian oceanography and the fleet was largely a civilian creation. In their customary reluctance to cast oceanographers in any warfare role other than advisers
twice removed, naval line officers left this task to ambitious and creative civilian veterans. After four years of war and frequent interaction with the navy,
many pivotal wartime translators and their apostles easily crossed the cultural
divide into senior civil service jobs as research scientists or administrators.

Thus, as the first postwar decade ended, a significant partnership and more productive dialogue slowly developed because ocean scientists working for the
navy understood their civilian colleagues, had little trouble communicating,
and had even less difficulty recognizing research significant to the navy.

In this very effective way, many of oceanography's translators became permanent insiders, living simultaneously and comfortably in two cultures. They
proceeded to define the nature of the postwar relationship between the navy
and the civilian oceanographic community, effectively offsetting the near incommensurability that historically handicapped this potentially productive
cross-cultural dialogue.

While the value of officers converted to the need for a deeper appreciation
of the ocean combat environment did not diminish, naval oceanographic policy and practice rested with civilians. Roger Revelle continued to command attention in the navy's higher ranks. Hydro, BuShips, and the ONR remembered
him for Operation Crossroads and his pioneering initiatives with the ONR
Geophysics Branch. Gordon Lill, Revelle's successor at Geophysics, and his
deputy Arthur Maxwell, a Revelle student from Scripps, sponsored oceanography in conjunction with the ONR's Naval Analysis Group and Naval Applications Group. During the first postwar decade, Revelle, Lill, and Maxwell controlled virtually all of the naval appropriations available for oceanographic
research in the United States. Their close personal links to Scripps and Woods
Hole, as well as their own views on science education and the navy's needs
made them critical to a naval-civilian dialogue that served oceanography both
as a discipline and a vital component of the national defense.

While the ONR Geophysics Branch emerged as the most outstanding il-
lustration of the personal dynamics critical to the growth of oceanography, Bu-
Ships and Hydro also made their marks as important players. Supplementing
the Office of Naval Research, the bureau supported ocean science in a more
humble way under the direction of Comdr. Butler King Couper, a pioneer in
Richard Fleming's Oceanographic Division at Hydro. Couper sustained his
strong ties to WHOI and also had many close friends among both Ewing's
team at Columbia and the Revelle-Eckart group at Scripps and the MPL. In
Hydro's case, Richard Fleming, Mary Sears, and Thomas Austin advanced
ocean research through the Oceanographic Division as a natural part of the
navy's mission. Lamar Worzel at Lamont, Gordon Hamilton at the SOFAR
station in Bermuda, or Brackett Hersey working out of Woods Hole easily
communicated plans, problems, and methods to Couper, Fleming, Austin,
Lill, or Maxwell. They knew each other very well and had a high mutual expectation based upon experience, trust, and a shared cultural perspective.[1]

Elizabeth Bunce, a WHOI scientist, female pioneer in oceanography, and
wartime explosives expert, frequently drew attention years later to the signifi-

cance of these relationships and the critical civilian role. She recalled Allyn Vine speaking so loudly on the telephone during frequent conversations with his friends in the Pentagon submarine community that simply yelling out the window toward Washington, D.C., with the same intensity would have placed a lighter burden on the institution's communications budget.[2] Whether it was a Revelle student making critical funding decisions at the ONR, Allyn Vine constantly on the telephone to the Pentagon from Woods Hole's Bigelow Laboratory, or the presence of Frederick Hunt and Columbus Iselin as naval advisers on the CUW, a tight web of professional friends and familiar personalities governed research projects, priorities, and funding in naval oceanography.

At times, it seemed almost incestuous. Woods Hole did not receive naval funds because of institutional stature or reputation. Revelle, Lill, and Maxwell sponsored ONR projects suggested by Woods Hole because they knew Iselin, Vine, and Brackett Hersey. Lill and Maxwell knew that Iselin consciously sought projects to enhance his own institution, to help oceanography develop, and to assist in the national defense. Not only did his wartime track record speak well for his reliability, but the WHOI director was also a personal friend. Iselin knew better than most that the navy's peculiar needs could serve a variety of ambitions.[3]

Gordon Lill shared these assumptions. When the University of Rhode Island needed a new director to bring its oceanography program to an acceptable level of maturity, John Knauss received the assignment. While Knauss had the qualifications, the appointment actually came to him because of his Scripps connection, friendship with Arthur Maxwell, his work at the ONR, and Gordon Lill's assessment of his abilities. Lill wanted the expansion of the Rhode Island program to succeed as part of his intention to ring the country with navy-funded oceanography schools. Unknown to Knauss at the time, he was Lill's personal choice. The same dynamic had earlier ensured the selection of Richard Fleming at Hydro, only in that case the Sears-Iselin component of the web brought that navy opportunity to the future head of University of Washington's oceanography program.[4]

They all knew each other and worked together in a very informal manner to achieve personal, institutional, and national goals. Regardless of whether the National Academy of Sciences and the nation's universities dictated collective cultural habits or they emerged from a combination of the Naval Academy, the Pentagon, and combat, by the end of the first postwar decade personal relationships among individual scientists shaped the dialogue between two historically reluctant bedfellows. Oceanographers effectively communicated with each other across the cultural gulf, giving voice to both naval and civilian interests and creating a more effective partnership. This solution placed the civilian scientists straddling both cultures in a position of unprecedented influence and significance. This was the Vaughan-Gherardi interwar common practice increased by several orders of magnitude.[5]

Coming of Age: *Albacore, Nautilus,* and Nobska

Just as it achieved a critical mass, this increasingly productive partnership drew motivation and sustained purpose from two landmark scientific and technical achievements. The CUW, BuShips, and the scientists at the David Taylor Model Basin in Carderock, Maryland, developed an entirely new submarine hull form that opened the way for truly extraordinary submerged speed. In 1954, the diesel-electric, teardrop-shaped experimental submarine USS *Albacore* (AGSS-569) exceeded thirty knots submerged, a pace most submarines had difficulty maintaining on the surface. After a guest ride on the *Albacore* in the summer of 1956 during a routine New England research site inspection tour, Comdr. Charles Bishop of the ONR's Undersea Warfare Branch commented: "The motto of ALBACORE is 'Praenuntius Futuri,' meaning 'Forerunner of the Future.' It is well chosen. A ride in her gives a new perspective of submarine operations."

In addition, the navy commissioned its first nuclear submarine, the USS *Nautilus,* on 17 January 1955. This vessel's ability to dive well beyond the 450-foot limit of World War II–vintage American fleet submarines and remain almost indefinitely in the deep ocean generated a series of disturbing new questions for those concerned with undersea warfare.[6]

While the navy celebrated these innovations, incoming CNO Adm. Arleigh Burke wondered as the summer of 1955 drew to a close if the navy could defend against such vessels. What if the Soviets mastered the new architecture and made similar progress with nuclear propulsion? This quandary represented a much-magnified reprise of the German Type 21 threat.

By 1955, the track record of the oceanographic community led Admiral Burke to approach the CUW for advice. The people he encountered confirmed the value and continuity of the translator's role, the importance of the relationships forged in war, and the significance of the more effective postwar dialogue. Nine years after its creation, the CUW could still draw upon the expertise of Gaylord Harnwell, Columbus Iselin, and Frederick Hunt, as well as more recent additions like Charles Wiebusch, director of underwater systems development for Bell Laboratories, and Harvey Brooks of Harvard's Department of Engineering and Applied Physics. At Iselin's suggestion, the navy and the CUW agreed to hold a summer study in the tradition of Projects Hartwell and Lamplight focusing on the antisubmarine warfare problems posed by the deep-diving submarines of great endurance that kept Admiral Burke awake at night. While Lamplight occupied the summer of 1954 by studying the extension of North American air defense seaward via the navy, Project Nobska would extend its gaze below the surface to examine the implications of advanced design and technology on undersea warfare. Named for the lighthouse that marked the Massachusetts coast at Woods Hole on Cape Cod, Nobska placed oceanography and the ocean environment on the strategic center stage.

While advanced submarines posed the problem, for the oceanographic community they also represented an opportunity. Unlike the circumstances surrounding Project Hartwell and the Low Report in 1950, the potential submarine threat projected for the 1960–70 timeframe could stay deeply submerged for a period limited only by its crew's endurance. The navy would have to thoroughly understand the deep-ocean environment or risk disaster. Oceanography thus emerged more strongly than ever as a discipline central to national defense strategy, and as such would immediately benefit from enhanced federal resources. Nobska offered oceanographers a chance for the greatest possible professional visibility. It also set them on a path to more complete integration with other scientists, naval engineers, technical specialists, and leading naval officers by offering a greater opportunity than ever to influence national planning.

Besides, Nobska was necessary—and Admiral Burke knew it. Combining the new Series 58 "teardrop" hull design characteristic of the *Albacore* with the nuclear endurance of the *Nautilus* would provide the navy with an amazing vessel. The progeny of this marriage would move faster, dive deeper, and remain there longer than any submarine in history. Indeed, nuclear power transformed American submersibles, previously dependent upon batteries and surface air, into the true submarines about which John Holland and Jules Verne had only dreamed in earlier years.

As Burke moved into the CNO's office, BuShips and the David Taylor Model Basin worked with General Dynamics's Electric Boat Division to create just such a submarine: the USS *Skipjack* (SSN-585). The new CNO realized that the Soviets would have no choice but to address the possibility of a ship like *Skipjack*. If they proved successful in the coming decade, could the U.S. Navy provide adequate defense? What scientific options and technological investments offered the best hope for meeting this challenge? Burke needed answers.

In the early autumn of 1955, the admiral and representatives of the navy bureaus provided the CUW with a classified briefing on the complete scope of undersea warfare. Familiar with the committee's activity, Burke used the meeting to appeal for its help in addressing future undersea threats emerging from new designs and propulsion technologies. On 18 October 1955 he solicited the opinion of Rear Admiral Furth, the chief of naval research, on a proposal to ask the CUW to conduct a study of the research and development aspects of navy ASW. Furth supported the idea and listed for the CNO the variety of subjects the study might address.

On 2 December 1955, Admiral Burke returned to the National Academy of Sciences with the bureau representatives in tow, only this time they met with the CUW to discuss the implications of recent submarine innovations. Although the undersea warfare problem took center stage, the correspondence between the CNO and NAS president Detlev Bronk indicates that Burke demonstrated a keen interest in long-term scientific advice to help the navy address its technical and operational problems. In addition to an ad hoc study group examining the ASW aspect of recent major submarine innovations, Bronk understood that the admiral also wanted to establish a permanent ASW scientific advisory team. This could mean either creating a new activity or enhancing the role of the Naval Research Advisory Committee and the CUW to the point that they would have rapid and complete access to naval ships, commands, and systems both to advise properly and to carry out significant long-term studies.[7]

The CUW informally decided upon the Project Nobska approach to the problem of advanced submarines and undersea warfare when it met at Harvard University on 5 January 1956. The committee opted for a summer study like Hartwell and immediately detected three areas of likely emphasis. Committee members suggested that participants in the project would have to appreciate the present state of undersea warfare and technical capability. They would also be required to project future circumstances and solutions based upon present capabilities and draw conclusions about the lines of research that offered the best hope of preserving and extending present advantages into the next decade.[8]

From the beginning, most advocates realized that this study might have a decisive effect on the naval approach to ASW. Chairman Eric Walker of Pennsylvania State University and his colleagues certainly realized that the *Nautilus, Albacore,* and nuclear weapons had initiated a revolution in undersea warfare. If Nobska explored this subject writ large, the project could provide direction for navy laboratories, commands, and personnel involved in substantive long-term studies in addition to a firm foundation for the advisory group sought by Burke. Walker and his colleagues took immediate action. The study would require detailed planning, an effective leader, and a favorable site for participants' families. They committed themselves to select a director before February, the scheduled time for their next meeting.

While the committee approached other more prominent science administrators like Vannevar Bush to direct the summer study, the combination of experience, subject knowledge, and reputation in a single person combined with a suitable site remained briefly elusive. This conundrum lasted as long as it took Columbus Iselin to volunteer for the job, offering his institution as the ideal gathering place.

The oceanographic community had a chance to prepare and reflect on the problems of warfare in the deep ocean just before the summer began. The CUW and ONR hosted an unclassified symposium on undersea warfare at the NAS in Washington between 29 February and 1 March 1956, and the roster of speakers read like a who's who of the ocean and earth sciences. Columbus Iselin presided over the event, which featured research reports on the ocean environment by the best in the field. Maurice Ewing reflected on radiological studies in the investigation of ocean circulation, and Woods Hole's Gulf Stream expert, Henry Stommel, spoke on deep-ocean circulation. Other papers by Hersey, Vine, Fleming, Fuglister, Isaacs, and others examined the ocean bottom, light penetration, ocean-model experiments, cooperation among the various oceanography disciplines, acoustics, and many other subjects of immediate interest to the ONR and the academic community. While these subjects had proven important to varying degrees during World War II, they now commanded the navy's attention by emerging as critical factors in the navy's mission.[9]

Iselin convened the Nobska summer study on 18 June 1956 with roughly two weeks of briefings held by various naval activities and commands for the sixty full-time participants. The ONR provided $275,000 to the NAS to cover the costs of operation through 15 September, when the project headquarters on Church Street in Woods Hole would close and move back to Constitution Avenue in Washington. At that point, the NAS staff would assemble the various parts of the final report for presentation to the navy and classified publication.

It made sense for an experienced translator to direct the study because success depended entirely on effective integration of naval officers with scientists from a variety of disciplines. This task seemed second nature to the WHOI director. In his preliminary appraisal of Nobska's task, Iselin asked questions that no lone professional, technical, or scientific community could answer. A variety of disciplines and warfare specialties representing a number of distinct professional cultures would have to compose a coherent set of policy recommendations for the CNO. Their collective informed vision would also have to project problems and solutions into an uncertain future. Bridge-building and cultural translation was the order of the day, and none of Iselin's questions had easy answers: "What are the most critical weaknesses of well-operated and well-designed true submarines? How can we exploit these? What sorts of ASW capabilities is our existing research and development program likely to achieve for ships, submarines and aircraft during the next ten years? What basic economic considerations are involved in undersea warfare? What are the most promising techniques for detecting, tracking, identifying, and destroying a fleet of say fifteen advanced design enemy submarines, such as might appear during the next ten years?"[10]

These questions and a score of others went straight to the fundamental policy issues of commerce protection, investment in research and development, evaluation of present capability, and threat assessment over time. Addressing these matters together with the navy made this gathering the kind of forum that Iselin and other oceanography leaders had earnestly sought since V-J Day. Project Nobska offered an opportunity both to do oceanographic research and to influence effectively the policy that determined research investment.

During Nobska, ocean scientists would work very closely with a wide variety of naval, scientific, and technical specialists, use their knowledge of the physical world to evaluate the nation's undersea defense, and project suggestions for improvement or enhancement ten years into the future. If Admiral Burke proved good to his word, this risky business would immediately and fundamentally influence naval investment in undersea research.

Most importantly, Nobska participants were able to take their advocacy of oceanography's potential and the promise of its combat applications far beyond the translator's art. Since 1940, key players like Iselin, Eckart, Hersey, Revelle, and Vine had devoted much of their time to heterogeneous engineering. They constantly worked to define oceanography, to advertise its value, and to find a place for it within the university community and the federal government. Oceanography came of age with Nobska, and its practitioners took the first step beyond these uncomfortably repetitive fundamental chores. Faced with a Soviet undersea threat, the CNO himself had publicly sanctioned oceanography as the most comprehensive way of effectively appreciating the navy's natural medium. Acceptance of oceanography as a strategic and tactical asset by America's most senior naval officer required all of his juniors to pay closer attention. Burke wanted not only his civilian counterparts' scientific insights, but also their informed vision born of imagination that would make Nobska projections into the future a foundation to build upon.

This was the kind of recognition and integration sought by ocean scientists since the inception of the Vaughan-Gherardi common practice some thirty years earlier. Each of the Project Nobska study groups demonstrated a critical mix of experience and knowledge to make its work as useful as possible to the navy's mission. Oceanographers working at the Whitman estate in Woods Hole, which WHOI had leased for the project, met with policymakers, industrial managers, fellow scientists, naval officers, engineers, and administrators to add their perspective to the deliberations. Nobska technical group T-4, whose focus was detection, offered a good example. What did they have to detect? Submarines combining the characteristics of the *Nautilus* and *Albacore:* nuclear propulsion, advanced hydrodynamics, and quiet operation. Group leader Persa Bell and his colleague Raphael Dandl both came to Woods Hole from the premier American nuclear laboratory at Oak Ridge, Tennessee. Their facility had trained Capt. Hyman Rickover and the navy's first team of nuclear engineers, many of whom helped make the *Nautilus* a reality. The panel also had a large contingent from the navy laboratory community. Stanley Peterson hailed from USNUSL in New London, William Finney and Harold Saxton from the NRL, and Julius Hagemann from the Mine Defense Laboratory. Robert Morse of Brown University and John Coleman contributed their expertise in physics, and the latter also offered his experience as secretary for the CUW at the National Academy of Sciences. However, the core of the detection problem lay with oceanography and ocean acoustics. Thus, to meet the challenge of detecting and identifying submarines with *Nautilus-Albacore* capabilities, T-4 required Brackett Hersey, Columbus Iselin, William Richardson, Allyn Vine, and Frederick Hunt, together with Charles Wiebusch and Winston Kock of Bell Laboratories and Margaret Sturm of Raytheon. The questions posed by Project Nobska took these engineers and scientists out of their customary professional environment and provided a common naval and oceanographic focus for their talents.

Ocean scientists also contributed to other critical aspects of the Nobska experience. Paul Fye, Willard Bascom, and the MPL's Alfred Focke worked with the group pondering weapons effects and limitations. Fred Spiess and Victor Vaquier, both from Scripps, were assigned to the navigation and communication group and the systems development team working on ways to contain submarines and to enhance continental defense. Allyn Vine joined those debating the nature of ships best suited for undersea warfare. Iselin himself sat at the table with Bascom, Fye, Paul Nitze of the Johns Hopkins School of International Studies, Capt. Ralph Williams of the ONR, and CUW secretary George Wood to discuss the implication of political trends on undersea warfare. During the course of the summer of 1956, this collective effort helped to define oceanography and its importance to the navy, to the research community, and to industry.[11]

Predictably, the Nobska final report spent considerable time reflecting on the undersea warfare problem from a purely oceanographic position. The conference participants urged the CNO and his staff to look at the problem from a Russian geographic and environmental perspective. The Nobska participants wanted Admiral Burke to place himself in the shoes of his adversary by imagining the difficulties involved in operating American submarine and ASW forces from a base in the Hudson Bay. Soviet submarines lived four hundred miles inside the two-hundred-fathom curve. The geography alone would compel Soviet oceanographers, about whom the West knew very little in the mid-1950s, to spend more time investigating the relatively shallow water close to their own shores. Given the location of their bases, it would take them considerable time to move to and from deep water, contending with ice all along the way. The Nobska participants strongly suggested, therefore, that the navy target Soviet icebreakers in the event of war. Their fleet would require insight into shallow water acoustics, an understanding of sea ice, experience with under-ice submarine operations, and secure access to relatively ice-free ports. In any case, in 1956 they did not yet have the proper ships for deepwater work, and the meager amount of time they spent in other than coastal waters limited the extent to which they could immediately employ oceanography as a national defense asset.

From this estimate, Nobska suggested that the CNO could draw some important strategic oceanographic conclusions. The participants had an excellent idea of the environmental forces that shaped Soviet planning. The United States needed a better sense of polar operations and a more complete knowledge of that unfriendly northern geography as a combat environment. The group also discussed the limits as well as the advantages of the BQR-4 shipborne sonar as well as the SOSUS. Both systems provided effective passive listening out to as much as four hundred miles in the case of the early SOSUS. The Nobska report suggested going further by exploring other acoustic phenomena such as surface mixed layer, bottom reflection, convergence-zone paths, depressed sound channels for which WHOI's Roy Rather had already designed variable-depth sonar (VDS), and echo-ranging systems to work in shallower regions of the world ocean. In the general section on oceanography, the conference reporters emphasized further American research in shallow water, ambient noise, marine geophysics, and biology. Naturally, marine biology played an important role in understanding ambient noise in the ocean, which constantly plagued sonar operators. The geophysical aspect played a role in calling attention to reflection, reverberation, bottom bounce, and many other aspects of the detection problem. As the Nobska report suggested, "Experience to date indicates that almost any aspect of oceanography, either physical, chemical, geological, or biological has some bearing on undersea warfare."[12]

The panel on detection, classification, and countermeasures took this statement even farther and provided an instructive comparison: "Generally speaking the oceanography is such that a good probability of detection at very long ranges will be encountered over about two-thirds of the deep ocean areas, and the naval tactician must learn to live with and exploit the topography of the ocean bottom just as the Army tactician employs mountains, rivers, and other features of land topography."[13]

The Nobska report emphasized sonar as still the best possible detection and defense method, but the recommendations on this subject took a turn many in the navy did not expect. With the very great promise of passive sonar on board ships and submarines as well as fixed listening and deployable forms such as SOSUS and VDS, active sonar systems had become the less-desired poor cousin. Submariners avoided active sonar because introducing sound energy into the water—the familiar active sonar pinging—immediately betrayed a boat's position and gave the enemy an opportunity. At best, a submarine commander would use a single ping to confirm target bearing and range immediately before torpedo launch. To detect an adversary without being discovered offered the highest probability of success in combat. This helped explained eagerness of the undersea warfare community to develop passive systems that exploited the ocean environment to increase assured listening ranges. The first boat to make a sound became dangerously vulnerable.

Nobska suggested an alternative way of thinking. Passive sonar development should continue at the best possible speed, including shipborne systems as well as SOSUS and VDS. However, what did the next decade hold? After all, Nobska's task was to explore potential and project at least one decade into the future. The study group on detection suggested that passive systems too quickly became vulnerable to jamming by explosive means. A well-executed saturation of part of the SOSUS system by depth explosives could effectively neutralize listening for a few critical seconds. Furthermore, the navy had to assume that the Soviets would work as hard as the United States in the future to quiet their relatively noisy submarines. Nobska scientists made a potent argument that the advantage afforded by assured passive detection would slowly evaporate as the adversary improved his vehicles.

Given the various recently revealed characteristics of the ocean, Nobska urged the navy to revisit active sonar — if for no other reason than to provide an alternative to passive listening. However, from the Nobska perspective, active sonar also possessed a variety of additional unexploited possibilities. It would not surrender as easily to acoustic jamming by explosives, it could enhance the accuracy of fire-control systems, and, together with the latest IFF devices, would make enemy identification much easier. The Nobska report also suggested that the tactical use of active sonar to jam individual submarines could force a Soviet commander to employ his own active sonar while attacking, making him immediately vulnerable to detection and counterattack.

Active sonar appealed to the Nobska participants because it had a series of predictable features that passive sonar lacked. If research and development produced transducers with sufficient power and directivity, an active system could reach far deeper at lower frequencies to search out submarines that would inevitably seek deep refuge as ship materials and architecture improved over the next decade. While its additional power would reach great depths, the sound energy introduced by an active system could also use convergence zones and the shallower sound channels that exist in higher latitudes to reach extraordinary ranges. The navy needed oceanographic surveys and research projects to develop this scenario. Would transducer-generated sound remain coherent enough over long ranges to make these capabilities possible? Studies at sea needed to focus on prototype transmitters and receivers, reverberation and ambient noise characteristics, tests in both shallow and deep water, and the identification by oceanography of regions that make convergence zones, sound channels, and good reflecting bottoms available. By the time Project Nobska was concluded, oceanography and the undersea warfare mission had finally become two sides of the same coin.

Of equal significance, Project Nobska enabled the navy to open doors, initiate debates, and cement significant relationships. On 25 September 1956 the NAS reported on Nobska to Admiral Burke, the naval operations staff, and the navy laboratories community at a briefing in Washington. Four weeks later, the NAS made a formal presentation to a larger federal and scientific audience. The ONR, the undersea warfare community, and naval analysts immediately descended upon the two-volume, blue-covered report and extracted from it a series of fifty-one recommendations the CNO marked for immediate action. Although fourteen proposals never made the list—and Captain Rickover resisted any suggestions regarding nuclear propulsion that conflicted with his own views—Burke largely kept his word. In return for his investment, the summer study faculty generated many important proposals, including one that demonstrated the potential of marrying the realistic promise of smaller nuclear warheads to the development of SLBMs. As Project Nobska ended, Burke set in motion the final development process necessary to create the Polaris missile system and the fleet ballistic missile submarine.[14]

As essential as Polaris became to the national defense, the Nobska final report revealed a new and equally important role played by ocean scientists in naval planning and strategy. Of the fifty-one final report recommendations scheduled for action by the ONR and Admiral Burke's office, the first eight pertained to physical oceanography, geophysics, and underwater sound—particularly active and passive sonar. He designated various agencies and offices within the navy to fund the work, monitor its progress, or accomplish the tasks at their own facilities. While not necessarily an indication of absolute priority over other Nobska proposals, Burke's list confirmed the ASW focus of naval planning and the crucial role in that process for ocean scientists.

Eagerly digesting the Nobska recommendations did not mean that the navy could always avoid indigestion. The varied background of the Nobska participants and the advisory purpose of the entire project precipitated some practical and cultural clashes with those in the navy responsible for operations at sea, strategic and budgetary planning, and operations analysis. Nobska's mandate permitted the participants to avoid the shackles of the legislative, bureaucratic, procurement, and budgetary processes. Admiral Burke and Detlev Bronk wanted the project faculty to consider every option in addressing the high-performance undersea threat.

Some agencies within the navy felt very keenly their restriction to the absolutely practical, and in varying degrees resented the flexibility, freedom, and influence granted to the Nobska participants. In one naval activity's view: "It is always possible to propose promising new avenues for exploration in the quest for large improvements. The difficulties arise only in reaching decisions as to relative worth, in terms of development cost and ultimate integration into a fleet capability, or in fitting all the recommendations into a hierarchy of priority within some given level of overall effort reference as [the Nobska Final Report] does not attempt to cope with these real difficulties."[15]

Select evaluations of the summer study thus proved very critical, suggesting that any planning and projecting not strongly linked to budget, current fleet requirements, and the ripple effect of redistributing resources had only marginal benefit.

This clash reflected very different perspectives on shared responsibilities. While Nobska participants had the freedom to explore advanced ASW issues widely and without constraint, those practiced in achieving the possible in the Pentagon, on Capitol Hill, or in the fleet viewed much of the activity at Woods Hole as an exercise without context executed in a vacuum. In this case, different methods and expectations applied to the same ASW issues produced select areas of intense disagreement.

Armed with Admiral Burke's support and considerable influence, Project Nobska explored possibilities and tried to initiate new ways of thinking, while some activities within the navy felt both challenged and bridled. Nobska veterans, accustomed to the burdens of working for a governmental institution, understood the dilemma of the naval administrator. The latter, however, did not entirely understand Nobska.

The most telling of these disputes occurred between the leaders of the Nobska effort and both the navy's Operations Evaluation Group (OEG) and Rear Adm. Thomas S. Combs, deputy CNO for fleet operations and readiness. Taking aim at one of the most uncomfortable Nobska recommendations, OEG director Jacinto Steinhardt contested, along with Comb's command, the apparent Nobska preference for active sonar as well as its critical assessment of passive sonar and SOSUS.

Both Admiral Combs and the OEG believed that Nobska scientists underestimated the potential of SOSUS, while overestimating its vulnerability. In commenting on the Nobska briefings in the autumn of 1956, Steinhardt went beyond his formal assessment and vented his frustration with Nobska views on SOSUS. In a memorandum to the CNO's office on the jamming issue, Steinhardt asserted: "The quotation from the NOBSKA Report quoted in paragraph 2 is not the only instance in which wholly wreckless and unsupportable statements were made in that Report."[16]

In this instance, Steinhardt's choice of words demonstrated that Nobska's evaluation of passive listening touched a nerve that may have had little to do with SOSUS or sonar. Driven in part by the constraints daily placed upon the OEG by the bureaucratic world of the politically practical and currently achievable, select Nobska recommendations seemed far too cavalier, capricious, and politically uninformed. As relative newcomers to the process of research and strategic planning within the navy, ocean scientists appeared naive and inexperienced by OEG standards.

Doubtless, part of the OEG-Combs reaction stemmed from the admission of oceanographers and other marine scientists to the CNO's inner circle of serious naval policy and planning advisers. With all of its flexibility and diversity, the summer study had invaded a sphere normally reserved for the anointed— for naval insiders and select contract consultants. Part of this negative Nobska critique thus stemmed from a natural inclination to defend hard-won advisory roles and political influence. In their critique, Admiral Combs and Steinhardt argued that Nobska's active sonar recommendations reflected a lack of experience. They were convinced that Iselin and his fellow scientists did not appreciate the technical and tactical ripple effect of new technologies. To them, the Nobska group lacked both experience with the ripple effect of new innovation and an understanding of operations at sea.[17]

The suggestion that Nobska participants had minimal experience with integrating new technology and science into the fleet does not withstand scrutiny. An examination of the research and applied results of the NDRC's work during the war demonstrates that Columbus Iselin and many of his Nobska colleagues had gone through extensive and often frustrating experiences while heterogeneously engineering both a place for their science in naval warfare and an awareness of the great strategic and tactical potential locked up in the navy's natural habitat. Furthermore, some of the scientists and engineers at Nobska had participated in the discovery of the deep sound channel, convergence-zone sonar, variable-depth sonar, sonar array engineering, and other aspects of SOSUS and underwater listening. Just a few weeks before Nobska began in Woods Hole, a CUW survey of active sonar programs within the navy discovered that roughly half of the fiscal year 1955 acoustics research budget contributed in some way to this type of detection. The survey reviewed programs underway at Woods Hole, the USNUSL, the NRL, and other laboratories either within the navy or under contract. Thus, few people knew better the possibilities and limitations of the systems the navy currently had in use or under development.[18]

A more careful reading of the final report demonstrates that Nobska did not reject passive acoustic detection. The summer study group strongly urged its continued development. Furthermore, various branches within the ONR had funded active echo ranging at both Hudson Laboratories and Woods Hole at least since the initiation of low-frequency acoustics research in support of

Project Michael in 1951. Five years later at Woods Hole, Project Nobska cast active sonar in the role of a promising supplement to passive detection.

Nonetheless, active sonar represented an often-unacceptable risk to the undersea warfare community. To introduce sound into the water immediately betrayed the user's position. This applied equally to surface ASW vessels, submarines, or shore stations with remote sound transducers. Revealing your position invited destruction at the hands of a competent adversary. Consequently, very few active naval officers favored the use of powerful on-board active sonar as a primary search tool.

Naval leaders other than Combs, Steinhardt, and concerned submariners objected to particular Nobska recommendations for very different reasons. For example, Rear Adm. Hyman Rickover, in charge of the BuShips nuclear propulsion project largely responsible for the success of *Nautilus,* argued against the Nobska call for smaller reactor plants and a separate activity for long-range nuclear research and development. Rickover was not interested in losing control of naval policy governing the innovations in nuclear propulsion generated by his team at BuShips Code 1500. It also made little sense to the politically savvy admiral to divide the federal budget assets currently available for nuclear propulsion between short and long-range programs. Congress would find it far too tempting to deny funding to one or both if conflict arose between the two.

The CNO reacted differently to the summer study results, as did some of his most influential submarine and ASW advisers. Burke's commitment to Nobska and his instructions to carry out or seriously explore virtually all of its recommendations demonstrated that the advanced submarine threat had made it strategically essential to understand the deep-ocean environment. Rear Admiral Lawson P. Ramage, representing Burke at Nobska, helped revise research priorities once the final report emergedand the major bureau and laboratory chiefs followed suit. Rear Admiral Albert G. Mumma and his staff at BuShips surveyed each Nobska recommendation and on 22 October 1956 reported on programs under way or pending that would address each suggestion.

While some recommendations remained at the bottom of the priority list and perhaps suffered a lack of resources, Mumma worked closely with Rear Adm. Armand Morgan, his assistant chief for ship design and research, in an effort to place each recommendation in the context of BuShips plans. They identified areas needing further attention, some areas that required new programs, and tried to ensure that worthwhile research did not suffer for lack of priority or visibility. In their report to Admiral Burke, the BuShips team took each recommendation in turn and demonstrated how the idea already formed part of the bureau's plans or how future plans would carry out the suggestions.

The controversial Nobska active sonar proposals offer a case in point. Mumma and Morgan did not agree with the position taken by Combs and

Steinhardt, but neither did they allow the emphasis to shift so completely that passive sonar developments like SOSUS would suffer or fall behind. Developing an active sound source or transducer powerful enough to ensonify distant and deep targets emerged as the major obstacle to progress. Mumma insisted that BuShips would stay the course with active sonar, exploring any powerful and portable sound sources, including explosives, with sufficient directivity to have the desired effect. He and Morgan strongly suggested to Burke that a proper balance between active and passive developments would yield the best results.

The reports filed by ONR inspectors after their road trips frequently demonstrated the effect of Nobska in the most immediate way. On 21 December 1956, Commander Bishop filed a report with the head of the ONR's Undersea Programs Branch after spending 12 and 13 December in New London visiting the USNUSL. After discussing both the operation of the BQR-2B sonar on board the USS *Darter* (SS-576) and the sound and communication systems destined for the USS *Thresher* (SSN-593), Bishop offered considerable detail on the development of special high-power sonar transducers for experiments in bottom-bounce sonar techniques. According to Bishop, "funds for this were requested in the CNO Nobska implementation request." Admiral Burke kept his word. The navy put the Nobska results under a microscope, examined them, evaluated them, the CNO demanded constructive debate, and then took action.[19]

Even with the conference complete and the CNO occupied, the minutes of the April and May CUW meetings reveal that the committee felt "an important continuing responsibility" to the navy on the issues of advanced submarine technology and ASW. Their sustained interest came from a desire both to see their initiative carried through and to continue playing a role in policy formulation. The latter particularly occupied the oceanographers as newcomers to policy planning at the highest levels within the navy. To encourage this interest, in 1957 the CUW organized a group of original participants to revisit the basic questions posed at Nobska. They wanted to discuss the extent of the navy's plans to implement Nobska's recommendations and the reasons for BuShips's reluctance to accept the project's suggestions regarding nuclear power. These Nobska veterans wanted to clarify any recommendations that seemed to disturb various naval activities, leaders, and policy advisers. This small-scale Nobska reprise took place between 5 and 7 August 1957 at the Naval War College in Newport, Rhode Island. The officers, engineers, and scientists who gathered there reviewed new information that the 1956 Woods Hole group did not have, discussed a proposal for a possible full-scale follow-up study, and, most importantly, sought careful clarification of the Nobska recommendations designed to ease the navy's fears and promote constructive interaction on the critical issues.

The ONR funded the effort and an oceanographer, Columbus Iselin, once again emerged as the preferred director.[20]

The Project Nobska review demonstrated the determination ofthe CUW and invited scientists and technical experts to ensure that the navy understood their suggestions. Criticisms of the sort leveled by Jacinto Steinhardt awakened the Nobska veterans to the continuing need for the entrepreneurship that had served them so well in the past. The final report issued by the 1957 Nobska review group recognized the Washington political environment and the critical budget realities faced by naval officers and civilians required to make hard political, strategic, and fiscal choices in a very short period of time. As scientists and engineers, however, they also recognized the realities of their professional practice. Few significant advances would come quickly in the ASW and pro-submarine science and technology fields. In their comments, the review faculty stressed the basically scientific nature of the deep-ocean problem facing the navy and the need for both short- and long-range research and development planning. Nobska provided the best avenues of exploration, but "The Navy must find a way of providing men, ships and submarines, not only for the gathering of basic data, but for the testing and experimental evaluation of new devices and techniques *in the research and early development stages,* while flexibility in design and method of employment is still available."[21]

At Newport, the Nobska group revisited the need for smaller submarines and ASW ships in greater numbers to achieve the dispersion required for a nuclear war, reminding the navy of the expense and risk of a smaller fleet of large vessels. They also called for a dual nuclear research program to plan for the long term in the same way that Admiral Rickover planned for the near term. The navy obviously needed a smaller, more compact nuclear reactor that would permit greater power for less cost in a smaller ship or submarine. This set the Nobska personnel in direct confrontation with Rickover. Furthermore, while rejecting the impression derived by Steinhardt and Combs, the Newport meeting reiterated the strong Nobska support for SOSUS and low-frequency passive detection, suggesting only that low-frequency high-powered active detection systems showed great promise. Why not explore both alternatives? The increasing dependability of SOSUS and shipborne sonar also led to the suggestion that concentrating on submarine noise reduction at maximum speed should give way in priority to silencing and listening at slow speeds. The review team reiterated that only the navy could make these decisions. The Nobska group could only make recommendations and try to persuade.

In any case, the navy should keep its options open and place more faith in developments over years rather than minutes. The scientists and engineers involved in Nobska drew the attention of a navy increasingly obsessed with the tumultuous present to the importance of careful reflection on the future. Unlike immediately putting to sea to meet a threat, scientific research took time, and the payoff usually came gradually. Even SOSUS could not lay claim to solving the undersea problem of the Cold War. Only promising scientific and technical developments evaluated over the short and long term in the context of oceano-graphic insight seemed wise to Iselin's study group. Flexibility, variety, and di-alogue would help the navy most in any confrontation with the Soviets.[22]

Nobska provided Cold War naval oceanography with its own "big bang." With more effective communication between ocean scientists and naval offic-ers now directly influencing the formulation of policy at the highest levels, scientific activity multiplied exponentially and accelerated dramatically. The increasingly effective dialogue demonstrated by Nobska served those who wished to define oceanography as a discipline, addressed the need for patron-age to educate a new generation of ocean scientists, and attended a navy ab-sorbed by undersea warfare. In all of this activity one historic element remained constant: the most promising projects and achievements depended completely upon personal initiative and fruitful interaction within a remarkably small web of veteran entrepreneurs and translators.

Post-Nobska activity also brought within reach cherished goals that had eluded Harvey Hayes, Theodore Roosevelt, Jr., and interested ocean scientists after the Great War. By 1957 the world had changed considerably. With spon-sorship of intensified oceanographic activity now part of its official responsi-bilities and routine, the navy stood out as the most committed of all the federal agencies involved in the ocean sciences. With the election of John F. Kennedy as president in 1960, the White House began to view oceanographic initiatives as part of its "Great Frontier" and looked to the navy to take the lead in coor-dinating what would become the National Oceanographic Program. Thus, the brass ring that political and scientific apostles of oceanography could not reach in 1924 seemed within the navy's grasp as the tumultuous sixties began. This promising recipe lacked only one vital ingredient.

The Last Ingredient

In the summer of 1956, money, talented people, effective communication, and compelling research brought the navy and the oceanographic community to within one step of a federally sponsored national oceanography program. While Nobska's success demonstrated the effect of combining naval need and patronage with an effective cross-cultural dialogue, the importance of focus and priorities also came to the fore. For years, postwar oceanographic leaders lamented the disparate nature of oceanographic progress. Many of the ques-tions posed at Wardman Park in 1949 still begged for answers. Science served the navy, but did this work also serve science? What of the need for carefully crafted programs of research and training? How would working for the navy better integrate the component fields of this relatively young interdisciplinary science? Who should determine research priorities, with their professional, political, and economic consequences? If the national oceanographic effort already possessed the fluid dialogue demonstrated at Nobska, it also needed its focus, its strict priorities, and its determined, single-minded leadership.

As the Nobska experience came to a close, both civilian and naval leaders took measures to provide direction and developmental planning for oceanography in the United States beyond the priorities of the 1956 summer study. The CUW's executive secretary, John Coleman, drafted a proposal while at Nobska to create within the Academy's NRC a new Committee on Oceanography. Unlike the ambitious but primitive first steps taken by the Lillie group thirty years earlier, this new committee would have to bring order, direction, and priorities to a swiftly growing and relatively popular field of study. Although very healthy in terms of interest and patronage, oceanography's growth seemed random at best, with priorities defined more by Cold War fears than a strategy designed to serve both the navy and civilian scientific concerns. Coleman's plan, which changed very little as it moved through the NAS for comment and revision, called for an insightful review of the general state of knowledge and scientific activity with an eye toward setting and supporting clearly defined priorities.[23]

At the end of the first postwar decade, American ocean science activity far outstripped that encountered by Henry Bryant Bigelow when he wrote his appraisal of oceanography for the Lillie Committee in 1931. American oceanography had barely learned to walk, but as 1960 approached it regularly set time and endurance records. Thus, in addition to looking at the state of the science, Coleman also wanted the new committee to coordinate planning among oceanography's primary patrons with an eye toward the most efficient utilization of personnel and equipment on the most pressing problems in the field. Since the latter still seemed hard to define because of oceanography's interdisciplinary nature, the committee plan also called for a strategy to integrate better the component fields to achieve increased disciplinary coherence. The new NRC committee should also provide for a for discussion and exchange of ideas as well as a way of assuring that American oceanography achieved the best possible representation at meetings abroad, in research at sea, and in the classroom and laboratory.

Following on the heels of Nobska, it became obvious to the ONR that such a committee could offer a central consulting authority of some stature to simplify and promote the development of oceanography as a national asset. Reacting quickly to an informal suggestion from Coleman and Bronk, Rear Adm. Rawson Bennett, then serving as the chief of naval research, approached his counterparts at the AEC, the Fish and Wildlife Service, and the NSF. He asked them to join the navy in formally proposing to the NRC the creation of a committee on oceanography. On 9 August 1956, Bennett wrote to Bronk: "In recent months there has been an increasing demand for advice on oceanographic problems of great magnitude. Many of these questions are of broad scope and long range, having far-reaching effects on the safety and benefit of mankind as well as considerable influence upon the foreign policy of the United States. They often require the concerted action of oceanographers and scientists in related fields. At present there is no established means through which the oceanographic institutions can act as a unit . . . we feel that there is an urgent need to establish a group which may be called upon for advice on current oceanographic problems and, in addition, will provide adequate planning coordination and direction of oceanographic research."[24]

For the navy, benchmarks like the Type 21 submarine, the Low Report, and Projects Hartwell and Nobska made it clear that investment in oceanography needed to increase. Just as they had in 1924, other federal agencies wrote their own letters echoing Bennett's request and assuring the success of Coleman's proposal. While the quick popularity of the concept momentarily surprised some, upon reflection the need for a committee with authority and insight guaranteed that the shock proved short-lived. William Thurston, executive secretary of the Division of Earth Sciences at NRC commented to Coleman on 21 August: "Inquiries within the Division of the Earth Sciences show there is a lot more interest and enthusiasm for an advisory group in Oceanography than I had realized, particularly if it is something more than a preen-and-screen committee."[25]

Working through the NRC's Division of Biology and Agriculture and its chair, Frank Campbell, John Coleman received permission from Detlev Bronk to solicit expert opinion from within the academic community on the need for such a committee. Those contacted responded with support and enthusiasm. Princeton's Harry Hess, a well-respected geophysicist and naval reserve officer, said he thought the creation of such a committee seemed long overdue. Paul Weiss of the Rockefeller Institute for Medical Research in New York City and former chairman of the Division of Biology and Agriculture thought that direction and focus implemented in an impartial manner would make a critical difference: "The need for such a committee exists and is emphasized by the many internal stresses and frictions that have developed in the field of oceanography between the several disciplinary branches composing this field. An impartial group of scientists of clear vision and perspective could render a real service and see this steadily growing field through some of its current infant diseases of internal dissention."[26]

Sensitive to the occasionally intense competition between Scripps, Lamont, and Woods Hole, as well as traditional divisions between ocean disciplines, Weiss felt the need to support oceanography in its adolescence.

Coleman completed his proposal for the NAS Committee on Oceanography (NASCO) on 23 November 1956 and it landed on Detlev Bronk's desk on 4 December, shortly after the Thanksgiving holiday. The NAS president quickly acted on the proposal. He instructed the Division of Earth Sciences to call a meeting in Washington, D.C., to initiate the process of establishing the committee. The division invited eighteen scientists to NAS headquarters to discuss the nature, roles, and membership of NASCO. Those invited included Maurice Ewing, Harry Hess, Columbus Iselin, Daniel Merriman of Yale's Bingham Laboratory, Roger Revelle, John Isaacs, and Athelstan Spielhaus. The meeting resulted in further correspondence, discussion and the establishment of a committee with an approved annual budget of $70,000, an executive secretary, and a membership of nine. Bronk asked CalTech's Harrison Brown to chair NASCO with the help of Richard Vetter, on leave for two years from the ONR, as executive secretary. The membership consisted of Revelle, Ewing, Iselin, Fritz Koczy of the University of Miami, former AEC commissioner Sumner Pike, Gordon Riley from Yale, Milner Schaeffer of the Inter-American Tropical Tuna Commission, and Athelstan Spilhaus.

The oceanographers assisting with the planning and those who served on NASCO could not overstate the importance of this group to the discipline's welfare and productivity. The disciplinary concerns voiced by Weiss, the fear of government control expressed by Iselin and seconded in correspondence to the Academy by Daniel Merriman, and the seemingly uncontrolled exponential growth of the discipline all demonstrated the need for sound advice on priorities, direction, and programs to serve both science and its customers. With the navy, other federal agencies, and the oceanographic community engaged in a productive dialogue, NASCO provided the missing ingredient needed to satisfy the concerns expressed at Wardman Park and Nobska. As a group experienced in science and management, NASCO's members proceeded to suggest changes that would bring order to oceanography's Cold War–driven awkward randomness made even more complex by popular interest in the oceans. With so many of the Wardman Park participants and Nobska scientists serving on NASCO, the concerns voiced eight years before regarding the effect of naval patronage and the need for greater attention to disciplinary development would certainly ascend to the top of the priority list.[27]

16

Listening, 1946–61

It appeared certain that sound from a shallow source in deep water must focus again at the same depth. . . . I decided to make some ray tracings for such a geometry and see what would turn up. . . . I cannot now recall but I believe I chose a mid-Pacific sound velocity and obtained a caustic pattern that *repeated every thirty miles*. We were some excited.

—JOHN BRACKETT HERSEY, reflecting on his discovery with ALLYN VINE of acoustic convergence zones, unpublished memoir ca. 1960 [emphasis added]

The trick to getting Caesar on the air is we never ate lunch.

—CAPT. JOSEPH KELLY, Project Jezebel program officer, to DAVID ALLISON, Navy Laboratories historian, 5 July 1983

SUBMARINE INNOVATIONS PROVIDED THE MAJOR THRUST and emphasis behind naval funding for oceanography in the postwar world. Those very able and imaginative Soviet partners to a wartime alliance gone sour had acquired truly revolutionary undersea technologies developed by the Germans. Without a more intimate knowledge of the ocean, the U.S. Navy might find itself at the mercy of faster and more lethal submarines with little hope of responding effectively. Discussions within the ONR, Hydro's Oceanographic Division, and the NAS Committee on Undersea Warfare thus quickly focused on making the ocean an ally against a potentially potent Soviet undersea threat.

Given the nature of the challenge, those projects involving underwater sound proved by far the most important. After Maurice Ewing left Woods Hole to design his own program at Columbia, George Woollard kept WHOI involved in seismic refraction and reflection. Woollard came to WHOI in 1942 from Lehigh as one of Ewing's colleagues in physics and remained until 1947 when he took over the Geophysics and Polar Research Center at the University of Wisconsin. Employing many of Ewing's techniques, Woollard continued to use his ships in pairs, the scientific staff of one vessel heaving different concentrations of explosives overboard and the team aboard the other vessel recording the echo return times. These data not only provided information on depth, but the explosive sound signals also penetrated the sediment on the ocean bottom to reveal the geologic characteristics of the layers below.

While this seemed to have little connection with naval operations, BuOrd expressed interest and support for very practical reasons. Data from Ewing and Woollard provided important information on detonators and the behavior of explosives at great depths. For this reason the bureau furnished surplus explosives and instrumentation from the Naval Ordnance Laboratory (NOL) at White Oak, Maryland. Lamar Worzel, then at Columbia University with Ewing, and Woollard and Allyn Vine of Woods Hole employed these assets in 1946, taking seismic refraction profiles across the continental shelf on research cruises publicly sponsored by the Geological Society of America.

The Bureau of Ordnance also continued funding WHOI's wartime "project seven" research conducted by explosives experts Paul Fye of the University of Tennessee and Arnold Aarons of the Stevens Institute of Technology. Experimenting with various types of explosive charges, the Fye-Aarons group, which included female oceanographic pioneer Elizabeth Bunce, employed Woods Hole research vessels in exploring seawater chemistry, cavitation, shock waves, submerged explosive illumination, photography, and long-range detection. This work continued through the early 1950s, with Aarons, Fye, and their graduate students working on explosive phenomena from some of WHOI's smaller research vessels.[1]

From the outset, other projects seemed more immediately applicable to the navy's operational concerns. During the last year of the war, the Naval Technical Mission in Europe, led by Commodore Henry Schade, discovered a wide variety of ongoing research projects with potentially fruitful applications to Allied naval problems. These ranged from underwater sound to submarine quieting to rockets to extraordinary advances in submarine design and propulsion. While some of these discoveries certainly added to the Allied arsenal, others revealed disturbing vulnerabilities.

Since submarines modeled on the Type 21 would move faster and with greater stealth, better location and identification techniques became more critical than ever. In the minds of submarine veterans, the ferocity of the naval war beneath the surface between 1940 and 1945 forever bound together sonar development, improved echo ranging, precise target identification, effective weapons, and a better understanding of the ocean environment. American submariners thus continued to consult with Scripps and Woods Hole, sought out experienced UCDWR personnel assigned to the MPL and NEL, and conversed frequently with those at the USNUSL. Their needs and advocacy guaranteed the continued collection of BT and SBT records at Scripps, Woods Hole, and the NEL as well as support for physical and geophysical ocean research.

To understand the navy's undersea warfare problems, scientists needed a closer acquaintance with the appearance and performance of the newest submarine designs. Shortly after the war ended, Allyn Vine and William Schevill spent time at the Portsmouth Naval Shipyard closely examining U-3008. They wanted to become familiar with this Type 21 U-boat captured from the Germans, defining its sound characteristics, exploring possibilities for further quieting, and evaluating the vessel's diving habits and buoyancy requirements with an eye toward future American designs with enhanced capability. One of these enhancements, the schnorchel, drew air from above the surface through the same extended mast that expelled diesel emissions. This appendage permitted U-boat commanders to run submerged on diesel engines while recharging their batteries, the usual source of submerged power. Since battery replenishment left a thermal trace attributable to diesel exhaust, detection by heat and the schnorchel mast's surface wake was a high-priority item for BuShips well into peacetime. Next to the possibility of actually sighting the schnorchel wake, the presence of a thermal scar or irregularity on the surface detectable by a heat scanner proved intriguing. The relatively young science of operations analysis, developed during the war under NDRC contract by Philip Morse and his colleagues at MIT, helped evaluate many of these phenomena and studied the best possible applications for emerging technologies and weapons systems. Scholars labored on these problems as the navy's Operations Analysis Group (AGUE).

In addition, evaluating German technology in 1944–45 drew attention once again to the importance and possible utility of low-frequency sound. The deep sound channel exploited by the SOFAR system took advantage of the low attenuation of sound and the long wavelengths at frequencies ranging from thirty to 150 cycles. The Germans began working with low-frequency sound as far back as 1927 and created the highly sensitive *Balkon* sonar array, based upon the *Gruppenhorchgerät* (GHG) ship-mounted hydrophone system found on board a variety of captured vessels, including the U-3008 and the battle cruiser *Prinz Eugen*. The low and slightly protruding shape of the bow blister designed to accommodate the GHG array on U-boats gave the *Balkon* its name. The Bureau of Ships actually had the entire *Balkon* system removed from U-3008 and installed in the USS *Cochino* (SS-345) for experimental testing in 1947–48.

The installation of the Pacific SOFAR rescue system offered further opportunities to explore low-frequency sound behavior at depth. At Woods Hole, Woollard and Columbus Iselin split their full schedule of sound research into BuShips projects B-16 and B-17. The latter concentrated on WHOI's aspect of SOFAR and the support promised BuShips during site selection and the installation of Pacific cables and hydrophones. Everything from bathymetry to ocean-bottom configuration and geology to explosives and detonators to the design and testing of unique electronic systems fell to Woollard and Vine, Worzel before he left for Columbia University, and another wartime scientific veteran, R. J. McCurdy.[2]

Woods Hole's B-16 project demonstrated the sheer variety of concurrent, sharply focused, and particular oceanographic investigations sponsored by naval activities after the war. While SOFAR seemed to have certain clear defense applications, the B-16 research sought to flesh out a picture of the ocean by concentrating on special oceanic attributes. Woollard directed research into bottom reflectivity in shallow water and the effect of surface reflection on short-range signals. Sound attenuation studies went on in tandem with investigations into the ocean's thermal microstructure. The latter became increasingly important as oceanographers closely studied the BT data collected during and after the war. They discovered that many minute variations in temperature never emerged from the sweeping general measurements taken by BTs and SBTs. These instruments provided only a broad evaluation of temperature variation as a submarine submerged or the surface-deployed instrument descended to maximum depth. This larger picture proved satisfactory during World War II. However, as sound became an increasingly important oceanographic tool after the war, practitioners needed much greater familiarity with the subtle and discrete temperature variations that BTs failed to detect during the war. Indeed, neither Spilhaus nor Vine ever intended the wartime instrument to demonstrate that level of sensitivity. After 1945, studies of thermal microstructure sought to determine variation as precisely as possible, finding temperature differences and irregularities in constantly moving fluid layers analogous to the often blending shades and occasionally sharp contrasts of watercolor brush strokes. With knowledge of temperature variation down to the smallest scale, refraction and reflection experiments, and Fathometer soundings, surface and submerged target determinations exhibited increased precision and held greater meaning. To complete this picture, Woods Hole personnel also studied sound reverberation and the results of target strength tests on different submarines, including Type 21s.[3]

On the West Coast, the MPL followed suit. Its 25 November 1946 contract with BuShips stipulated studies of the physical principles governing the generation and propagation of sound in the sea and the creation of a solid theoretical foundation for understanding those principles. Led by Carl Eckart, the MPL also studied signal transmission and recognition. Russell Raitt, the MPL's senior research associate, explored ocean-bottom reflection in much the same way as WHOI's Woollard. In September, 1946, Leonard N. Liebermann left

Woods Hole for the MPL, taking his thermal microstructure studies to the West Coast. In addition, MPL director Eckart himself contributed fundamentally to the theory of attenuation, absorption, and signal processing.

Ewing, Worzel, and Liebermann departed Woods Hole in 1946, as did another person from the BuShips projects, Donald Wilson, who took a teaching position at the University of Florida. George Woollard left for the University of Wisconsin the following year. Iselin knew that Doc Ewing would continue to work out of Woods Hole until Columbia acquired a research vessel. Nevertheless, to keep the WHOI's sound program active, in May, 1946, Iselin asked Brackett Hersey to understudy Woollard as the latter prepared to leave. Fresh from the navy and with oil company experience, Hersey joined Allyn Vine to direct Woods Hole's underwater sound effort, employing acoustic methods in research on reflection, refraction, SOFAR, and sound propagation.[4]

Dividends

By the beginning of 1947, naval investments in oceanography began to show results with myriad practical naval applications. Beginning its experimentation with missiles by making test launches over the Pacific from the West Coast, the U.S. Army Air Forces entered into negotiations with the navy to employ the Pacific SOFAR system to locate the remains of spent rockets. The navy decided to take on the task, but insisted on a new listening station closer to the northern missile firing ranges. A short-range missile launch site already existed at Point Magu in California, and the army envisioned another in the Pacific Northwest, plus two more in Alaska. One of the latter would send its test missiles into the Gulf of Alaska, while the third site would employ the open Pacific in the direction of Unimak in the Aleutians.

Along with the four original SOFAR stations being built along the West Coast and in Hawaii, the Navy Department argued that it would need only one more station in Alaska to provide complete coverage of the army's missile testing program. Moreover, discussions with Maurice Ewing convinced the navy and Col. Michael Duffy of the Army Air Forces's Watson Laboratories that installation of the four new SOFAR stations in the Pacific would make even the planned survey work for a fifth station in Alaska unnecessary. The familiar ocean sound characteristics of the Gulf of Alaska region and the current geometry of Pacific SOFAR surveillance could determine the best possible location for the fifth station. Using SOFAR in reverse, the behavior of sound in the gulf would virtually point to the best possible location for listening.

While this method would certainly serve the navy and Army Air Forces personnel, it also suited Ewing well. The wily Columbia geophysicist used the $95,000 allocated for the planned Alaska listening station surveys to complete preliminary investigations for a SOFAR network in the Atlantic. This grid would cover the New York–Bermuda–Puerto Rico triangle and offer extraordinary research opportunities for Woods Hole and the staff of Ewing's program at Columbia University.

Commander Roger Revelle at BuShips's Code 940 managed the funds and tracked the program for the navy. As Revelle watched, Ewing proceeded to lay the foundation for a very productive SOFAR system based at Bermuda that would open in 1949. Eventually the station served the army, the fledgling air force, the navy, and the Lamont Geological Observatory. The army called this missile location effort Project Torrid, and BuShips expected Columbia's preliminary report on the Atlantic range by 15 January 1947.

A large portion of Woods Hole's 1947 responsibilities supported the creation of Ewing's Bermuda station, the development of the Atlantic SOFAR range, and the insatiable curiosity of submariners intrigued by new possibilities for acoustic detection and transmission. Such concerns dominated B-16 and B-17, as well as the newer project H-50—all funded into 1947 by BuShips, Hydro, and the ONR. These research efforts concentrated on bottom reflection, ocean-bottom topography, the transmission properties of the deep sound channel, hydrographic surveys in the Bermuda vicinity, and target strength measurements, as well as observations on submarine quieting.[5]

On the West Coast, oceanographers explored further the deep scattering layer and explored sound convergence zones, knowledge critical to the effective use of SOFAR and shipboard sonar. Charles Eyring, R. J. Christensen, and Russell Raitt of the MPL and Scripps first discovered the deep scattering layer during a UCDWR research voyage at a point near Guadeloupe Island off Baja California in 1942. This dense, roughly horizontal layer of shifting marine life had plagued sonar operators for years and fooled Fathometers into reporting shallow bottoms at fifteen hundred feet, as opposed to the reality of twelve thousand or more. After the war, Raitt and his colleagues discovered that millions of small sea creatures moving in a diurnal cycle fashioned this acoustic phantom. The entire layer would remain in the depths during the day to avoid predators, ascending closer to the surface in search of food after dark, only to scatter and reform in the depths by morning. In his early postwar H-50 labors on sound reflection from the bottom, Woods Hole's Brackett Hersey confirmed the Raitt discovery with similar observations in the Atlantic Ocean.

Hersey and his WHOI colleagues also discovered a physical feature of the ocean second only in significance to the deep sound channel itself. As Hersey recorded the event years later in an undated handwritten memoir:

> During that summer [of 1946] my old friend Allyn Vine from
> Lehigh days returned to Woods Hole from the Bikini Atom Bomb

tests. . . . In our very first conversation (I believe) we shared the results of some acoustics experiments I had done as a mining officer
west of Oahu . . . in 1944 and some thoughts he had about possible
significance of Ewing and Worzel's SOFAR experiments to acoustic
systems operated at shallow depths. It appeared certain that sound
from a shallow source in deep water must focus again at the same
depth. . . . I decided to make some ray tracings for such a geometry
and see what would turn up. The computations of ray paths were
made using Ewing's wartime formulas and a few days work produced a set of the now familiar patterns of caustics—starting at considerable depth, going to the surface and back down in a not-quite
symmetric pattern. . . . I cannot now recall but I believe I chose a
mid-Pacific sound velocity and obtained a caustic pattern that repeated every thirty miles. We were some excited. . . . Vine and I together with a new friend and colleague Bill Schevill put together a
report (classified SECRET) describing the effect and presenting a
prediction of sound intensity as a function of horizontal distance
from the source.[6]

The regularity of the sound repeatedly rising to the surface promised shallow submarine or surface-ship detection over hundreds of miles and well beyond the horizon. The naval application of this discovery became immediately
apparent.

Not every voyage or project undertaken by Woods Hole, Scripps, or Columbia had the quest for naval applications as its primary goal. Scientists usually conducted naval research at sea concurrently with experiments designed to
satisfy their personal inquisitiveness. This practice of piggybacking extended
well back into the interwar period and had the advantage of satisfying personal
curiosity while promoting naval interest, critical professional familiarity, and
careful focus.[7]

Michael, Jezebel, and Caesar

The Pacific SOFAR system, Ewing's ambitions to install a hydrophone array in Bermuda, and the Hersey-Vine discovery of convergence zones led the
submarine community to look even more closely at the strategic and tactical
possibilities presented by underwater sound transmission in the low-frequency
range. In conjunction with the CUW, Frederick Hunt, veteran director of Harvard's demobilized underwater sound laboratory and in charge of that University's continuing work in underwater sound called Project 10, applauded their
curiosity. The navy supported and promoted these efforts in 1949 by establish-

ing Submarine Development Groups 1 and 2, based at San Diego and New London respectively, to explore new discoveries in sound behavior from zero to three hundred hertz.

With the assistance of SubDevGru 1, the Naval Electronics Laboratory in San Diego employed the SOFAR stations installed for air-sea rescue at Point Sur and Point Arena to track submarines by listening for low-frequency ship noise. Submarine sounds originating at 450 to 500 feet quickly reached the deep sound channel, where they encountered the SOFAR system hydrophones at nearly 2,500 feet. The NEL's initial research with SubDevGru 1 proved the possibilities of the channel by extending detection ranges out to fifteen miles in the very early stages of research.

The promising preliminary results from the Pacific experimentation won the support of Rear Adm. Charles Momsen, the assistant CNO for undersea warfare, and the interest of Bell Laboratories director of research Mervin J. Kelly. With both Kelly and Momsen describing the low-frequency sound channel work as promising, the CNO, Adm. Forrest Sherman, began to take a more serious view of this kind of research.

Admiral Momsen had already consulted informally with the CUW on the best way to initiate a broad study of the undersea warfare problem that would provide the proper context for determining the value to the navy of the sound channel and its properties. The CUW suggested a study of ASW in the broadest possible sense, covering transport and cargo handling, vehicle and weapons systems, and submarine defense. Momsen and Rear Adm. Thorvald A. Solberg, the chief of naval research and head of the ONR, discussed the matter with Kelly in New York City on 27 February 1950. Also in attendance at the meeting were J. B. Fiske, Kelly's assistant, and Julius Stratton, the MIT provost. Given the navy's view that Soviet submarines posed the most critical future naval threat, Stratton proposed that the Navy Department and MIT work jointly. He offered his institution as the best possible site for the broad ASW study suggested by the CUW and nominated Jerrold R. Zacharias as chair. At Momsen's request the CUW wrote to Admiral Sherman formally recommending the study, and Stratton went to MIT president James R. Killian, Jr., for approval of his offer to host the project. With CUW support and Killian's endorsement via a 23 March 1950 letter to the CNO, Momsen's proposal won Sherman's approval.

Admiral Sherman and the MIT president agreed to call the study Project Hartwell. Lore has it that the name derived from Hartwell Farms, a favorite watering hole some twenty minutes from the MIT campus. The project had precedent in both approach and method: "Mr. Killian pointed out that such a group would desire in no way to duplicate any evaluation of current methods and plans already undertaken by agencies of the Navy, but would explore possibilities of new methods and their potentialities to emphasize salient features of old ones. The method of attack would be to utilize a small, highly skilled group, which could be brought together for the summer."[8]

Science and the navy sought to sustain their common dialogue, credited with bridging the gap between the two cultures during World War II, and apply it to the issues raised by new potential threats, revolutionary undersea technologies, and recent revelations about the ocean. The early Cold War environment offered a unique political context and provided ready political support.

After a series of spring conversations to refine plans for the project, a tentative program scheduled a series of general briefings to begin on 5 June at the Pentagon. After a week, the venue for the briefings would shift to the Navy Underwater Sound School at Key West, then after seven more days to the USNUSL in New London, ending on 23 June. The ONR's Undersea Warfare Branch assigned Comdr. William H. Groverman as project liaison officer and asked him to arrange the June events. Zacharias would then direct the work of the group at MIT through 26 August with a final progress review between 28 August and 2 September. The contract for MIT's services and those of the participants received required approval from the Joint Research and Development Board, and then negotiations commenced between the Navy Department and MIT's Division of Industrial Cooperation.[9]

The preliminary gathering of participants took place on 27 April in the National Academy of Sciences building during the annual meeting of the American Physical Society. Admiral Solberg made it very clear in his remarks to the group that the navy fully supported the project and expected to draw on its conclusions to shape future planning. The caliber of talent sitting in the audience doubtlessly strengthened the admiral's faith in the Hartwell process. The group of thirty-three participants included Luis W. Alvarez from Berkeley, Harvard's Frederick Hunt and Harvey Brooks, Merle Tuve and Lloyd V. Berkner from the Carnegie Institution, Vice Adm. Edward Cochrane (Ret.), then serving as the head of MIT's Department of Naval Architecture and Marine Engineering, Carl Eckart of Scripps and the MPL, Winston E. Kock of Bell Laboratories, Charles C. Lauritsen from Caltech, and the father of operations research, MIT's Philip Morse.

In the ensuing discussion, scientists and naval officers alike realized that the project goals went far beyond just killing submarines and enhancing undersea detection. Project Hartwell would focus on "the problem of overseas transport and the protection of fast carrier task forces. It was made clear also that this project is not simply an evaluation group; the navy looks to it for guidance on future long range plans; its results can be used to initiate new programs, new groups."[10]

Before the preliminary meeting broke up, Admirals Solberg and Momsen encouraged the Hartwell recruits to remain in Washington to attend the CUW's fifth undersea symposium scheduled for 15 and 16 May 1950. A number of them did and thus experienced what many came to call the "bombshell" report, delivered by physicist Frederick Hunt. Reflecting on these events, one of the navy's pioneers in underwater sound surveillance, Capt. Joseph Kelly, commented that Hunt "hit practically everything right on the nose. The use of the deep sound channel, the concentration on noise below 500 Hz (this is where [Dr. Harold L.] Saxton [of the NRL] disagreed with him), build an array of at least 20 wavelengths (we made it 40), orient the array vertically (we did it horizontally), use magnetic recordings to do post analysis (we did it in real time)."[11]

Hunt vividly demonstrated for his audience the combat potential locked up in Ewing's 1937 discovery of the deep sound channel. Hunt's declaration, based upon his own wartime research and Ewing's discoveries, far exceeded the startling effect of Harvey Hayes's SDF on the navy and scientific community in 1924. Together with Norman A. Haskill of the air force's Cambridge Laboratories, Hunt looked to the general problem of sound transmission in the atmosphere for direction as he came to terms with the riddle of underwater acoustics. As Hunt described it: "almost all sounds generated at any depth in the deep seas are transmitted to all ranges beyond 5 miles with intensity that falls off no more rapidly than the inverse first power of the range. In terms of the transmission anomaly . . . this result represents the welcome bonus of . . . 60 db at ranges of the order of 500 miles, beyond which you begin to gain additionally due to the convergence of great circle transmission paths. On the basis of this I feel that I can say to my colleagues in sonar: have heart, boys, we're back in business again."

At the conclusion of his comments, Hunt brought the audience and the navy into modern Cold War antisubmarine warfare: "Sounds queer doesn't it; but not as queer as the fact that the time interval corresponding to echo-reception from a range of 1500 miles is just 60 minutes. But for a small-interconnected network of fixed, shore-based, listening stations the sweep rate would be an ocean per hour!"[12]

The groups into which the Project Hartwell scientists were divided advised the navy on a wide variety of scientific concepts and technologies for protecting the overseas transport of troops and materiel from submarine attack in case of war. Recommendations came forth on hunter-killer submarines, radar, enhancement of long-range detection, low-frequency sonar,[13] helicopter-deployed sonar, atomic depth charges, air-deployed homing torpedoes, and a wide variety of other countermeasures and methods of detection and identification. Indeed, Hartwell defined for the next five critical years the primary avenues of scientific research in undersea warfare sponsored by the navy.[14]

While the Hartwell scientists had subsurface warfare under their microscope, the world political situation changed considerably. On 25 June 1950 the
Korean War erupted and the United States entered that conflict at the head of
a United Nations force. Once again the determination of a nation at war guaranteed that work on undersea warfare would receive careful attention and all of
the required resources. Just as the Korean conflict help save the ONR's funding
and research programs, there was now little doubt that the navy would listen
with even greater interest to the conclusions reached by Jerrold Zacharias and
his colleagues. The climate was right.

What the navy heard proved truly exciting. The Hartwell group disclosed
that low-frequency research suggested the feasibility of *assured* detection ranges
of nearly five hundred miles in the case of diesel-driven snorkeling submarines.
Below five hundred hertz, sounds more readily penetrated to the deep sound
channel, making listening over hundreds of miles possible. According to the
final report: "This frequency range is particularly important for submarine detection, because there is evidence that there are strong peaks in machinery noise
both from Diesels and battery operated submarines at these frequencies even
when propeller cavitation is absent. The low frequencies are also known to have
a very low attenuation. . . . The need is very great for information on: (a) how
such sounds propagate; (b) whether correlation exists in time and also in space
for sound waves at these frequencies; (c) the question of ambient and self noise
versus target noise; and (d) how large can one make listening arrays that will
give the expected increase in gain with size."[15] This statement alone largely determined the focus of naval oceanography and the programs at Woods Hole,
Scripps, the MPL, the University of Washington, and other institutions for the
foreseeable future.

To a certain extent, the navy had already begun the kind of focused research
suggested by Project Hartwell. In an effort to evaluate the submarine as a target, the Navy Department initiated Project Kayo in 1949 to solve the problem
of employing submarines to detect and destroy other submarines. In an effort
to make the best use of the growing familiarity with sound behavior in the
ocean, the navy assigned four boats to SubDevGru 2 in New London, together
with the services of Submarine Division (SubDiv) 11 at Pearl Harbor. Their
mission was to study everything from self-noise to long-range detection to submerged evasion to actual combat with other submarines. The submarines assigned to these commands would work with navy laboratories, other submarine and surface groups, and private oceanographic institutions to explore
the submarine versus submarine scenario. As early as March, 1949, for example, Brackett Hersey's WHOI acoustics team worked with the USS *Amberjack* (SS-522) on passive sonar assured detection ranges, noise characteristics,
and acoustic signatures. He recorded audible signals from the submarine on an

office-recording machine, the Gray Audograph Electronic Soundwriter, for ease of storage and playback.

With Capt. Roy Benson at the helm, SubDevGru 2 had at various times during 1949 the services of the USS *Toro* (SS-422), *Corsair* (SS-435), *Tusk* (SS-426), *Cochino* (SS-345), and *Halfbeak* (SS-352). The *Halfbeak* took *Cochino*'s place when the group lost the latter to an on-board electrical fire that precipitated a battery explosion and generated hydrogen gas during arctic exercises off Norway.

A conference called for 10–11 January 1950 at New London explored the priorities the Project Kayo participants should follow, the most important of which came from a veteran translator, Allyn Vine of WHOI. The Woods Hole physicist proposed using the assets available to SubDevGru 2 and Pearl Harbor–based SubDiv 11 in very specific ways. Extending detection ranges took first priority, followed by both passive listening rather than echo ranging at depths greater than twenty-five hundred fathoms and thorough exploration of the possibilities afforded by the deep sound channel. More specifically, he recommended new submarine-mounted hydrophones to afford better listening at close range, investigation of the convergence zones postulated by the WHOI acoustics group, longer periods on station to simulate picket and chokepoint operations, underwater communication, and careful research into the best depths and positions for listening. Searching for a way to keep its planning along the most promising tracks, Kayo submariners established a postconference formal planning group on the advice of Vine, WHOI's William Schevill, and Columbia's Lamar Worzel.[16]

In July, 1949, just as Kayo and the organization for Hartwell began, the ONR funded a new Columbia University sound-channel listening station on Bermuda. The navy already had a small SOFAR facility at Cove Point devoted to air-sea rescue. Finally sensing the astonishing potential of Ewing's 1937 discovery, Admiral Solberg invested $190,000 to explore it more fully, with an additional $10,000 in aid coming from the New York Navy Yard and BuShips. Eager to explore sound-channel characteristics and the limits of low-frequency, long-range detection, Maurice Ewing, along with graduate students Worzel and Gordon Hamilton, placed a deepwater geophone and a hydrophone in 650 fathoms of water 5.5 miles from the beach on a bearing of 110 degrees true from Bermuda's Saint David's Lighthouse. Ewing also had several shallow hydrophones placed at thirty fathoms 1.5 miles from the beach at 125 degrees true. He employed a thirteen-conductor armored cable for the shallow hydrophone, and a four-conductor armored cable ran from the island to the deep instruments. While a leak in the deep hydrophone's case quickly took that instrument out of the picture, all of the other devices worked well.[17]

Contracts signed by the ONR and WHOI assured the latter's support for

Ewing's effort on Bermuda. *Atlantis* went to sea under Hersey's direction to provide explosive sound sources to test the Columbia University installations. In the summer of 1949 alone, the Woods Hole group spent three months both in the Atlantic Ocean over the Mid-Atlantic Ridge and in the Caribbean. This process of *Atlantis* participation in the initial Atlantic SOFAR experimentation continued through the early summer of 1952.

In addition to the WHOI activity, the USS *San Pablo,* a small seaplane tender converted to oceanographic survey work in 1948, put to sea for a three-month summer cruise the following year to explore a substantial portion of the Atlantic. The ship covered an irregular route from England to Dakar in North Africa, then repeated the process on the ocean's west side from Argentia to Barbados. All the while, the Bermuda sound station at Saint David's Lighthouse monitored the SOFAR bombs dropped by the vessel. Columbia University's contract renewal report to the ONR confirmed the great potential of Ewing's discovery and seemed to fulfill the promise suggested by Frederick Hunt's "bombshell report": "Preliminary information from the station indicates that they are able to hear 1 lb. sofar bombs at ranges up to at least 1000 miles and from azimuths of about 090 to 220 degrees true. . . . By combining the shallow and the deep instruments, a very good azimuth to Sofar drop has been obtained. . . . In April [1949] a cooperative experiment was carried out with personnel of the Woods Hole Oceanographic Institution and with Submarine Development Group Two. Sound level signatures of a submarine in various conditions of operation were obtained, with emphasis on the low frequencies."[18] The *San Pablo*'s task mixed with other Hydrographic Office priorities and continued through 1953 while the Bermuda station tested its equipment and refined its procedures.

Gordon Hamilton, one of Ewing's Columbia students who later became head of oceanography for the ONR, took over as the sound-station chief in Bermuda. Hamilton applied the station's sound-channel listening capability to a variety of defense and scientific tasks, thus demonstrating the increasing utility of effective undersea listening. He later recalled:

> The station was set up primarily, or initially, to develop methods of surveillance for underwater nuclear weapons tests, the concern being that Russia might fire some tests off Novaya Zemlya in the arctic . . . we put in an underwater hydrophone east of Bermuda out about 25,000 feet of cable, where in those days the only sensor available to drive a cable of that length with no underwater electronics was a geophone—that is, an oil company exploration instrument—and we tied various experiments to that phone for the next twenty years. We recorded it continuously, which meant we were recording the T-

phases, that is, the earthquake phases which translate to underwater
acoustic signals from quakes around the Atlantic basin. We [also]
recorded the whales as they moved through the Bermuda waters in
the spring each year, primarily the 20 Hz ones . . . that geophone we
put in off Bermuda in '49 and '50 was used to first demonstrate that
you could hear a distant submarine snorkeling on the surface using a
general radio sound analyzer to search for the resonant schnorchel
acoustic peak somewhere around 80 Hz, . . . it was [started by] Joe
Worzel from Lamont and a man named—a captain [Roy Benson]
from submarine development. SUBDEVRON 2.[19]

With funding from both the navy and air force, Ewing employed the sound
station to explore the potential of the deep sound channel. The entire Bermuda
operation acquired a uniform identity under the aegis of the navy and Columbia's
Department of Geology. By the autumn of 1950 the Columbia University Geophysical Field Station, as the facility became known, absorbed both the Cove
Point installation and a project conducted by the old wartime sonar analysis
group that had become affiliated with Brown University as the Research Analysis
Group. Defense tasks and research work focused on the utility of the deep sound
channel for detection and tracking. The Columbia scientists also wanted to learn
more about the influence of ocean-bottom topography on underwater sound
travel. As was usually the case in a Ewing operation, the instruments served their
intense master without reprieve and recorded data around the clock.[20]

When Hartwell ended, both the CNO and the scientific community better
appreciated the possibilities of low-frequency underwater sound and the deep
sound channel phenomenon. In the autumn of 1950, Mervin Kelly of Bell Laboratories entered into discussions with Admiral Sherman that resulted in an
ONR contract with AT&T's Western Electric division on 12 December. This
arrangement provided for a thorough research program in underwater sound
with an emphasis on the detection and classification of low-frequency sound radiated from submarines. Bell Telephone immediately began a program very
similar to that administered by Ewing and Hamilton on Bermuda. Bell's first
laboratory, complete with cable-borne deep-ocean listening devices, opened in
1950 on Sandy Hook and monitored New York Harbor traffic with hydrophones installed three miles offshore in only forty-two feet of water. The
company planned to set up a southern laboratory on Bermuda, but in June,
1951, opted for Eleuthera, the site of Ewing's wartime operation.

Shortly after the contract signing, Bell submitted a momentous report out-
lining the general details of a new low-frequency analyzer. This device presented the possibility of both submarine detection and classification. Called
Low-Frequency Analysis and Recording (LOFAR), this technique and its

hardware emerged from research conducted by Hartwell veteran Ralph Potter, the director of sound transmission research at Bell Laboratories, together with his colleague David Winston. Shortly before the Hartwell group met at MIT, Potter and Winston developed a method of acoustically fingerprinting low-frequency sound signals in a way similar to that used for the analysis of human vocal patterns. In cooperation with SubDevGru 2 in New London, the two scientists explored the unique regularities and rhythms of snorkeling submarines. Motor, engine, and propeller operation in these boats produced a natural rhythm in the detected signal that formed the basis for determining submarine sound signatures. With LOFAR, Bell Labs could both detect submarines at great distances and identify the target as a particular vessel or class of vessel. This system finally revealed the significant operational applications captive in Ewing's 1937 discovery of the deep sound channel. In one of the most significant historical events in postwar oceanography and submarine warfare, the navy received the first LOFAR apparatus from Bell Labs at Murray Hill, New Jersey, on 2 May 1951.

As the significance of LOFAR emerged, a committee commissioned by the NAS and the original Hartwell group met at Columbia University. Its members lobbied for the creation of a single scientific entity to investigate the defense applications of underwater sound as discussed in their summer study report of the previous year. Concerned about the project's recommendations, ONR chief scientist Emmanual R. Piore summoned a group of Hartwell veterans, science policy makers, and naval leaders to Washington to explain the meaning of the Hartwell proposal that called for a comprehensive overhaul of the navy's underwater sound program. The 15 November 1950 meeting included Piore; Alan Waterman, former ONR chief scientist and future director of the National Science Foundation; the president's science adviser, MIT's James Killian; and representatives from BuShips, the Bureau of Aeronautics, ONR, and the office of the CNO (OpNav). The Hartwell participants included Frederick Hunt and Carl Eckart as members of the CUW's Special Panel on Low-Frequency Sonar.[21]

After lamenting Eckart's early return to California with influenza, Hunt recalled that the Hartwell group had asked the navy to focus on six or seven primary themes. These included sharing gains in sonar systems thinking, information transfer between service laboratories, and research on passive listening and low-frequency sound transmission. The group explored several laboratory and administrative models for enacting the Hartwell proposals before urging the creation of either a central university laboratory under direct ONR control or a research facility operated by Bell Laboratories under contract to ONR. At first, a BuShips representative wanted to treat the problem as a new project with its own contract, personnel issues, and budget estimates, but the Hartwell participants would have none of it. They viewed the effort as a research task, a laboratory matter. They argued that a central laboratory should reside at Columbia, funded by $10 million in navy seed money.

Not every scientist supported this option. Hunt, for example, thought that the CUW and Hartwell should avoid involvement in the implementation, administration, politics, and development of summer study recommendations. He preferred the option that the navy considered most seriously: a direct contract with Bell Laboratories to carry out the Hartwell recommendations on underwater sound. Hunt informed Gaylord Harnwell, deputy chairman of the CUW, that Admiral Sherman and Mervin Kelly of Bell Labs had already discussed the possibility. A proposal to enhance Ewing's program at Columbia by creating a centralized implementation laboratory in New York City with a field station in Bermuda had emerged from the CUW in August, 1950, but it never seriously rivaled the Bell Labs option.

Although the vision of a single central underwater sound laboratory was never realized, the CNO saw possibilities in the centralizing concept and provided a more modest amount to finance what became known as Columbia's Hudson Laboratory. The revelation of Bell's low-frequency analyzer in December, 1950, doubtless led the CNO, ONR, and BuShips to live with two entities rather than one. They did not want to interfere with the remarkably quick and productive work at Bell Telephone or retard the momentum Hartwell veterans initiated at Columbia. Thus, next to Ewing's Bermuda operation, two official projects existed to explore deep-ocean sound and surveillance. Contracts for Project Michael governed the Hudson Laboratory activity, supporting scientists who wanted to work in low frequencies. The code name Project Jezebel was assigned to the commercial effort at Bell sponsored primarily by the ONR and BuShips. Captain Joseph Kelly, the first BuShips project officer for Jezebel, described well the situation and its immediate consequences: "So I mean they let them both go off together. It didn't mean they loved each other. . . . I got this feeling from people like Dr. [Robert] Frosch who was from Hudson Labs, but that's the way the world went at the time. There was competition for ships' time, there was competition for money, and there was competition for glory, you know, of getting the job done."[22]

In May, 1951, LOFAR went to sea aboard the USS *Halfbeak* in the vicinity of Guantanamo Bay, Cuba. The device recorded the boat's signature both on board and at Bell's Sandy Hook facility, confirming the effectiveness of the system and its greater possibilities in deeper water. With valuable analysis emerging from data taken in forty-three feet of water, the sound channel depths off Eleuthera or Bermuda held great promise.

It took Bell barely a month to obtain British permission for a Bermuda installation. In July, three hydrophones went into the water at 40 feet to compare with the installation at Sandy Hook, two more went down to 960 feet, and a last penetrated to 4,000 feet, the axis of the Atlantic deep sound channel. By October, the data Bell had received and discussions with the CUW brought a general vote of confidence from the National Academy of Sciences.

The following month, BuShips and Bell Laboratories entered into a development contract, NObsr-57093, for the production and deployment of equipment using LOFAR technology for detecting and classifying submarines. The effort expended under this contract ran along two avenues. The first involved basic research into ambient sea noise, target acoustic output, and the transmission of sound underwater. The second concentrated on the development and production of detection and analysis equipment. The entire effort pointed at a greater understanding of the acoustic nature of the deep ocean and the creation of SOSUS.

The effort to deploy Michael and Jezebel products was given the code name Project Caesar. The navy and Western Electric began preparations for installing Caesar's first-generation listening devices and stations in 1952 under the phase code name Caesar I. The first Caesar station—Station Charlie, built in 1954 at Ramey Air Force Base ninety miles west of San Juan, Puerto Rico—began effective listening in February, 1955.

Charlie was soon followed by Stations Baker and Item, which operated from San Salvador and Grand Turks Island—both in the British West Indies. Station Fox listened from Shelbourne, Nova Scotia, while Dog and Easy directed their attention southeast and north from the shores of Bermuda, sharing expertise and occasionally space with Ewing's SOFAR station on that island. Along with the Bermuda facilities, Stations How (Nantucket), George (Cape May), and Able (Cape Hatteras) started listening by the end of the year. As knowledge and experience increased, the CNO's request for six stations continued to grow in number over the next dozen years through Caesar installation phases IIA, IIB, III, and IV. The navy initiated Caesar IV in 1961.

The easy relationship between the people involved in Jezebel and Michael proved critical to the rapid installation and effective operation of the Caesar stations. Captain Kelly, who served as the Jezebel project officer for twenty years, clearly recalled the value of shared professional experience to scientists, naval officers, and industrial personnel. Most of the essential navy personnel working on the underwater surveillance project came from the same warfare specialty. Although some of their habits might raise eyebrows in today's climate, present-day systems managers would envy their economy of effort and effectiveness. Kelly noted that

> All of these people were submariners believe it or not. At that time, the chief of the Bureau of Ships was Admiral Leggitt. Leggitt had a

Captain Lewis, Jack Lewis, who was a former shipmate of [sub-
mariner and Captain, later Admiral] McCain's. So they all knew each
other, and there was a rapport that you couldn't believe. . . . The
Chief of the Bureau of Ships did things that no chief would even
think about doing today. I mean about arbitrarily giving one shipyard the contract [to refurbish a needed cable-laying ship] and telling
the rest of them to go to hell, and allowing Western Electric to enter
into the contract for the [cable-laying] machinery and putting it on
the installation contract because we considered it nothing but a set of
tools. I mean, that's the way he looked at a ship, just the same as if we
went out and bought a welding torch or something like that. My
boss during those days, Commander Ernie Schwab, he was a submariner too. He had been at sea with both Jack Lewis and McCain,
so they got along. Then we had Howie Thompson, who was a submariner, who had the 466 (ONR) billet. It was a closed community,
and they got along fine together. . . . Now the other thing, and I'm
not bragging, I had this similar rapport with the Bell Labs and the
Western Electric people. I was probably a little more acceptable to
them as being the Naval officer coming up there, because I had been
in industry.[23]

The system obviously rested on professional competence, personal relationships, and a high level of trust. While the latter certainly invited abuse, the navy,
industry, and the scientific community have since discovered the high cost and
woeful inefficiencies of suspicion and doubt.

The basic tools used in the surveillance effort rested on the ocean bottom.
This first generation of American postwar ocean surveillance grids used fifty-
pound moving coil hydrophones measuring 4.75 inches in diameter and
twenty-six inches long. Each of the early arrays carried forty of these listening
devices attached to a type of armored cable called 21 Quad, which accommo-
dated forty-two pairs of wires along its entire length.

Few of the listening stations were exactly alike. The type of installation varied according to surface and ocean-bottom conditions, and the same criteria
determined cable lengths for the arrays. The latter always remained an important consideration for project technicians and scientists because extraordinarily
long cable journeys reduced the intensity, quality, and usefulness of the detected signals. From the nineteen nautical miles of the Station Charlie cable,
Caesar I configurations went to a maximum length of 116 miles for the Station
How installation at Nantucket. These LOFAR arrays focused on the five to
150 hertz bandwidth, passing on any signals detected to a console that recorded
the contact via a continually moving stylus placing horizontal strokes across a

moving track of electrosensitive paper. The entire project derived fundamentally from the studies of sound transmission in the ocean depths performed in the Atlantic by Columbia University and Woods Hole, drawing on research at sea by Ewing, Woollard, Vine, Worzel, Hersey, and the naval survey teams working on board the *Atlantis* and naval survey vessels. In the end, all of these developments rested on an observation made in 1937 without instruments by an imaginative and able scientist who placed his ear to *Atlantis*'s rail.[24]

Between Ewing's Bermuda station; the Bell Laboratories activity on Eleuthera; the development work at Western Electric; the research at Hudson Labs, the NEL, Woods Hole, and Scripps; and the flow of funding from BuShips and the ONR, listening had finally reached the stage Ewing had imagined it would not quite two decades earlier. For the early cold warriors at sea, the timing was perfect.

The advent of SOSUS demonstrated that current oceanographic knowledge, growing daily and exponentially, offered the navy an array of attractive ASW possibilities in instrumentation and environmental intelligence. However, the rate of innovation and discovery seemed to alter frequently the current wisdom regarding the most effective method of detecting and tracking submarines. With ships at sea and productive institutional and university programs under contract to the navy, flag officers needed greater program coordination, planning, and informed advice.

As a direct result of a sonar conference held in New London between 15 and 17 September 1948, the chief of naval research, Admiral Solberg, created the Underwater Sound Advisory Group (USAG), to consult with the CNO and ONR on the best avenues of current research and the most promising ways to address the problems of undersea warfare. The USAG received its charter as a permanent advisory body in January, 1950, during the chairmanship of the NRL's Harold Saxton. This group also took responsibility for the laboratory community's classified in-house professional publication, the *Journal of Underwater Acoustics*. This journal offered a necessary and welcome vehicle for peer review, publication, and professional dialogue among those working in this highly secret field.

The CNO also created a temporary committee on underwater sound research programs, christened CORPUS, which began to evaluate programs for naval sponsorship in 1949. The ONR funded early CORPUS evaluations of seemingly worthy projects proposed directly to the CNO, those emerging from ONR contract programs, and still others offered to BuShips and Hydro. J. P. Maxfield of the NEL chaired CORPUS, which also included underwater sound expert John Ide from the USNUSL and Harold Saxton. The committee evaluated particular projects with an eye toward formulating a long-range research plan for the CNO's consideration. Both the USAG and CORPUS drew their members from the naval laboratory community acting in a part-time advisory capacity.

Integrating a flexible research program with strategic planning and tactical implementation presented a formidable challenge. For example, one of the criticisms of USAG planning offered in March, 1949, by OEG director Jacinto Steinhardt emphasized the committee's tendency to restrict its focus to particular projects, and thus losing sight of the varied and complex undersea warfare context. How did the naval and civilian efforts fit into the total picture of undersea warfare? Was this consideration a truly determining factor in the navy's decision to support specific lines of investigation?

While Steinhardt's remarks exhibited a measure of validity, the communication between cultures had improved dramatically and scientists had more influence on strategic planning than ever. The OEG had a tendency to credit scientists with particular knowledge, but to discount their comments on the larger picture, citing a lack of comprehensive familiarity with the demands of naval warfare. The scientific community certainly had much to learn, but the OEG also feared the diminishment of its influence in the face of civilian scientific expertise. In this environment of gradual change and transformation, effective collaboration still depended upon personal relationships, entrepreneurship, and initiatives by veteran translators from both cultures. Given the enhanced level of collaboration at the end of the first postwar decade, translation and persuasion by oceanographic leaders proved very effective. In addition, the ONR's constant sensitivity to the fleet's strategic and tactical requirements gave oceanographers regular insights into the navy's immediate direction and preferences. Many scientists thus started using the ONR's priorities as a barometer of naval patronage.

In this spirit, Columbus Iselin called for a session on sonar and underwater sound at the 1950 ONR/CUW-sponsored undersea warfare symposium to provide information both on the state of the art and the navy's requirements. John Coleman, the CUW executive secretary, noted that scientists inquiring about the symposium's program wanted fewer presentations exploring specific technologies in favor of a greater indication of where their research might fit into current naval planning. They wanted to know the extent to which they could depend on ONR funding in the future and those research fields applicable to the navy's current concerns and obligations.

Reacting in a way reminiscent of Iselin's concern for the growth of oceanography as a discipline, the CNO perceived postwar naval oceanographic developments as random responses to the particular, rather than part of a coordinated program. Upon becoming CNO in 1949, Admiral Sherman wanted to ensure that the navy sponsored ocean science with true tactical and strategic significance. At his request, the ONR asked the USAG's Floyd Firestone to draw up a comprehensive twenty-five year underwater sound research plan to cover that particular aspect of naval oceanographic work. The CNO's concerns mirrored those expressed in the same year by civilian ocean scientists at the Wardman Park discussions. As the academics awaited Carl Eckart's final report, the CNO asked Firestone to submit his findings in the spring of 1950.[25]

Sherman also initiated Project Kayo in 1950 in an effort to evaluate the significance of offensive submarine operations to the ASW problem. Underwater acoustics research revealed a quiet and stable submarine as the most effective ASW system. If the primary focus shifted from surface ASW to a submarine-versus-submarine scenario, stealth would become even more critical to success and survival. The quietest vessel with the longest listening range would have the advantage in a sea battle.

To help it cope with the sheer volume of oceanographic research undertaken in the areas of highest priority, the Navy Department asked Brown University's Research Analysis Group to make a single picture of it all. Analysts at Brown integrated data on underwater sound and detection received from navy contractors as well as related naval activities. The CNO's staff needed to know what it all meant and how it fit together.[26]

In a presentation made to the CUW's sixth undersea symposium in early May, 1951, BuShips's Capt. William Pryor focused attention on the need to co-ordinate myriad projects and significant scientific results, to establish effective communication between scientists and the fleet, and to insure that "good housekeeping" reigned over the entire process. Pryor had long since qualified as a veteran in the navy's laboratory and ocean science community. As commanding officer of the USS *Semmes,* he worked with Columbus Iselin before World War II on the discovery of the afternoon effect, and he had spent considerable time as head of the Naval Research Laboratory. While his talk provided information on the current state of sonar and different types of systems, including the ship-deployable, variable-depth sonar developed at Woods Hole by Roy Rather, Pryor constantly emphasized his conviction that current research had transported the navy to a takeoff point in the field of underwater sound. If "good housekeeping" permitted laboratory staff, scientists, and technicians to apply themselves completely to acoustics research as well as instrument design and operation, extraordinary advances lay just ahead in listening and detection.[27]

Ocean sound transmission seemed to hold the key to undersea warfare, and in 1952 the chief of naval research, Rear Adm. Calvin M. Bolster, added two new committees to the array of navy advisers. He intended not only to tap the expertise of the committee members but also to promote cooperative research across institutional boundaries.

These new advisory groups, called the Deep Water and Shallow Water

Propagation Committees, would provide a regular forum for the exchange of current research information in a way that would make it readily available to concerned scientists and naval planners. Woods Hole's Brackett Hersey chaired the Deep Water Propagation Committee, which was dominated by East Coast institutions working on SOSUS and acoustics for submarine and ASW application. Others serving on the committee included Robert Frosch of Hudson Laboratories, J. C. Steinberg of Bell Laboratories, Lamar Worzel from Lamont, and the lone West Coast representative, G. H. Curl from the Naval Electronics Laboratory at Point Loma. Unlike the USAG and CORPUS, this group obviously depended heavily on members of the ocean science community working outside the navy laboratories, encouraging intellectual and project collaboration.

Organized to apply the same talent and methods to the acoustic challenges presented by the shallows, the Shallow Water Propagation Committee mixed naval and private affiliations. Frosch and Hersey were also on this panel, only the former occupied the chair, serving with a selection of scientists from Bell, the NRL, the NEL, Hudson, the ONR, and Lamont.

Project Jezebel demonstrated the extent to which this energetic exploration of underwater acoustics had already begun unmasking the secrets of the ocean and the warships hiding in its depths. After 1954, the ocean surveillance system that had begun as projects Jezebel, Michael, and Caesar grew into a system of eighteen deepwater arrays reaching from Barbados to Argentia. The navy and Bell Laboratories also added ten shallow-water arrays covering the banks south of Newfoundland. On the West Coast, nine deepwater arrays covered the Pacific shore from San Nicholas Island to Vancouver Island, with two others farther west terminating at Adak. In 1958 the commanders of the Atlantic and Pacific systems set up their respective SOSUS control centers at Norfolk, Virginia, and Treasure Island, California. The following year they upgraded the system, adding digital spectrum analyzers and new display consoles to multiply the low frequency ranges covered by the system.

A new crop of listening stations thus supplemented the first generation Caesar installations. X-ray began listening from Point Sur, Mike from Eleuthera, Love went on line in Barbados along with Sugar at Cape Hatteras and Jig on Antigua. By 1960, the system received the balance of the Caesar IIA and IIB stations and the Caesar III series. Uncle took its place on San Nicholas Island, Tare at Centerville, and Zebra at Coos Head, joining William at Pacific Beach South, Yoke at Pacific Beach North, King at Argentia, Victor at Shelbourne, and the series III installations at Eleuthera, Shelbourne, and Bermuda. The navy expended roughly $51 million on Jezebel research during the first SOSUS decade, with an additional $375 million to actually create the network. The individual arrays averaged between $10–12 million, with annual operating costs of slightly over $13 million. More than 75 percent of the latter went to support the twenty-two hundred personnel assigned to the program.

Just as the wartime underwater sound program depended upon instruments like the BT, the early SOSUS program generated demands for a variety of unique instruments and materials to expedite underwater communication. The expectations of the Caesar installations included not only the reception of immediately significant sound data, but also system longevity. The SOSUS provided a powerful incentive for the further refinement of armored coaxial cable beyond the standards used by international telephone companies for transoceanic communication. The multiple sound signals drawn into the system by each hydrophone and transmitted via cable also required coordination for proper analysis. With each ocean-bottom array employing forty listening devices, only the proper correlation of numerous independent signals would enable LOFAR to determine the signatures of possibly hostile vessels. In 1955, Victor Anderson, teaching at Harvard University on loan from the Marine Physical Laboratory in San Diego, developed the Delay Line Time Compressor (DELTIC). This critical device provided the first practical method of correlating and processing individual signals from the forty hydrophones serving each Project Caesar shore-monitoring station. Three years later, back at the MPL, Anderson brought digital technology to this same problem by creating Digital Multibeam Steering (DIMUS), which formed or assembled acoustic beams. Rather than emerging from the Caesar installations like a blast from Buck Rogers's sidearm, DIMUS formed beams from multiple incoming acoustic signals detected simultaneously. The system fashioned them into a single data stream on the way to a Caesar station LOFAR console, like so many divergent lanes of highway traffic merging into one. Forming a single, meaningful acoustic beam allowed Caesar personnel to view the entire target puzzle in an instant, with all of the radiated sounds coordinated in proper relative proportion. Vernier displays also permitted a six-to-one compression of the signals to make variations and regularities in the beam stand out more boldly for better interpretation.

The signals recorded by Caesar technicians and the art of rendering the proper interpretation emerged from regular work at sea. The beginning of the second postwar decade witnessed the advent of nuclear submarines and the further refinement of conventional propulsion systems equipped with snorkels for submerged diesel operation. While the sound transmission work of Ewing, Hersey, Vine, Wollard, Raitt, and others constantly enhanced the naval and civilian appreciation of sound behavior in seawater, the technology of undersea warfare made threat identification ever more difficult and sophisticated. Submarines radiated noise from many different sources, including diesel engines, nuclear-driven turbines, propellers, shafts, gears, pumps, electric motors, transformers, hull vibration, and the sound of water flowing along the hull as the ship passed through the water. Coordinating all of these detected sounds in a way that made detection and identification possible, and accomplishing this end at much greater assured ranges, remained a perpetual goal of those constantly refining the Caesar system.[28]

17

A Closer Look, 1955–60

We did not bother to inform Washington . . . because we felt they might lose their nerve.

—CAPT. DON WALSH, USN (Ret.) on repairing *Trieste* before the world record 23 January 1960 deep dive, 20 June 1995

A S COLUMBUS ISELIN DISCOVERED BEFORE THE ATTACK ON Pearl Harbor, opportunities are made—they rarely are the product of serendipity. Furthermore, they are often discovered in some of the most unusual places.[1] For example, the origins of both naval deep submergence work and the historic dive to the deepest point in the world ocean began in the spring of 1955 in Brown's Hotel in London's Mayfair District. Attending an ONR-sponsored deep-diving symposium, Scripps-educated marine geologist and ONR London representative Robert Dietz encountered Jacques Piccard, son of Swiss physicist and inventor Auguste Piccard. Trying to overcome the obstacle of language and the younger Piccard's aversion to prying journalists, Dietz managed to introduce himself as a naval scientist who saw great possibilities for both scientific investigation and naval service in Auguste Piccard's experimental deep submersible *Trieste*. For the next few hours, an American geologist and a Swiss engineer extended the navy's dialogue with civilian science to include the abyssal depths of the world ocean.

Determined to interest the navy in *Trieste*'s potential, both Dietz and Piccard offered presentations on deep submergence and the vehicle's capability at an ONR-sponsored Symposium on Aspects of Deep Sea Research conducted 29 February to 1 March 1956 in Washington. After Jacques Piccard lectured on the nature of the *Trieste* and bathyscaph operation, Dietz spoke on the practical scientific and naval applications of deep submergence, discussing other vehicles and the dives made aboard the Piccard invention to date. As a direct result of their appearance, seventy-five oceanographers passed a resolution applauding Piccard's efforts and proclaiming the bathyscaph's research potential. They also called for a national deep-sea research program. One of the enthusiastic advocates was Woods Hole's Allyn Vine. In just eight years his wife, Adelaide, would christen Vine's own submersible concept—funded by the ONR and operated by WHOI—which bore an abbreviated version of his name: *Alvin*. In another initiative, dated 26 April 1956, and doubtless sparked by Dietz's connection to Scripps, Roger Revelle formally asked the ONR to fund an effort by his institution and the NEL to evaluate the *Trieste* for oceanographic and naval use. Although the ONR realized the significance of the proposal, Gordon Lill had different plans.[2]

In February, 1957, the ONR's Geophysics Branch signed a contract with the Piccards to fund further refinements and eventually acquire the *Trieste*. Arthur Maxwell, Gordon Lill's assistant for oceanography, negotiated the arrangement with Jacques Piccard that set the stage for a series of dives in the Tyrrhenian Sea primarily designed to satisfy American curiosity about the vehicle while permitting some European participation. The ONR wanted to address the basic questions of utility and significance. The emphasis of the summer diving program thus relied upon firsthand human viewing to effectively contrast personal observation at great depths with the traditional methods of retrieving benthic objects and sea life for examination aboard ship. The bathyscaph's role as a scientific platform proved just as attractive to the navy as to those scientists eager to see the unreachable ocean floor for the first time. As an instrument for underwater sound research, the *Trieste* could well prove ideal. The vehicle's ability to hover with minimal self-noise made it an excellent transducer and hydrophone platform at any depth and in any geographic location. Only the fragility of the instruments mounted externally suggested possible limits, and many devices designed by Bell Laboratories and Ewing's Lamont group had already exhibited their durability. The possibility of using the *Trieste* in submarine rescue work and in the development of deep-diving technologies and methods for warship applications also attracted the submarine community, eventually leading to its support further development.

In April, 1957, a team of evaluators from the navy and American scientific institutions journeyed to the southern shore of the Bay of Naples, to the Navalmeccanica Shipyard at Castellammare di Stabia, where the *Trieste* lay in dry dock. To explore those aspects of the submersible most interesting to the navy, the ONR formulated a research program focusing on acoustics with small measures of biology, geology, and physical oceanography to provide as broad a basis for evaluation as time would permit between June and September. Gordon Lill and Arthur Maxwell invited a variety of naval activities and research institutions to send representatives. These included the ONR itself, led by Maxwell, Comdr. Charles Bishop, Robert Dietz, and five others. In addition, the Geophysics Branch also invited representatives from the USNUSL, NEL, Columbia University's Hudson Laboratories, and Woods Hole. After receiving approval from the ONR's Deep Water Propagation Committee, the research program began with orientation dives for scientists and American and Italian naval officers. Bishop initiated the first series of dives with Jacques Piccard and University of Milan geologist A. Pollini beginning on 24 June. The chairman of the CUW, CalTech's Harrison Brown, made a point of visiting Capri while on vacation in Europe in 1957 and participated in the *Trieste*'s summer evaluation program.

The success of these dives led the ONR to request a classified CUW conference on the bathyscaph early the following year. Already contemplating the possibility of acquiring or leasing the *Trieste,* the ONR wanted further discussion at the NAS before reaching a final decision. Richard Vetter, the CUW executive secretary, scheduled the meeting for 20 January 1958 at the Academy's headquarters in Washington.

This gathering immediately confirmed the enthusiasm expressed at the February, 1957, ONR meeting and suggested that the feeling had become contagious. *Trieste* veteran and NEL scientist Andreas Rechnitzer later recalled that those in attendance almost immediately followed the lead of a senior oceanographic veteran and seasoned translator who wanted to bring the vehicle to the United States as the basis for an American deep-submergence program. According to Rechnitzer, "before the meeting started, Columbus Iselin, who was then the director of Woods Hole said, 'Gentlemen, before we start, I have something to say.' He says, 'You know I haven't been prepared to endorse deep submersibles, but I've changed my mind. I think we ought to buy it and bring it to the United States.' Well, it changed the whole tenor of the meeting. I mean, before we even got started."[3]

Reflecting on the 1958 gathering in a letter to Vetter, Robert Dietz observed that any doubts among oceanographers evaporated after the success of the 1957 summer research trip. An American program now seemed likely given support from yet another critical source. As Dietz pointed out: "in spite of the unanimous adoption of this resolution [in 1957] many oceanographers retained mental reservations. These now seem to have disappeared, perhaps partly due to last summer's successful *Trieste* dives. And more importantly, the attending naval officers expressed an operational usefulness for manned submersibles which can operate at least to great depth if not all the way to the bottom. It seems to me quite likely that the Navy soon may feel the need for an experimental deep submarine and may go ahead with such construction."[4]

Jacques Piccard expressed the same conviction in his published collaboration with Dietz recounting the *Trieste* story. He also recalled the brief trip to Washington for the CUW conference and the essential and obvious support offered by naval officers, especially the submariners. The effect of the mature dialogue between the navy and the ocean science community openly displayed at Nobska and encouraged by Admiral Burke certainly manifested itself in the case of the *Trieste*. Piccard recalled that he "sensed excitement in their voices when discussing the problems of abyssal diving. Submariners now seemed to feel that the bathyscaph's problems were theirs as well. Her future development was pertinent to their own operations."[5]

Piccard's last sentence identified the significant development that virtually guaranteed the *Trieste*'s future. Although the bathyscaph did not even faintly resemble an operational naval vehicle, officers eager to go to sea in command of submarines and ASW vessels looked beyond the *Trieste*'s ungainly appearance and found something valuable, something directly tied to their mission and welfare. The navy-scientific community dialogue on oceanography had come a long way.

The ONR began negotiating with the Piccards in the spring of 1958. Arthur Maxwell represented the ONR and Charles Carpenter acted as the navy's purchasing agent in Italy. From the beginning, the ONR intended to push bathyscaph technology to the limit. Although the *Trieste* itself comprised roughly $267,000 of the total price, the ONR had budgeted more. Lill and Maxwell wanted a vehicle that could probe the full extent of the world ocean as "inner space" in the same way the space program had begun to explore the heavens. Ocean science needed the technology and capability, the submarine community wanted to probe the ocean more deeply, and both needed greater public attention to command the necessary resources.[6] Looking back, Maxwell recalled that

> the Navy decided that they ought to buy the TRIESTE. I went over there to Italy. Charlie Carpenter was the purchasing agent there. He was there, and we sat down with Jacques Piccard, and we negotiated a deal on this. We talked not only about getting the TRIESTE as it was, but buying a new sphere that could go to the deepest part of the ocean, getting his engineer, he sent Giuseppe Buono. Getting Piccard to be part time on it and so on. We get this all settled at night and he says, "I'll meet you in the morning." He'd go home and talk to his father, Auguste Piccard, and he'd come back and we'd start all over from scratch. We spent a week on that.

> We ended up buying the—we bought it for steel, when you look at it. I think we paid about $1 million for it, but it also bought the new sphere, [which] sent it [Trieste] down to depth, it bought Piccard's services, and the TRIESTE. All of this was about $1 million, and it got the Navy started in the deep-submersion business.[7]

Maxwell's $1 million bought an impressive package. The services of Jacques Piccard as primary pilot included an agreement that also reserved for the inventor's son the right to sit at the *Trieste*'s controls on any special or unique dive. This would ensure that Piccard would pilot the *Trieste* if the navy authorized a dive into the Challenger Deep, the deepest point on the Earth's surface. Unfortunately, the crew selected by the navy to operate the *Trieste* had no knowledge of the contractual promise made to Piccard.

The new personnel sphere purchased for the submersible with the initial navy investment came from Germany's Krupp Works in Essen, famous for nearly two centuries as one of the world's premier armaments manufacturers. In late 1958 and early 1959 Krupp built this second sphere in three sections, imitating the general dimensions of the first. Six years earlier the Italian firm of Terni fabricated the first sphere in hemispheres to an exterior diameter of seven feet two inches. The Krupp model weighed in at thirteen tons out of the water and eight when afloat, with a wall five inches thick reinforced to seven inches at the portholes. Fitted to the *Trieste* in 1959, Krupp built the new sphere to resist ocean pressure at thirty-six thousand feet or nearly seven miles. Lill's office had obviously set its sights on the Challenger Deep.

In July, 1958, naval authorities and the Italian shipyard staff loaded the *Trieste* aboard the USS *Antares* (AK-258) for the trip to San Diego and the Naval Electronics Laboratory. With clement weather all year around and deep water only a few miles offshore, San Diego seemed the ideal place. In addition, southern California had the most significant navy presence on the West Coast. The naval base could provide repair facilities as well as the expertise of SubDevGru 1, with scientific assistance coming from the NEL, Andreas Rechnitzer's home base.

When the *Antares* arrived in San Diego and the technical support personnel reassembled the bathyscaph, the discussion within the lab and the navy turned to the most effective use of the *Trieste*. As Christmas, 1958, approached the deepest point in the world ocean, the Challenger Deep in the Marianas Trench, began to emerge as the most significant immediate goal of the deep-submergence group. To many within the navy it presented a goal just as compelling and remote as that sought by the National Aeronautics and Space Administration (NASA) and its astronauts. Indeed, many within the deep-submergence community hoped for a level of general enthusiasm for ocean exploration that might generate a foundation of popular support for the creation of a "wet NASA."

The navy had officially entered the deep-submergence business. For Gordon Lill and the ONR staff this step brought the navy closer to a more complete knowledge of its own natural medium. While diving to the deepest point in the world ocean had a circus-like side to it, from Lill's point of view the attention given to a sensational accomplishment would help ensure that the navy could sponsor and facilitate research or conduct its own investigations anywhere in the world ocean. The *Trieste* was an extraordinary instrument that would reveal much about the nature of the ocean, from spectacular bioluminescence to the character of the bottom geology to the effects of tremendous pressure.

After a series of biological and experimental dives beginning in December, the *Trieste* team took the vehicle out to Guam in 1959 to prepare for the deep dive. Project Nekton, as Andreas Rechnitzer dubbed it, received only qualified approval from the CNO. Admiral Burke approved the effort on the condition that it would remain secret and under navy control. Lieutenant Don Walsh, the *Trieste* officer in charge, was assigned to carry the project request to Washington. He found support for deep submergence, but little enthusiasm for the deep dive and virtually no one willing to approve the project. He thus found himself before Admiral Burke. After a series of embarrassing aborted rocket launches, the admiral did not care to see any more sensational front-page stories devoted to navy technical failures. Given assurances of success from Walsh, the admiral took a chance on the fourteen-member naval-civilian team then on Guam making preparations to set an unbreakable record.

Guam, the nearest naval base to the Challenger Deep, served as the project's base of operations. The first two test and scientific dives took place without difficulty and provided technical and observational data as well as biological information, especially on bioluminescent marine life. With Jacques Piccard at the controls on dives, Lt. Lawrence Shumaker, the assistant officer in charge of the *Trieste,* and Andreas Rechnitzer, the project's chief scientist took turns serving as copilot.

The third Nekton dive, *Trieste*'s sixty-first, proved truly frightening and drew attention to the hazards of the project and the determination of those involved. Auguste Piccard had designed the *Trieste* as a large underwater balloon. The vehicle's entire upper portion served as a float—essentially a large compartmentalized steel bag with a wall one-fifth of an inch thick. Piccard then filled his float with hundreds of gallons of aviation gasoline. The latter provided a source of buoyancy that was cheap and plentiful in any modern navy. When completely filled, the float did not present any danger of explosion due to the internal absence of oxygen. Only during process of filling or emptying the float did true danger occur. Since gasoline compresses with the increasing pressure of depth and deprives the vehicle of up to roughly ten tons of buoyancy, the elder Piccard attached two cylindrical drums to the underside of the float—one mounted forward, the other aft. Flanking the passenger sphere, these drums, which had conical downspouts at the bottom, contained metal shot in the form of tiny spheres somewhat like the BBs used in an air rifle. The volume of shot in each drum served as ballast that the pilot could gradually drop simply by turning off the electrical current holding them all fast in a magnetic field. In the event of an emergency, the pilot could drop the contents of both drums—and the containers themselves if need be—initiating a barely controlled ascent.

Attached to the bottom of the float amidships, the passenger sphere provided ample room for two individuals. Krupp manufactured the new sphere in three parts as opposed to the Terni model's two halves. The company that had produced armor plate and guns for generations of German warships did not lack the skill for hemispheres, but it did lack the tools. The German firm's relationship with the Nazis during the war led the postwar allies to deprive it of many important machine tools and technical capabilities. The sphere used in Nekton thus consisted of a central band with two nearly hemispherical caps to complete the piece. A resilient epoxy held the components together.

Jacques Piccard and Andreas Rechnitzer piloted *Trieste* on dive sixty-one. The deep-submergence team selected the Nero Deep, not far from Challenger, to test the submersible for the impending deep dive. After reaching the bottom at 18,150 feet, the pair experienced a rude awakening during their ascent. Rechnitzer recalled hearing

> these explosions, and my eyes immediately shifted to that depth gauge, because I thought for sure we were going to go right back down again if we had lost our buoyancy, because we were pretty close to neutral anyway. If you lose one of the tanks, why, down you go.
>
> Well, we stayed there, and I said to Piccard, "I think maybe you should go topside and see what could have happened," but it takes fifteen minutes to empty the entrance tube [to the passenger sphere]. That's done with compressed air. So we had to wait that fifteen minutes. That was a long fifteen minutes this time. So when it's empty of water, then you can open the 400-pound door and go through the 15-inch hole, get up into the ladder and go on up topside.
>
> Well, he called back down. He said he didn't see anything wrong in there. So, bingo, I was out of there and slammed that door, and it wasn't until we returned to the harbor the next day and opened it up, we found several gallons of water inside.[8]

The epoxy seal holding the three sections of the Krupp passenger sphere together had given way. As long as the *Trieste* remained deep, the ocean pressure sustained the seal's integrity. When shallower depths gradually released the pressure on the sphere during its ascent, the sphere began to leak, revealing difficulty at the component seams. Upon examination back on Guam, the *Trieste* group discovered that temperature rather than pressure had caused the seal to rupture. Rising rather quickly from a deepwater temperature of 37 degrees Fahrenheit into the tepid 85- to 90-degree surface water of the South Pacific caused some epoxy seal sections not of the same mass to expand at different rates, which produced a shearing effect.

A quick and careful repair seemed the only option. But how could the support group accomplish that on a small, western Pacific island? They removed the sphere from the float at the naval base on Guam and proceeded to take it apart. All three sections came away for cleaning and the removal of the remaining epoxy and paint. Led in their repair effort by master machinist Chief Jon Michel, the team put the sphere back together by applying to the precisely fitting contact surfaces a combination of grease and white lead called lubriplate, along with a Permatex-like gasket compound on the external edge of the sphere. They covered the seams with strips of rubber to create an external lap seal.

Achieving a precise alignment and maximum effective contact between the three components presented a difficult problem thousands of miles from the naval repair facilities at San Diego. Fortunately, Chief Michel had served aboard the destroyer tender USS *Prairie* (AD-15) and recalled that one of the shops on board had two hydraulic jacks. Two days after a radiotelephone call to the repair officer on the *Prairie,* which was tied up at Yokosuka, Japan, Michel had his jack. His former shipmates crated the tool and shipped it to Naval Air Station Agaña on Guam. It performed admirably for a field device applied outside a controlled environment. The repair team removed the observation window roughly opposite the hatch and fixed the jack outside the curve of the hemispheres with the arm of the tool passing through the inside of the sphere from window to hatch. With the jack sustaining both the alignment and the precise surface contact between parts, Chief Michel attached a series of six bands around the circumference of the sphere to hold it fast. He added four bands of his own design to a pair that served the Krupp sphere when it came from Germany. When the technicians removed the jack to restore the window and hatch, the sphere's three components seemed as one, perfectly sustaining their alignment.

Shortly after completing the repairs, Lieutenant Walsh held what they called a "green table"—a brainstorming session staged around a table covered with a green cloth. They now had to decide among themselves if the submersible could make the dive into the Challenger Deep. Rechnitzer asked Chief Michel if he felt confident enough about the repaired sphere to make the dive himself. The chief replied in the affirmative, reminding all present that the alignment seemed perfect and the increasing pressure of depth would further tighten the seal. He felt strongly that a leak seemed unlikely.

All at once, the entire team wanted to make the record-breaking dive. As commanding officer, Lieutenant Walsh had ultimate responsibility for the mission's success or failure and the right to make a critical crew decision. Given the circumstances, he preferred to have the CNO place Rechnitzer, a naval reserve officer, on active duty so he could make the dive as the navy representative with Piccard. That request made its way up the chain of command and made perfect sense. As the Project Nekton director and chief scientist, Andreas Rechnitzer's scientific expertise would perfectly complement Piccard's vehicle knowledge. Other members of the team thought that Walsh could easily function as pilot and Shumaker had also demonstrated his ability to effectively control the vehicle. Was Piccard's participation necessary? Jacques Piccard then stunned everyone by claiming that his contract with the ONR stipulated that he would go on any record-setting or unusual dives. When Rechnitzer queried Washington, Admiral Burke confirmed Piccard's claim and selected Walsh to accompany him.

At eight in the morning on 23 January 1960, in rough conditions hovering between sea state four and five, Jacques Piccard and Lt. Don Walsh started for the deepest point in the world ocean. Only the support group on board the USS *Wandank* (ATA-204), an auxiliary naval ocean tug, and Andreas Rechnitzer, Chief Michel, and Lieutenant Shumaker aboard the USS *Lewis* (DE-535) witnessed their submergence. As the vehicle descended, Rechnitzer received word from the NEL to discontinue the project and stop the dive. The motive for the order to abort still remains uncertain. The possibility exists that word of the need to perform major repairs on *Trieste* after its previous dive found its way back to San Diego. According to Chief Michel, who watched as the project's chief scientist read the NEL message to abort, Rechnitzer put the message in his pocket and continued to listen as Walsh and Piccard reported their progress toward the bottom on the underwater telephone. Turning back was out of the question.

Don Walsh watched as the surface light faded in the porthole and the only indicator of movement became the bioluminescent marine life "falling" upward as the *Trieste* descended 35,800 feet into the Challenger Deep. He later recalled this defining moment in his life:

> I got on the underwater telephone and called Larry [Shumaker] topside—this was a sound-powered telephone, no wires or anything—and told him that we were on the bottom, because I had looked at the depth gauges as we were going down and Jacques was at the window, because what happens, in mid-water, if you turn off the outside

lights, and if you have any bioluminescence, then you can see your
rate of vertical movement, because it's like snow falling backwards.
You get these lights coming up this way. When they stop, you stop.
When you start going like this, you'd better get busy, because otherwise you're going to go all the way back to the surface. You can valve
off a little gas if you're fast, and then resume your descent.

But when you get near the bottom, we had a fathometer, which
would give us about 100–150 foot—we weren't sure because of the
pressure on the transducer at that depth, but basically it would give
us some early warning the bottom was coming up. So then you start
to really slow down. Then you turn your lights on, because even
though you can't see the bottom, you can start to see the lume. The
water will start to sort of glow, because it's bouncing off [the bot-
tom] even though you can't see it. Then you really slow yourself
down. You make a nice gentle landing that way.

So he was doing that, I was watching the fathometer, calling off
the height above the bottom, because I could see the slope coming
up, and I was ready to talk to topside that we had got on the bottom,
and I was also reading the depth gauges. It was at 37,500 feet—
37,500, it can't be off that much. We were really around 36,000. But
we had no way of doing measurements then. We didn't have these
deep-sea sound recorders that could really see that far, not with that
precision. So it was admissible that maybe it was a couple thousand
feet deeper than anybody thought.

Finally, down at the bottom, 37,800 feet, so I called that up. Of
course, that was sent—Larry [Shumaker] . . . sent a radio message
over to the *Lewis. Lewis* sent it back to Guam; Guam sent it to Washington. The whole world knew, 37,800 feet. That got a great hoot of
derision from the [oceanographic] community. As it turns out, they
were right. We didn't know that for a few days afterwards.[9]

The depth reading sent to Washington did raise some eyebrows because
the number seemed far too high. When naval and university scientists questioned the figure, Rechnitzer soon discovered that Piccard had calibrated the
Fathometer on the *Trieste* for fresh water before the vehicle came to the United
States. The *Trieste* crew and scientists at the NEL assumed that he had more recently prepared the device to work in the expected medium, the much denser
saltwater of the South Pacific. Scripp's John Knauss and the ONR's John Lyman, a veteran of Fleming's office at Hydro, corrected the figure and arrived at
a precise record depth of 35,805 feet. Walsh had planned a second dive to equal
the first record to provide other members of the team with experience at the

ultimate depth. Unfortunately, a cracked porthole window prevented another journey to the bottom that might have permitted Rechnitzer or Shumaker to share the record for the deepest dive in history.

After the return to Guam, the navy flew Walsh, Piccard, Rechnitzer, and Shumaker to Pearl Harbor and then on to Washington for a meeting with President Eisenhower. After a meeting with several flag officers in Hawaii, Walsh received the Legion of Merit from the president, who in turn gave Jacques Piccard the navy's Distinguished Public Service Award. Although the navy and the oceanographic community now had access to any point in the world ocean, the CNO decided against too often repeating the risks taken in making the deep dive. He limited the activity of the *Trieste* and other submersibles in development under naval sponsorship to a maximum depth of twenty thousand feet. Still under Walsh's command, the *Trieste* dove to that depth near Guam with Rechnitzer and Shumaker in the spring of 1960. In more than 70 percent of the world ocean, the bottom lies at that depth or less. More to the point, this restriction kept submersible development focused for the foreseeable future on that slice of the ocean available to the submarine community and its adversaries.

After Project Nekton, the *Trieste* worked out of San Diego in support of geophysical and oceanographic research. The bathyscaph presented an ideal platform for deep-sea research in acoustics, ocean currents, bottom composition, and marine biology. Many scientists from the NEL, MPL, and Scripps requested and received time aboard the *Trieste* to further their research, especially if their funding came from the ONR. To the satisfaction of the often-neglected marine biologists, the bathyscaph devoted many dives to the study of marine life. Firsthand observation at depth permitted an appreciation of fragile, pressure-sensitive organisms that a net and examination on deck could never achieve. Besides, marine life held a greater importance than ever for the navy with the discovery of the deep scattering layer and its effect on sound transmission and bottom location. The *Trieste*'s other activities provided American submarine and ASW crews with a better appreciation of currents and the acoustic characteristics of deeper water now that submarines could regularly dive below the World War II limit of 450 feet. In addition, with Projects Jezebel and Michael still active and a familiarity with the deep sound channel and acoustic convergence zones, an intimacy with ocean layers beyond the submarine's current capacity seemed wise.

In April, 1963, a tragedy interrupted the *Trieste*'s scientific schedule. While performing deep-diving tests off Provincetown, Massachusetts, the submarine USS *Thresher* failed to surface. The navy immediately mobilized all of its resources in an effort to discover the vessel's fate. When it became apparent that the *Thresher* had exceeded its test depth and found its way to the bottom with all hands, the navy had to come to terms with a twofold disturbing reality. First, its most advanced nuclear submarine would never return—and neither would its highly skilled complement. The second reality proved almost as disturbing as the first. In this terrible case, the navy had no way of attempting a rescue, and it lacked a system designed to locate a submarine in distress at extreme depths. It took naval survey vehicles assisted by scientists and ships from Wood Hole and Lamont weeks to locate the debris trail of the broken and twisted submarine in the spring of 1963. Moreover, there was only one naval asset that had a chance of directly examining the evidence of the tragedy. The board of inquiry looking into the loss of the *Thresher* needed as much insight as possible if its conclusions were to have a constructive meaning for the future safety of the submarine force. The CNO therefore called Lt. Comdr. Donald Keach, who had relieved Walsh as the officer in charge of the *Trieste* in July, 1962, to ask if the deep submersible could help in the *Thresher* search. Keach and his team responded immediately, but warned the CNO that the barely maneuverable *Trieste* would require fairly precise wreck coordinates before he could promise effective examination.

Finding debris scattered by implosion, explosion, and currents posed no easy task. In the process of locating the submarine's remains, Lamont geophysicist J. Lamar Worzel found himself dreadfully embarrassed when one of Columbia's cameras took a picture of its own mounting device, momentarily convincing everyone that the debris trail lay directly below. Not long afterward, the trail revealed itself to a magnetic anomaly detector and then to towed, still-image cameras. The team from Lamont on board the USS *Robert D. Conrad* (AGOR-3) photographed a piece of twisted metal piping, an upright compressed-air bottle, and bits of metal and insulation. Woods Hole's *Atlantis II* dredged up a battery plate of the kind used on board the *Thresher*. When Keach finally took the *Trieste* down after a quick shipboard trip from San Diego to the North Atlantic, he retrieved the artifact that brought certainty to a universal suspicion. In the "elbow" of a partially working manipulator arm hurriedly attached to the *Trieste* he trapped a piece of pipe that bore the legend "593 boat." When more dives revealed piles of hull plating and other scattered parts, the CNO, Adm. George Anderson, revealed the findings to the public and terminated the search on 5 September.

For the public and the families of the crew, the search answered the most fundamental questions about the disaster. While the Navy Department wanted to give the families time to cope with their loss out of the limelight and to offer members of the press a chance to find other stories to distract them, the submarine community had other ideas. In May, 1964, under the cover of a scientific effort sponsored by the ONR, the *Trieste* again came east from San Diego. In the months since the dives piloted by Keach and his executive officer, Lt.

George Martin, the Mare Island Naval Shipyard effected major alterations to the deep submersible, fitting it with a new float, moving the reinstalled Terni passenger sphere forward, providing more powerful maneuvering motors, and rechristening the result *Trieste II.* The virtually new vehicle worked in conjunction with the naval research vessel USNS *Mizar* (T-AGOR 11) to find the remains of the *Thresher*'s hull.[10] After *Mizar*'s sonar and towed cameras suggested the best dive points, Lt. Comdr. J. B. Mooney, the new officer in charge of the *Trieste,* made the critical dive on 18 August. On that day the navy took its first close look at the effect of deep-ocean pressures on nuclear submarines. Mooney later vividly recalled the scene he saw from the *Trieste*'s pilot seat:

> We went down into the area where we found some fairly heavy debris, and then our first indication that something was amiss was that there appeared to be some cracks in the sediment. So something had disturbed the sediment, that looked as though some big event had taken place there, and I guess it had, because it was where the hull had impacted.
>
> As we moseyed around, it appeared that we were among large piles of sand that had been pushed up, sort of disturbed, and sat down in one area just to look around, and could see out the front window of the bottom, maybe thirty feet away, and tried to release gas to get down that thirty feet, so that we could get closer to the bottom, but we could not move downward. . . . So we knew that there was something man-made that was there, so we turned 90 degrees and suddenly realized the reason that we couldn't get down on the bottom was because we were sitting on the hull of *Thresher.* One side of that hull was butted up against the side of the sand pit, and the other side of the hull had no sand around it at all. You could see down to where it was sitting on the bottom.
>
> It looked like it was a swirling impact. As it swirled, it sort of dug its own hole there. It wasn't just where it had hit and made the impact of the hull, it was where it swirled around.
>
> So we moved around on top of the hull and could see the safety track that's on top of the submarine where you have safety lines. We could see where a hatch had been. We could see where the messenger buoy had been. We could see a large piece of metal that stuck up, . . . the point of it sort of stuck back toward us, like . . . a beer can opener, like a church key. We, of course, with a three-sixteenth-inch hull wanted to stay fairly far away from that, so we gave it a lot of berth.
>
> We did attempt to see if there was any additional high-level [radiation] readings down on the side of the hull, so we moved off the

hull, eventually went down between it and the side of the sand pit. It was a tight squeeze. It turned out that we scraped paint off both sides of the gas bag of *Trieste* as we descended down into that hole, and down there, there were still no elevated readings.[11]

By 1965, when the navy transferred the *Trieste* from the NEL to the submarine force in San Diego under the control of ComSubPac, the experience of the *Thresher* catastrophe confirmed the strong and obvious link between the deep-submergence program and American submariners. A heightened general awareness of the value of deep submergence descended on the fleet. In the twenty-four months following the discovery of the *Thresher*'s hull, *Alvin* made its debut at Woods Hole while the navy contracted with General Dynamics's Electric Boat Division to build the *Alvin* clones *Turtle* and *Sea Cliff*. Originally christened *Autec I* and *Autec II,* these vessels initially worked in the navy and air force missile splashdown detection range off Bermuda. In direct response to the difficulty involved in finding the *Thresher* and the impossibility of rescue even if the vessel had come to rest at a depth its hull could withstand, the CNO initiated the Subsafe Program. The program's measures to improve safety included a navy contract with the Lockheed Missile and Space Corporation of Sunnyvale, California, to build two Deep Submergence Rescue Vehicles (DSRV) capable of removing twenty-eight people at a time from a stricken submarine at a depth of five thousand feet.

Although the deep-diving community achieved neither the "wet NASA" many hoped for nor the corresponding national support and funding, in 1965 the navy established a deep-submergence program in San Diego with SubDev-Gru 1 to integrate the new capability into fleet operations. The 1960 deep dive and the loss of the *Thresher* in 1963 reinforced the importance of deep-ocean access to operational security and success. Achieving extraordinary goals and suffering profound tragedy provided a context in which both scientists and seagoing officers could recognize the potential represented by vehicles like the *Trieste.* Developments would not proceed at the pace of the manned space program, but they would proceed. In 1965 the Navy Department built its deep-submergence program around the *Trieste.* Robert Dietz and Jacques Piccard had anticipated that step—not in 1965, but ten years earlier, while conversing over tea in the comfort of Brown's Hotel in London, far away from the Challenger Deep.

18

Coming Full Circle

HILE THE DEMISE OF THE AXIS IN 1945 BROUGHT A welcome return to peace, the latter proved a mixed blessing for the navy. Redefining its roles and missions in a world no longer held hostage by the Germans and Japanese, the navy almost immediately descended into a very significant domestic political battle over its part in the postwar national defense. The absence of an imminent external naval threat in 1945 combined with the birth of the U.S. Air Force to render the navy vulnerable during critical Capitol Hill debates on the defense budget, service roles and missions, and airpower. Only by 1950, with the successful "Revolt of the Admirals" in defense of naval aviation and the significant conclusions on undersea warfare reached by the Low Committee and Project Hartwell, did the navy finally began to see the faint outline of its future. As a result, a mature professional dialogue emerged between 1950 and 1956 that led the navy and the oceanographic community into both a permanent partnership and the federal agency alliance known as the National Oceanographic Program.

Oceanographers acutely felt the anxiety of this uncertain time. The war obliged civilian scientists and naval officers to cooperate and communicate with a professional intimacy rarely witnessed in the interwar years. Would this continue? Who would emerge as the navy's next primary adversary and what effect would that have upon naval strategy? How would the navy's patronage policy change with new strategic decisions? What role would oceanography play in the navy's postwar mission? Indeed, many scientists did not want to play a role, but returned instead to their university laboratories and prewar research projects. Others welcomed a continued partnership, but with reserve. The navy proved a very generous patron. However, what effect would naval investment have on oceanography's development as a discipline? Without proper management, naval funding might encourage oceanographic growth along varied project lines as opposed to a singular discipline with diverse interests and applications. Furthermore, would naval funding remain consistent enough to enable civilian oceanographic institutions to cope with the extraordinary postwar expense of operating at sea? Few things seemed certain.

Naval oceanography, with its direct tie to the fleet's operational requirements, attracted millions of federal dollars, but at the cost of considerable confusion. While research sponsored by the navy after 1940 paved the way for significant scientific and technical progress, the wartime experience and the massive infusion of postwar money seemed to exaggerate the lack of cohesion and identity in this multidisciplinary science. Oceanography clearly needed a defined research practice and heritage of its own, distinct from the traditional independent disciplines of biology, physics, geology, and chemistry. It also needed a multitude of instruction programs with a capacity for students beyond that of the existing graduate programs at the Universities of California and Washington. As opposed to other sciences, oceanography emphasized an integration of disciplines to study the dynamics of a constantly changing ocean system. Only continued patronage, the focus that comes with disciplinary identity, and effective research and training programs would attract talented people and convince universities to admit oceanography as a field of study on par with other scientific pursuits. Would the navy assist civilians in building a discipline and its identity while using oceanography as a tool in the national defense?

Soon after the surrender of Japan, the civilian oceanographic community took the initiative to sustain the relationship that proved so productive in the war against the axis. With the NDRC demobilizing, only an effective replacement could avoid a reprise of the 1924 ICO disaster. Roger Revelle's initiative, taken from his position within the navy's BuShipsCode 940, created the oceanographic component of Operation Crossroads and produced critical information on the effect of nuclear weapons as well as new data on Bikini Atoll and other island groups in the central Pacific Ocean. Revelle's excellent performance at Crossroads and his promotion to commander placed him in a position in 1946 to influence naval leaders. He also found himself coordinating the activities of and supervising scores of the best ocean scientists in the country, an image that he gladly and quickly cultivated. The oceanographic project at Crossroads kept naval officers and oceanographers working together on a project critical to the nation's future, sustaining for the moment the wartime cooperation with all of its possibilities.

Three critical agencies created after the Crossroads experience helped promote the wartime working relationship between civilian oceanography and the navy into an effective professional dialogue. Gaylord Harnwell's ambition for continued ocean research in conjunction with the navy prompted the creation of the National Academy of Sciences' Committee on Undersea Warfare. Populated almost entirely by veteran translators, the CUW assisted in planning programs to address what would soon become the navy's principle strategic postwar priority, antisubmarine warfare. It also placed many of the institution-based translators like Iselin, Ewing, Vine, Revelle, Ted Hunt and others in the quasi-official position of adviser to the federal government on undersea warfare.

In a similar way, the creation of the ONR and the Oceanography Division at Hydro in late 1946 placed within the federal service select scientists with wartime naval experience and a talent for cultural translation. These people knew the navy, the needs of oceanography, and the importance of their postwar activities to the creation of an effective dialogue. They built firm cultural bridges and brought considerable federal patronage to oceanography.

In any case, it was easier for scientists to communicate their needs to the federal patron if the person speaking for the government was a friend, colleague, and scientist or technician of stature. For example, Lt. Comdr. Butler King Couper, a friend of oceanography with very close Scripps and WHOI ties, also managed BuShips's modest postwar budget for work on ship fouling and underwater acoustics. The Couper case was not unique. Assisted by colleagues like Mary Sears and Columbus Iselin, who used their rank and experience to influence and persuade the navy's leaders, Roger Revelle and Gordon Lill found themselves at the ONR and Richard Fleming at Hydro, where they were in a position to control substantial budgets, ships, and other research assets.

Once in place, scientists like Fleming and Revelle could communicate easily with a counterpart at a university or research institution. They used the same culturally constructed vocabulary, appreciated purely scientific needs and concerns, and addressed more effectively and immediately the incommensurability that often occurred between academy-molded officers and university-shaped scientists. They truly had one foot in each camp and brought an understanding to the process of building a lasting postwar dialogue that no other person or institution could match. It proved even more significant that many of these individuals who could best understand this difficult relationship had in their control the bulk of the navy's money for sponsoring oceanographic research. The rapid progress of the next decade was not simply good fortune but the result of this enhanced and effective dialogue empowered by individuals who collectively possessed the scientific talent, administrative authority, and assets to act.

The promise of an enhanced and more effective dialogue proved both promising and disturbing. The navy's problems certainly drew attention, talent, and money, but what of oceanography's even more basic needs? These concerns, driven by the effect of the navy's awesome spending power and its changing postwar strategy, precipitated the 1949 meeting of the NAS panel on oceanography. This group is often mistakenly called the second NASCO, after the Lillie Committee of 1927, when it actually never had similar responsibility. *Oceanography 1951* emerged from both the panel's rather informal discussions at Wardman Park and the pen of Carl Eckart as a succinct statement on the varied and fundamental dilemma facing oceanographers in the first postwar decade. The Cold War era presented the mixed blessing of generous naval funding, the debate over research priorities and federal influence, excessive operating costs, the internal requirements of a discipline still defining itself, and the need for peer recognition and university training programs.

The war, scientific ability, and cultural translation skills determined the cast of characters that would address these problems and prove most influential during the Cold War. Veterans like Frank Jewett and Columbus Iselin initiated the process that led to Wardman Park. Revelle, Ewing, Eckart, and others with valuable naval experience contributed to *Oceanography 1951*'s coherent statement of needs. The mature cultural bridge-building skills of Iselin, Jewett, John Coleman, and others made Project Nobska possible, and by the end of 1956 ensured that the most effective translators were positioned as naval advisers at the highest political and naval levels.

These achievements complemented the work of postwar translators who had become civil servants—like Richard Fleming and Gordon Lill, who were already resident in Hydro and the ONR. In the very different world of 1950, the experience of war and twelve years of Franklin Roosevelt had placed the central government in a dominant and very influential role in American society and science. Furthermore, the Korean conflict and a Cold War foreign policy seeking to contain the Soviet Union ensured for the moment that Presidents Truman and Eisenhower would continue funding levels reminiscent of World War II. The ocean scientists of 1950 now had little reason to fear a reprise of 1924, when the ICO's plans failed to receive presidential or congressional support. Effective translation and a true dialogue placed civilian oceanography and the navy in a position to contemplate a coordinated national oceanographic program by the mid-1950s—thirty years after the NRL's Harvey Hayes and his naval supporters initiated the effort.

The energy and determination of American oceanography's return to an ocean at peace during this first postwar decade demonstrated the immediate effect of the translator's art enabled in a new way. Rather than depending upon the NDRC, now just a memory along with the war, the navy-oceanography relationship very soon rebuilt itself on a new foundation. After Crossroads, a combination of Hydro, the ONR, BuShips, and the CUW provided the navy with channels for patronage and a productive dialogue with civilian science on oceanographic research. They defined projects, provided assets, granted extraordinary amounts of money, and promoted oceanography in a vigorous way as an important component of the national defense.

The rapid resumption of research at sea after 1946 demonstrated the postwar vitality of the navy-oceanography relationship. While the internal disciplinary needs of oceanography only slowly received attention, officers and scientists pursued naval projects and institutional goals with great energy. Maurice Ewing took the first postwar step. He journeyed to the Mid-Atlantic Ridge in the summer of 1947 sponsored by Woods Hole, Columbia University, and the National Geographic Society. After returning Ewing's company to Woods Hole, WHOI's *Atlantis* once again put to sea, this time in December, 1947, on its Med cruise, *Atlantis 151*—a project as important for classified naval work as for the fundamental oceanography of the Mediterranean Sea. The following year, Brackett Hersey initiated his acoustics research for the navy sponsored by Woods Hole, BuShips, and the ONR. This program continued well into the 1960s, providing critical insights into the deep sound channel, bottom-bounce sonar, convergence zones, and submarine sound signatures for submariners and ASW officers. Project Cabot resulted in innovative work on the Gulf Stream in 1950, the same year that Roger Revelle, as newly appointed associate director at Scripps, led that institution on MIDPAC in cooperation with the NEL. The latter introduced the staff at La Jolla to regular, independent, deep-ocean research.

Regular polar investigations also resumed during this early postwar flurry of activity with Operations Highjump and Nanook II in 1947 and later with Operation Skijump in 1951. These first careful forays into the polar regions initiated the career of one of the navy's great polar scientists, the NEL's Waldo Lyon, who eventually traveled submerged aboard the USS *Nautilus* under the North Pole in 1958. Indeed, when the USS *Skate* (SSN-578) actually surfaced at the North Pole on 17 March 1959, Comdr. James Calvert conducted a funeral service for the recently deceased Sir Hubert Wilkins. That pioneer's plans to travel to the pole under the ice in the company of the young Harald Sverdrup on board the O-12 failed nearly thirty years before when the submersible lost one of its dive planes.

Indeed, much of the navy's early postwar work at the poles supported a global project that served the national defense, helped allay international tension, and dramatically increased knowledge about the Earth. Conceived on the model of the two Polar Years held to involve many countries in the exploration of the Arctic and Antarctica, the International Geophysical Year presented great possibilities for the earth sciences and especially for oceanography. Planning began in 1950 for eighteen months of concerted international research into the earth sciences beginning in 1957. From the beginning, the organization and execution of the IGY program lay with civilian scientists from many countries. Even the American IGY committee reflected its purely civilian nature. The National Academy of Sciences took the lead in organizing the U.S. program, and in 1953 asked the National Science Foundation to approach Congress for the necessary funding and take charge of program administration. When Congress responded to the NSF request with an appropriation of $39 million, the U.S. program was initiated.

From the beginning, the navy took part in aspects of IGY administration and bore a heavy logistical burden. This seemed logical. The U.S. program called for very ambitious studies of ocean currents, the polar regions, gravity and magnetism, and Earth observations from a new satellite. The latter became very important from a political standpoint, as did science in general, after the Soviets launched Sputnik in 1957. All of this work required mobility on the high seas and the ocean transport of staff and equipment. While competition with the Soviet Union accounted for a large part of the political and military commitment, these years also demonstrated the positive effect of the navy's improved and productive dialogue with the ocean science community. The period 1950 through 1958 spanned the Low Report, Project Hartwell, the creation of the ONR and Hydro's Division of Oceanography, the CUW, and Project Nobska, and would witness the creation of second NAS Committee on Oceanography, as well as the ONR's Ten Years in Oceanography (TENOC) program. Under CNO Adm. Arleigh Burke, the navy made a commitment to the ocean sciences in a way that promised to serve both the navy's mission and the needs of the ocean science community.

In this environment, the navy's support of the IGY seemed only natural. Gordon Lill of the ONR Geophysics Branch served as chairman of the U.S. IGY oceanography panel. The balance of the committee's membership seemed almost predictable: Ewing, Fleming, Iselin, Revelle, Coast Guard Rear Adm. Edward Smith as the retiring Iselin's replacement at Woods Hole, Hydro's John Lyman, and Dale Leipper from Texas A&M's oceanography program, which the ONR had helped to create. In addition to university and institution vessels, the navy provided much of the ship time required for research at sea, especially in the Arctic and Antarctic. The submarine force also welcomed scientists on board select vessels to take IGY gravity measurements while submerged, an activity reminiscent of Vening-Meinesz aboard the S-21 in 1928, and of the 1932 Navy-Princeton Expedition that employed the services of the S-48. The navy even managed the joint IGY satellite program with the army and air force that gave birth to the Vanguard multistage rocket and launch vehicle. The latter emerged from pioneering research conducted by John P. Hagen at the Naval Research Laboratory.

In 1957, the navy and the NAS formalized a part of this network of professional and personal friendships, information, patronage, and scientific practice that characterized both the development of oceanography in America and the navy's role in that process. As the Nobska summer study concluded in 1956 and in tandem with preparation for the IGY, the Academy created the second NASCO. With the identification of Soviet submarine potential as the primary national maritime threat by both the Low Report and the Project Hartwell summer study, submarine warfare and ASW became the navy's top priority. This aspect of naval warfare demanded as great an intimacy as possible with the unforgiving ocean environment. The submariner operating in the depths swiftly recognized the ocean as both his greatest potential ally and most deadly enemy. Many officers learned their first environmental lessons during the war at sea between 1940 and 1945 from those very translators who moved into Hydro and the ONR after the war. At first populated almost entirely by veteran translators, NASCO continued to influence oceanographic policy nationally and within the navy for the next two decades.

Directions

Immediately upon its creation in 1956, NASCO's members realized the need for a carefully crafted, comprehensive plan for American oceanography that went beyond *Oceanography 1951*'s definition of issues. Circumstances required concrete planning and program promotion. As the committee embarked upon the composition of its multivolume report *Oceanography 1960–1970*, the ONR realized the need to work with the committee and to prepare the navy for NASCO recommendations.[1]

As the NASCO report took shape, Gordon Lill communicated his reluctance to extend early ONR sponsorship for the initial draft to committee secretary Richard Vetter and the secretary of the NRC's Division of Earth Sciences, S. Douglas Cornell. The document itself had little to do with Lill's concern. As head of the ONR's Geophysics Branch, he knew all too well that NASCO would not necessarily set priorities and make recommendations in terms of naval need. Rather, the committee would naturally dwell on the needs of oceanography as a significant science and an emerging discipline. He agreed with Hydro veteran Richard Vetter that Admiral Burke would only accept guidance if it seemed to place the navy's interests first. On 6 March 1958 Cornell brought to Detlev Bronk's attention a comment by Vetter that naval leaders "are not inclined to ask for anyone's advice from outside." While this flew in the face of the civilian scientific role at Crossroads, the critical civilian presence at numerous summer studies including Nobska, the activity of the ONR, and the navy's role in requesting the creation of NASCO, the statement still contained a measure of truth. The Navy Department's internal process of evaluating and implementing civilian scientific recommendations demonstrated the continued importance of the translator as catalyst and facilitator between the two cultures. The debate over Nobska's interest in active sonar and the reactions of Jacinto Steinhardt and Admiral Combs to the summer study's recommendations in this category provided a case in point. As the 1960s dawned, there still remained the occasional unwillingness by those intimately involved in operational matters to consider that scientists outside the Navy Department might have a comprehensive grasp of the strategic, tactical, and technical significance of applied oceanography in naval warfare.

Some within the NAS appreciated Lill's reluctance. He had no desire to volunteer his domain within the ONR as the first sponsor of the forthcoming NASCO report before briefing the CNO on the committee, its goals, and their significance for the navy. While the navy had joined other agencies in requesting the creation of the committee, this did not mean that it would immediately endorse or heed the group's recommendations. Taking Lill's point, committee chair Harrision Brown and NAS president Bronk wrote to Admiral Burke on 8 May 1958 explaining their perspective on NASCO's purpose and priorities and requesting naval assistance in making the committee's work productive for all concerned.[2]

In constant communication with old friends and colleagues on the committee as deliberations opened, Gordon Lill knew from the beginning the substance of NASCO's ambition for American oceanography. In conjunction with his ONR Geophysics Branch colleagues Arthur Maxwell and Feenan Jennings, Lill prepared a report entitled "Project TENOC: The Next Ten Years in Oceanography" to present the essence of the still unfinished NASCO program in a form familiar to the navy and in harmony with its mission priorities. On 7 October 1958 the ONR presented the TENOC report to the CNO over the signature of Rear Adm. Rawson Bennett, the chief of naval research. Sounding very much like Harvey Hayes thirty-five years before, Gordon Lill noted in the document's preface that every naval weapons system and all of the specialized programs recommended by recent studies like Project Nobska had to recognize the determining influence of the ocean environment. In the opening paragraphs Lill and his colleagues asserted that oceanography formed "the broad platform of knowledge which must be strengthened in order to support the applied problems facing the Navy." Without a sustained oceanography program, the navy's plans would fail on all fronts. It had to come to terms with its natural environment.[3]

Success would require trained people and a significant budget increase. If, for example, the ASW active-sonar project dubbed Artemis received full support over the 1958–63 period, it could easily occupy every qualified oceanographic scientist and technician and every facility on the Atlantic seaboard. There were only between six hundred and seven hundred qualified people available. This had to change. Furthermore, the ONR's Geophysics Branch expected a budget of only $5 million for the ocean sciences if current fiscal year 1960 projections held true. Lill's budget usually provided nearly half of the Scripps research budget and 70 percent of the money flowing to Woods Hole. His 1957 investments included the purchase and refinement of the deep submersible *Trieste* that effectively created the navy's deep-submergence program.

The success of Auguste Piccard's bathyscaph set the stage for advanced submersible technology and the discoveries made by research vessels like the *Alvin, Sea Cliff,* and *Turtle,* and by the remotely operated vehicles (ROVs) that eventually brought Robert Ballard's "telepresence" to millions of television viewers in the 1980s and 1990s. The ONR's sponsorship of submersible technology also helped advance deep-sea photography. Access to the depths drove basic single-frame photography a la Ewing and Jenkins to Super 8 or 16-mm film by the time the *Trieste* searched for the *Thresher,* and more recently to Hi-8 video.

Through the ONR, Hydro, and BuShips, by 1958 the navy funded 80 to 90 percent of all U.S. oceanography—including instruments, ships, and vehicles—playing the role of catalyst, patron, and participant.

Each civilian institution depended on the navy absolutely and felt its limits acutely. The navy spent $1.5 million on studies like Projects Hartwell and Nobska to learn that only more money, facilities, and training programs would adequately serve the national defense and the needs of civilian science. Even Vice Adm. William F. Raborn, first head of the navy's Polaris missile program, argued for a better grasp of oceanography in support of submarine-launched ballistic missiles. Lill did not hesitate to use this high-profile program and its environmental requirements to make his point to the CNO in the TENOC report.

Lill, Maxwell, and Jennings then added some spice to the argument by drawing the CNO's attention to the recent Soviet emphasis on oceanography. Invoking the Soviets early in the Cold War was sure to elicit a response: "Our mastery of the seas is being seriously challenged through the submarine menace and through science by the Soviet Union. We have been unable to make a good estimate of their scientific ability, but their efforts are enormous and they have obviously decided that they must excel in oceanography. The Soviet initial phase is a big show under the auspices of the International Geophysical Year, a scientific activity made to order for their political purposes."[4]

In the conclusion of the report's preface, the three authors also took pains to link disciplinary development in oceanography with the national defense by commenting on the navy's "unique opportunity to supervise and foster the proper development of oceanography." Lill wanted the navy to pay very close attention indeed to oceanography's development. The TENOC report discussed the eleven existing research programs from Miami to Oregon and from Woods Hole to the SIO, as well as the ONR's estimate of the minimum expansion needs in each case. While drawing Admiral Burke's attention to the kind of work performed at each institution, the TENOC report also indicated the current portion of each institution's budget funded by the ONR. In each case, a considerable ONR investment clearly related the research at these ocean science centers to the navy's operational priorities and indicated their considerable dependence upon naval support for survival.

Using Scripps as one example among many, Lill and his colleagues closely associated existing research to potential and significant naval applications. The staff at Scripps studied waves, tides, geomagnetism, circulation systems, and chemical analysis techniques, developed instruments like buoy systems, and took soundings, samples, and feature observations. Commenting that the link to the navy's mission seemed clear, the ONR suggested that the navy spend $1.5 million on a new building, $12 million on two new SCB-185 research ships of roughly 1,000 tons displacement, and $5 million on a single 2,000- to 3,000-ton vessel. If Scripps obtained these assets, Lill and his colleagues reminded the CNO, the institution could more constructively contribute to the knowledge of the environment required by the new sound surveillance system currently under construction by the navy. On-line and listening by 1955, Projects Jezebel, Michael, and Caesar, later dubbed SOSUS, would benefit greatly from research into oceanic circulation, marine acoustics, the development of ultraquiet platforms, as well as the instruments, ships, and deep-ocean vehicles that Scripps could help develop and operate. The TENOC report envisioned the Scripps budget increasing over ten years to triple its 1959 level.

The ONR's vision for the future held similar ambitions for Woods Hole, Lamont, Hudson Laboratories, the new ocean-science programs at the University of Miami and Texas A&M University, as well as other older laboratories like the University of Washington's.

In their determination to place these ideas before Burke, Lill, Maxwell, and Jennings intended that the TENOC program should pave the way for the

NASCO report. In the process, their efforts became a virtuoso performance in cultural translation. The authors and Admiral Bennett wanted to prepare the political and professional environment within the navy for the research and infrastructure proposals as well as level of financial commitment NASCO wanted from the CNO. Gordon Lill and his colleagues gave the ideas and the program suggestions they shared with the NAS committee a naval flavor that would ensure a favorable reception when NASCO submitted its final report to the CNO. The ONR's endeavor ensured that Burke and the naval hierarchy found nothing new or unfamiliar in NASCO's *Oceanography 1960–1970*. Instead, it came to them as the product of a mature professional dialogue.

The endorsement of the TENOC report Admiral Burke sent to Bennett and other naval commands on 1 January 1959, before the appearance of NASCO's *Oceanography 1960–1970* report, demonstrated that Lill's efforts had met with success. The CNO fully supported TENOC. The clarity and effective presentation of the issues by the ONR Geophysics Branch suggested massive expenditure for facilities, personnel, and research but also persuaded the navy's highest-ranking officer of the pressing need. On the subject of the forthcoming NASCO report, Burke reacted in precisely the way Lill had hoped: "while that portion of the Committee on Oceanography report now under preparation pertaining primarily to the navy may differ somewhat from the TENOC (Ten Years in Oceanography) program proposed by reference (c) [the Chief of Naval Research], the latter paper clearly states enough of the problem, its urgency and needs, to serve as a program that would have the endorsement of the National Academy of Sciences."[5]

It was clear to Lill that in giving his support the admiral also demonstrated an appreciation for the scope of the TENOC program. Admiral Burke realized that navy money would create laboratory buildings, office space, and education programs to set new training standards and to establish oceanography as an important specialization in universities and professional schools. For Burke, the commitment fit the mission and addressed the anticipated threat. Since 1950, ASW stood out as one of the navy's highest priorities. In his response to the TENOC report, Burke placed the ONR's requests for oceanography in the category of an ASW priority. Although he realized that his budget had limits, this priority provided as absolute a commitment as possible to oceanography as defined by TENOC and NASCO: "The Chief of Naval Operations endorses the TENOC program as presented . . . and will support it within budgetary limitations in the navy's research and development, shipbuilding, and military construction programs. Since military oceanography is research in direct support of anti-submarine warfare, a commensurate priority within the Navy is assigned."[6]

When NASCO issued its report in 1959, the navy had no difficulty with its provisions. The TENOC program clearly became a cultural bridge useful to both civilian scientists and naval officers as they planned and executed oceanography's future both as a discipline and a component of the national defense.

By 1960, oceanography's fortunes had truly come full circle. Ever since Wardman Park, many working scientists and administrators in federal agencies began meeting informally to share knowledge and resources. Only in this way could they address the responsibilities assigned to their agency in a manner adequate to the challenge. Beginning in 1950 the Informal Ocean Discussion Group (IODG) began a series of occasional meetings to share information and discuss mutual problems. Evolving into the Coordinating Committee on Oceanography (CCO), these agency leaders, sensitive to the importance of oceanography, met as the opportunity permitted to develop a joint sense of federal needs, productive policies, important programs, and potential solutions. From the beginning, the navy played a leading role in this forum through the ONR.

Determined that their agencies collectively act to address pressing issues in oceanography, two of the civilian CCO agencies, the Atomic Energy Commission and the Bureau of Commercial Fisheries, joined with the ONR in the request that turned John Coleman's Nobska-inspired idea into a formal appeal to create NASCO in 1956. These links also enabled Lill, Maxwell, and Jennings to anticipate effectively the contents of *Oceanography 1960–1970* via informal conversations with their longtime friends and colleagues both in NASCO and the CCO before the multivolume report began to emerge from the NAS.

The influence of the CCO and NASCO, as well as the recommendations of *Oceanography 1960–1970,* suggested that great possibilities for success in federal ocean science lay in joint or interagency initiatives. As Capt. Steven Anastasion put it during his term as special assistant to the assistant secretary of the navy for research and development: "The logical conclusion was that there appeared to be several aspects of the various agency programs which might be developed within the framework of a national program. And this was, generally, the conclusion which evolved simultaneously at all upper levels of government, including the Office of the President, the Congress, and several departments."[7]

President Eisenhower had already created the administrative means to bring about the collaboration. In March, 1959, he created the Federal Council for Science and Technology (FCST), which provided the chief executive with a science adviser and a consultation committee on scientific issues of national significance. In May, MIT's James Killian, as science adviser to Eisenhower and chair of the FCST, reacted to the just-published NASCO report by suggesting that the FCST appoint a committee to examine the nation's oceanographic needs and to evaluate the conclusions of the NASCO effort. In July, James H. Wakelin, Jr., the new assistant secretary of the navy for research and development, assumed the chairmanship of the FCST oceanography subcommittee. At summer's end, Wakelin reported to the FCST and the president on his committee's deliberations. The members gave their wholehearted approval to the NASCO report and its call for ambitious educational, research, seagoing, and institutional programs. The subcommittee also called upon the FCST to create within its own council structure a permanent Interagency Committee on Oceanography (ICO) that would promote and guide federal ocean science initiatives. On 22 January 1960, Killian's council acted and the ICO became a permanent part of the Federal Council for Science and Technology. This was clearly the kind of result that the navy-led ICO had sought in vain more than thirty years earlier.

Also inspired by the TENOC and NASCO report deliberations, Congress addressed the importance of the oceans to the national welfare, beginning with the hearings chaired by Rep. George Miller in his House Merchant Marine and Fisheries Committee in March, 1959. This represented only one of a series of bills and resolutions that passed the House to strengthen the nation's ocean resources. With Sen. Warren G. Magnuson's sponsorship of Senate Resolution 136, passed unanimously in June, Congress expressed its desire to attain a greater knowledge of the ocean environment as a matter of federal policy. This encouraged further action within the executive branch.

The membership roster of the 1960 ICO proved even more ambitious than that of 1924, and some of the same agencies that had supported the post–World War I effort joined the new enterprise. The navy led the list of members, with Wakelin assuming the role of chairman, along with regular representation from the ONR, BuShips, the Bureau of Weapons, Hydro, and the army Corps of Engineers' Beach Erosion Board. The director of the Coast and Geodetic Survey represented the Department of Commerce and the interests of his own activity as well as the Weather Bureau and Maritime Administration. The Bureau of Commercial Fisheries director represented the Interior Department's concern with fisheries management, sports, mines, and the activity of the Geological Survey. The Coast Guard's chief of operations brought the Department of the Treasury to the table, and the assistant Surgeon General represented the Health, Education, and Welfare Department. Other regular participants included the AEC, NSF, and the Smithsonian Institution. The State Department, Bureau of the Budget, and NASCO also sent observers. In March, 1961, under the auspices of the Kennedy administration, the new science adviser, Jerome B. Wiesner, helped the ICO compose a formal mission statement and further solidified its position as a component of the FCST.

On 29 March 1961, President John F. Kennedy sent to Congress a request for appropriations to support the first National Oceanographic Program. With help from the ICO, he revealed a multiagency commitment to oceanography and the budgetary needs of numerous federal agencies coordinating their own efforts and programs while reaching out to universities and private ocean-science institutions. In his message to Congress, the president remarked: "Knowledge of the oceans is more than a matter of curiosity. Our very survival may depend upon it."[8]

By 1961 the navy and oceanography had truly come full circle. Harvey Hayes and Theodore Roosevelt, Jr., would have quickly recognized and supported these policies and plans. Although it took another world war and Cold War antagonism to provide a receptive political environment, thirty-seven years of naval initiative and great cultural sensitivity, as well as translation and entrepreneurship by civilian scientists, placed oceanographic priorities on the cabinet table in the White House, on the agenda of both houses of Congress, and on the Pentagon planning board.

The dialogue with the ocean-science community that enabled the navy to assume a leading role in the national program coalesced between 1950 and 1956. Energetic postwar research efforts combined with an ease of communication made possible by translators trained in war. Fueled by Hydro's new oceanographic programs, the ONR's funds, and the navy's ASW priority, as well as the determination voiced at Wardman Park to bring oceanography its own disciplinary identity, an official partnership became possible and desirable. Given Cold War imperatives, the dialogue characteristic of this partnership had an almost immediate effect at Nobska, with TENOC and the creation of NASCO, and in the expectations of *Oceanography 1960–1970*.

While there was still much to be done in terms of disciplinary development and further naval applications, oceanography had achieved national and professional recognition as both an innovative and legitimate scientific inquiry and a vital aspect of the naval warrior's art. The latter was now a composition rendered not only in gray steel images, but also in the ocean's blues—the navy's truly natural hue.

NOTES

Repository and Archive Abbreviations

A&M/OH: Cushing Library, Oceanography Oral History Collection, Texas A&M University, College Station

AM Phil: American Philosophical Society Library, Philadelphia, Pa.

AR/NHC: Operational Archives, NHC

BP:COO: Beginning of Program, NAS: Committee on Oceanography, Organization, 1948

BuShips Centcorr: Bureau of Ships Central Correspondence, 1947, RG 19, NA

BuShips Confidential Centcorr, (year): BuShips Confidential (declassified) Central Correspondence, RG 19, WNRC

BuShips Secret Centcorr, (year): BuShips Secret (declassified) Central Correspondence, RG 19, WNRC

BuShips Secret Gencorr, (year): BuShips Secret (declassified) General Correspondence, RG 19, WNRC

CAH/Ewing: W. Maurice Ewing Papers, Center for American History, University of Texas, Austin

CH/NHC: Contemporary History Branch Collection, NHC

CNA: Archive of the Center for Naval Analysis, Alexandria, Va.

Cominch: Records of the CNO–Headquarters COMINCH Confidential [unclassified], RG 38, NA

DTRC: Navy Laboratories Archive, David Taylor Research Center, Carderrock, Md.

General Board: General Board Studies, 1947–49, RG 45, NA

General Board S/F: General Board Subject Files, 1950–51, RG 45, NA

HO: Hydrographic Office

HO Gencorr: Navy Hydrographic Office General Correspondence, 1907–24, RG 37, NA

HO Gencorr 1924–45: Navy Hydrographic Office General Correspondence, 1924–45, RG 37, NA

InGeo/UT: Institute for Geophysics, Pickles Research Campus Warehouse, University of Texas, Austin

Insurv: Records of the Board of Inspection and Survey, RG 181, WNRC

Lasky: Marvin Lasky Papers, RC 21–5, DTRC

MCZ: Special Collections, Library of the Museum of Comparative Zoology, Harvard University, Cambridge, Mass.

NA: National Archives, Washington, D.C.

NA2: National Archives 2, College Park, Md.

NAS: Archive of the National Academy of Sciences, Washington, D.C.

NAS-Paris: Records of the National Academy of Sciences—Paris Report Series, RG 189, NA

Navoceano: Naval Oceanographic Office, RG 456 (National Imaging and Mapping Agency), NA2

NCB: Naval Consulting Board Correspondence Files, 1915–23, RG 80, NA

NHC: Naval Historical Center, Washington, D.C.

NOSC/TL: Technical Library, Naval Ocean Systems Center, San Diego, Calif.

NSB: Naval Studies Board Archive, NAS

NSWC: Naval Laboratories Archives, Naval Surface Warfare Center, White Oak, Md.

OD/WHOI: Office of the Director, WHOI

ONR: Records of the Office of Naval Research, RG 298, WNRC

OSRD/NDRC: Office of Scientific Research and Development/National Defense Research Committee Records, RG 227, NA2

SecNav Gencorr 1916–26: General Correspondence of the Secretary of the Navy, 1916–26, RG 80, NA

SecNav Gencorr 1926–40: General Correspondence of the Secretary of the Navy, 1926–40, RG 80, NA

SecNav Corr 1927–39: SecNav Confidential [unclassified] Correspondence, 1927–39, RG 80, NA

SIO: Archives of the Scripps Institution of Oceanography, University of California, San Diego, La Jolla

SOC: Submarine Officers Conference, Submarines/Undersea Warfare Division Files, AR/NHC

SONRD Gencorr: Office of Naval Research, General Correspondence of the Coordinator of Research and Development, 1941–45, RG 298, NA

Tharp: Private Collection of Marie Tharp, Oceanographic Cartographer, South Nyack, N.Y.

WHOI: Data Library and Archives, Woods Hole Oceanographic Institution, Woods Hole, Mass.

WNRC: Washington National Records Center, Suitland, Md.

Chapter 1. Selling Bellevue, 1914–24

1. Hydrography and oceanography are different. The former is closely related to physical oceanography, concentrating on physical conditions, boundaries, and currents. The latter is far more comprehensive and includes, among other things, the study of marine life, the physics and chemistry of the ocean, and the geology of the ocean bottom. In the U.S. Navy, the function of the hydrographer was to take his physical oceanographic data and apply that information to the composition of maps and sailing directions to improve navigation. Thus, the naval commitment to oceanography, which is documented and analyzed in this chapter, represents a new and more comprehensive mission for the U.S. Navy Hydrographic Office.

2. Henry Bryant Bigelow, *Oceanography: Its Scope, Problems, and Economic Importance,* 3–7; Susan Schlee, *On the Edge of an Unfamiliar World: A History of Oceanography,* 107 ff; Eric Linklater, *The Voyage of the Challenger;* and Eric Mills, "Problems of Deep Sea Biology," in *The Sea,* vol. 8, *Deep Sea Biology,* ed. G. T. Rowe.

Challenger also demonstrated the significance of government involvement for long-range scientific efforts.

3. Bigelow, *Oceanography,* 7–9.

4. Dean C. Allard, *Spencer Fullerton Baird and the U.S. Fish Commission: A Study in the History of American Science.*

5. Bigelow, *Oceanography,* 8.

6. Henry Bryant Bigelow, "A Developing Viewpoint in Oceanography," *Science,* 24 Jan. 1930, 84–89.

7. Bigelow, *Oceanography,* 8; Eric Mills, "From Marine Ecology to Biological Oceanography," *Helgoländer Meeresuntersuchungen* 49 (1995): 29–44; and box 193, HO Gencorr.

8. Submarine Signal Company to Hydrographic Office, #17518, 1 Apr. 1909; Schlee, *On the Edge,* 265; Bigelow, "Developing Viewpoint in Oceanography," 84–89; and box 193, HO Gencorr.

In 1916, the NAS, at President Wilson's request, established the National Research Council to facilitate research by composing committees of scientific and technical experts in many war-related fields. The goal was to overcome some of the cultural barriers that inhibited scientific participation in mobilization. While the NRC was technically part of the NAS, it operated independently, often causing responsibility disputes and questions about relative authority. This condition persisted until Frank Lillie assumed leadership of both agencies in the mid-1930s.

The following documents relate to the creation of the NRC by executive order and its official relationship to the agencies of the federal government: Hale to President Wilson, 26 Mar. 1918; Wilson to Hale, 19 Apr. and 8 May 1918; Hale to Wilson, 22 Apr. and 10 May 1918; J. F. Tumulty, secretary to the president, to Hale, 13 May 1918; Executive Board: Science Advisory Board, NAS. See also A. Hunter Dupree, *Science and the Federal Government: A History of Politics and Activities to 1940,* 327–28.

9. The term *prosubmarine* is used here to denote research undertaken to enhance the offensive and defensive capabilities of the submarine itself, as opposed to the ASW surface ships.

10. Dupree, *Science and the Federal Government,* 318–19; and Lloyd N. Scott, *Naval Consulting Board of the United States.*

11. Bush was a physicist working for J. P. Morgan's American Radio and Research Corporation of Medford, Mass. During World War II he directed the National Defense Research Council and later the Office of Scientific Research and Development. For further information on this device, which never detected a U-boat in operational use but was successfully tested in America and Great Britain and deployed on British submarine chasers see Vannevar Bush, *Pieces of the Action,* 72 ff.

12. NCB to SecNav Josephus Daniels, 22 and 23 May 1918; and Thomas Robins, NCB secretary, to Daniels, 18 Mar. 1918, box 47, NCB.

13. "Problems Assigned to the Board by Secretary Daniels . . . ," 26 Oct. 1915 and 10 Feb. 1917; Daniels to BuCon, 26 Oct. 1915; Daniels to NCB, 7 and 10 Feb. 1917; and editorial in the *Scientific American,* 11 Aug. 1917, box 47, NCB.

Daniels felt the pressure of editorial columns criticizing his lack of success with the antisubmarine campaign. Even the scientific journals and magazines took up the critical chorus. See also Schlee, *On the Edge,* 245; and Scott, *Naval Consulting Board,* 14–15, 67–83.

14. Robins to Daniels, 13 Apr. 1917, box 47, NCB.

15. For a general discussion of the underwater sound efforts of the Submarine Signal Company, the ancestor of the modern Raytheon Corporation, see *Submarine Signal Log* (Raytheon, 1963). The subcommittee was part of the Special Problems Committee (box 3, Lasky).

16. Schlee, *On the Edge,* pp. 246 ff.; Daniel J. Kevles, *The Physicists,* 118–24; Robert H. Kargon, *The Rise of Robert Millikan: Portrait of a Life in American Science,* 86–87; Scott, *Naval Consulting Board,* 14–15, 67–83; and *History of the Bureau of Engineering, Navy Department, During the World War.*

After the war, only the navy's underwater sound team under the direction of Harvey Hayes, transferred from the Engineering Experiment Station in Annapolis to the NRL in Anacostia, and the Submarine Signal Company remained in the underwater acoustics business.

17. After moving to the United States before World War II, Chilowsky was involved in the exploration of plastic materials to replace the Rochelle salt crystals originally used in his piezoelectric work before and during the Great War. (See note 18 for a definition of the piezoelectric effect.) In October, 1943, he contacted John Tate, head of NDRC Division 6, on this same subject. See Chilowsky to Tate, 29 Oct. 1943, box 61, SONRD Gencorr; and John Herrick, "Subsurface Warfare: The History of Division 6, National Defense Research Committee," chap. 5, NSB.

18. Langevin's device was a primitive transducer designed to send out a conical beam of sound from a surface ship or submarine with sufficient power to produce a return echo. This would enable the monitoring vessel to determine the location of the object causing the echo—whether it be iceberg, animal, ocean bottom, or submarine. The piezoelectric effect, upon which Langevin's work was based, is the process of generating an electric polarization in certain crystals, Rochelle salt for example, by applying mechanical stress.

19. Karl T. Compton was attached to the Research Information Service, sometimes called the Research Information Committee, in Paris. Later in his career he became the president of MIT and served on the NDRC during World War II.

20. "Report of the Conference on Detection of Submarines by the Method of Supersonics," report no. 161, 31 Oct. 1918; and "Report by Professor Morecroft on 'Historical Survey of Developments in the United States'," report no. 161A, 7 Nov. 1918, box 11, NAS-Paris.

21. Dupree, *Science and the Federal Government,* 319. See also Willem Hackmann, *Seek and Strike: Sonar, Anti-submarine Warfare, and the Royal Navy, 1914–54,* passim; Bush, *Pieces of the Action,* 71 ff; and Marvin Lasky, "A Historical Review of Underwater Acoustic Technology 1916–1939 with Emphasis on Undersea Warfare," *U.S. Navy Journal of Underwater Acoustics* 24, no. 4 (Oct., 1924): 559–601.

22. George T. Davis, *A Navy Second to None: The Development of Modern American Naval Policy,* 270 ff; and Robert Gordon Kaufman, *Arms Control in the Pre-Nuclear Era: The United States and Naval Limitation Between the Two World Wars.*

The agreed upon battleship ratio for Great Britain, the United States, Japan, France, and Italy was set at 5:5:3:1.75:1.75. The unit of measure was in increments of one hundred thousand tons.

23. C. E. McClung, chair, Division of Biology and Agriculture, to Henry F. Moore, Bureau of Fisheries, 22 May 1920, NRC Committee on Oceanography, 1919–23, NRC Division of Biology and Agriculture, NAS; and Rexmond C. Cochrane, *The National Academy of Sciences: The First Hundred Years,* 499.

24. Marc I. Pinsel, *150 Years of Service on the Seas: A Pictorial History of the U.S. Naval Oceanographic Office from 1830–1980,* vol. 1, *1830–1946,* 2–9; and Francis Leigh Williams, *Matthew Fontaine Maury: Scientist of the Sea.*

25. Rear Adm. Edward Simpson, Hydro, to BuNav, 23 May 1919, box 303, SecNav Gencorr 1916–26.

26. Asst. SecNav Theodore Roosevelt, Jr., to the secretary of commerce, 7 May 1919, box 303, SecNav Gencorr 1916–26.

27. Daniels to Stephen G. Porter, chairman, House Committee on Foreign Affairs, 5 June 1919; and Daniels to commander, U.S. Naval Forces Operating in European Waters, 7 June 1919, box 303, SecNav Gencorr 1916–26. The conference began on 24 June 1919.

28. Simpson to Daniels via BuNav, 26 Nov. 1919, box 303 SecNav Gencorr 1916–26. See also the enclosure with this document entitled "General Statement Relating to the Proposed Creation of an International Hydrographic Bureau."

29. SecNav Edwin Denby to the secretary of state, 5 Mar. 1921, 8 Sept. 1921, 12 Dec. 1921, and 8 Apr. 1922; Denby to ViceAdm. Albert P. Niblack, n.d., Feb., 1922; Roosevelt to the secretary of state, 2 Aug. 1922; and Denby to the secretary, International Hydrographic Bureau, Monaco, 26 Feb. 1923, box 303, SecNav Gencorr 1916–26. See also Schlee, *On the Edge*, 136–38.

Monaco became a center for oceanographic studies just before the Great War when Prince Albert I established the Oceanographic Museum of Monaco as a center for his research and that of other interested scientists.

30. Roosevelt to Pres. Warren G. Harding, 27 June 1923; memorandum by Rear Adm. Frederic D. Bassett for the SecNav, 28 Aug. 1923; Secretary of Commerce Herbert Hoover and Roosevelt to President Harding, 28 Aug. 1923; and Acting SecNav Theodore D. Robinson to the secretary of state, 2 Aug. 1926, box 303, SecNav Gencorr 1916–26.

Within the navy, Hydro was the primary interested activity, and within the Department of Commerce it was the U.S. Coast and Geodetic Survey. These organizations had the most to gain by participation and effective representation in the bureau. For the vote tally on Niblack's election to the presidency of the IHB, see IHB Secretary-General G. Spicer-Simson, Circular Letter no. 9-R/27-H of 1926, 15 Nov. 1926, box 239, HO Gencorr 1924–45.

31. Coontz was already well aware that Hydro had undertaken a detailed evaluation of the Pacific islands that might provide a strategic or tactical advantage to the fleet in the event of war with Japan. Each island in every island group was evaluated and placed in a system of priorities according to importance. This was entirely for the use of the fleet planners. See Hydro to CNO, Subject: War Portfolio, 7 Feb. 1921, box 145, HO Gencorr.

32. Hobbs to Coontz, 18 May 1921; Coontz to Hobbs, 23 May 1921; and McVay to Brig. Gen. W. D. Conner, U.S. Army, 2 June 1921, box 2191, SecNav Gencorr 1916–26.

33. Annual Meeting, 21–22 Apr. 1922, NRC Division of Geology and Geography, NAS.

34. Schlee, *On the Edge*, 250–51; and Pinsel, *150 Years of Service*, 110–11.

35. Barrows to Littlehales, 11 Jan. 1922; Parker to Barrows, 12 Jan. 1922; memorandum for the SecNav submitted by Herbert E. Gregory, 13 Jan. 1922; and Barrows to Capt. W. E. Parker, 18 Jan. 1922, general files, NRC Foreign Relations: Committee on Pacific Investigations, 1921–24; "Report of the Committee on Pacific Investigations," 21 Apr. 1922, Projects: Hawaiian Bird Island Expedition, NRC Foreign Relations: Committee on Pacific Investigations, 1922–23; and Barrows to Nelson, 6 May 1922, Reports to Division of Foreign Relations, NRC Foreign Relations: Committee on Pacific Investigations, 1922–37, NAS.

36. Nelson to Roosevelt, 5 May 1922, box 920, SecNav Gencorr 1916–26.

37. George W. Littlehales, "In Relation to the Extent of Knowledge Concerning the Oceanography of the Pacific," *Proceedings of the National Academy of Sciences* 2 (1916): 419–21.

38. Nelson to Vaughan, 12 May 1922, Projects: Hawaiian Bird Island Expedition, NRC Foreign Relations: Committee on Pacific Investigations 1922–23, NAS. Vaughan was at that time affiliated with the U.S. Geological Survey. In February, 1923, he became the director of Scripps Institution for Biological Research in La Jolla, California. In 1925 he changed the institution's name to Scripps Institution of Oceanography. It became the country's first great oceanographic institution.

39. Gregory to Nelson, 8 July 1922; and commandant, 14th Naval District to CNO, 19 July 1922, box 920, SecNav Gencorr 1916–26.

40. Roosevelt to Nelson, 20 May 1922; Denby to Wallace, 5 Feb. 1923; CNO to

commandant, 14th Naval District, 8 Feb. 1923; and Wallace to Denby, 8 Feb. and 10 Feb. 1923, box 920, SecNav Gencorr 1916–26.

41. Hayes to officer in charge, EES, 19 Feb. 1923; and Navy 1924–30, NAS—GOVT: AG and Departments, NAS.

42. Ibid. At virtually the same time, Germany's Weimar Republic was contemplating a very similar program with the University of Berlin that eventually culminated in the 1925–27 *Meteor* Expedition.

43. Ibid.

44. Halligan to BuEng, 19 Feb. 1923; Bassett to Denby, 14 Mar. 1923; BuNav to the Sec-Nav, 21 Mar. 1923; acting SecNav to BuNav, 4 Apr. 1923, Navy 1924–30, NAS—GOVT: AG and Departments, NAS.

45. Secretary of state to Denby, "Projects: Hydrographic Expedition—W. M. Davis," 22 Jan. 1923; and Denby to Barrows, 7 May 1923, Committee on Pacific Investigations, 1923, NAS. See also Davis to Denby, 30 Apr. 1923; Fenneman to Denby, 30 Apr. 1923; Bassett to Asserson, 3 July 1923; and Asserson to CNO, 26 Dec. 1923, box 2431, SecNav Gencorr 1916–26. For general naval records on the Pan-Pacific Congress, see file 253880, box 230, HO Gencorr.

46. Barrows to Gregory, 14 July 1924, Projects: Pacific Ocean Expeditions in Cooperation with the Navy Department, NRC Foreign Relations: Committee on Pacific Investigations, 1924, NAS.

47. Gregory to Barrows, 10 Mar. 1924, Projects: Pacific Ocean Expeditions in Cooperation with the Navy Department, NRC Foreign Relations: Committee on Pacific Investigations, 1924, NAS; and commandant, 14th Naval District to SecNav, 10 Mar. 1924, box 920, SecNav Gencorr 1916–26.

48. Projects: Pacific Ocean Expeditions in Cooperation with the Navy Department, NRC Foreign Relations: Committee on Pacific Investigations, 1924, NAS. For the private endorsements, see Thomas Wayland Vaughan (Scripps) to Barrows, 4 Apr. 1924; L. Steiniger (Smithsonian) to Barrows, 3 Apr. 1924; and William Bowie, Department of Geodesy (U.S. Coastal and Geodetic Survey, Department of Commerce) to Barrows, 1 Apr. 1924.

49. Nelson to Prof. Vernon Kellogg, NRC secretary, 3 June 1924; Projects: Pacific Ocean Expeditions in Cooperation with the Navy Department, NRC Foreign Relations: Committee on Pacific Investigations, 1924, NAS; and commandant, 14th Naval District to SecNav, 10 Mar. 1924, box 920, SecNav Gencorr 1916–26.

Chapter 2. The Hayes Initiative Bears Fruit, 1923-25

1. Robert V. Bruce, *The Launching of Modern American Science, 1846–1876* (New York: Alfred A. Knopf), 205 ff.

2. Hayes to officer in charge, EES.

3. Hobbs to Denby, 7 Jan. 1924, Projects: Evaluation of Proposed Studies, Committee on Pacific Investigations, 1924, NRC: Foreign Relations: Naval Oceanographic Expedition—W. H. Hobbs, NAS.

4. Bassett to Gano Dunn, NRC Executive Board chairman, 6 Feb. 1924, NAS.

5. Barrows, memorandum for members of the Committee on Pacific Investigations, 26 Feb. 1924; and Dunn to Bassett, 9 Feb. 1924, NAS.

6. Dunn to Bassett, 9 Feb. 1924.

7. Hydro responded to Denby's request by soliciting opinions in forty different professional scientific and engineering societies. See William H. Hobbs, "A Proposed New 'Chal-

lenger' Exploring Expedition," *Science,* 7 Mar. 1924, 237–38; and NRC: Foreign Relations: Committee on Pacific Investigations, 1924, NAS.

8. Davis to Lawson, 29 Feb. 1924, Navy 1924–30, NAS—GOVT: AG and Departments; Andrew C. Lawson, memorandum for HO, 17 Mar. 1924; Gregory (chairman, Pacific Investigations Subcommittee) to Davis, 26 Mar. 1924, Projects: Pacific Ocean Expeditions in Cooperation with the Navy Department, NRC: Foreign Relations: Committee on Pacific Investigations, 1924, NRC Division of Geology and Geography; and "Research in Oceanography, Annual Meeting of the NRC—Minutes," 26 Apr. 1924, NRC Annual Meeting, 1924, NAS. See also Andrew C. Lawson, "The Continental Shelf Off the Coast of California," *Bulletin of the National Research Council* 8, pt. 2, no. 44 (Apr. 1924): 3–23.

9. Dunn to Bassett, n.d., Feb. 1924; Stejneger to Barrows, 27 Feb. 1924; Barrows to Stejneger, 29 Feb. 1924; Vaughan to Barrows, Bowie to Barrows, Evermann to Barrows, and Merrill to Barrows, 3 Mar. 1924; Swingle to Barrows and Ritter to Barrows, 4 Mar. 1924; Vaughan to Davis, 6 Mar. 1924; Gregory to Barrows, 11 Mar. 1924; and Vaughan to Barrows, 26 Mar. 1924, NRC: Foreign Relations: Committee on Pacific Investigations, 1924; and Vaughan to Barrows, 1 Mar. 1924, general records, 1924–25, Division of Geology and Geography, NAS.

10. Standley to Eberle, 20 June 1924, box 231, ser. 255978, HO Gencorr.

11. Roosevelt, "Letter of the SecNav Inviting the Conference, Navy Department," 2 June 1924, box 2432, SecNav Gencorr, 1916–26.

12. Bigelow spent a great deal of time aboard Fisheries ships before World War I collecting samples for his work at Harvard's Museum for Comparative Zoology (MCZ). It thus was not unusual that the Department of Commerce should call upon him to act as a representative with academic and NAS credentials. See Henry Bryant Bigelow Papers, Special Collections, Library of the MCZ, Harvard University.

13. Littlehales, "Keynote Address," 1 July 1924, box 2432, SecNav Gencorr, 1916–26.

14. "Report of the Conference on Oceanography," 1 July 1924; Bassett to SecNav, 6 Oct. 1924; and Nelson, "Oceanographic Prescience," 168–78, box 2432, SecNav Gencorr 1916–26.

15. "Report of the Conference on Oceanography."

16. Bassett to SecNav, 6 Oct. 1924, box 2432, SecNav Gencorr 1916–26.

17. Ibid.

18. IHB Circular Letter no. 37 of 1922, 29 Sept. 1922; IHB Circular Letter no. 28-H of 1924, 30 Aug. 1924; and IHB Circular Letter no. 28-H of 1925, 3 Nov. 1925, box 239, HO Gencorr.

19. Lord to SecNav Curtis D. Wilbur, 5 Feb. 1925, box 2432, SecNav Gencorr 1916–26.

20. Bassett to Wilbur, 11 Feb. 1925, box 2432 SecNav Gencorr 1916–26.

21. Memo for Dr. Kellogg, Bi-Monthly Report (Dec.–Jan., 1924–25), 2 Feb. 1925, Bi-Monthly Reports, NRC Division of Geology and Geography; Annual Meeting, 25 Apr. 1925; and minutes of the Fifth Annual Meeting of the AGU Section on Oceanography, 30 Apr. 1925, Sections: Oceanography, AGU 1923–33, NRC Executive Board, NAS. See also Helen Raitt and Beatrice Moulton, *Scripps Institution of Oceanography: The First Fifty Years,* 106–107, 114.

Chapter 3. Disappointment and Persistence, 1926–30

1. By 1929, Hydro had signed track agreements with the major international steamship companies to ensure that the charts carried these tracks and that the identity of the ships traveling them was clear. See Cunard to Roberts, 31 May 1929; and Shonerd to commandant, U.S.

Coast Guard, 8 June 1929, box 138, HO Gencorr 1924–45. For examples of the latter, see Sirch to HO, 16 Jan. 1925 and passim (H1–15–307515); and Harlan T. Stetson to HO, 23 Feb. 1927 and passim (H1–15–54401), box 140, HO Gencorr 1924–45.

2. CNO to commandant, 15th Naval District, 26 Feb. 1926, box 845, SecNav Gencorr 1916–26; Crosley to American Geographical Society, 4 June 1926, box 145, HO Gencorr 1924–45.

3. *Submarine Signal Corporation v General Radio et al.,* no. 2450, District Court, D. Massachusetts, 14 F. 2d 178, 20 July 1926; and *Submarine Signal Company v United States,* no. C-318, U.S. Court of Claims, 61 Ct. Cl. 652, 15 Feb. 1926, Library of Congress Legal Research Division.

There were these challenges and yet another from a different quarter. Samuel Spitz, whose company designed and manufactured scientific equipment, inquired as to the nature of the SDF as he had patented a similar device. For all three cases see, Spitz to Acting SecNav, 1 Aug. 1922; reply, 14 Aug. 1922; Submarine Signal Co. to Navy Department, 5 June 1924; Wilbur to Submarine Signal Co., 8 July 1924; Hayes to SecNav (with enclosure), 9 and 25 June 1925; BuEng to SecNav (JAG), 19 June and 9 July 1925; Robinson to JAG, 1 Sept. 1925; U.S. asst. attorney general to SecNav, 4 Sept. 1925, 1 Oct. 1925; BuEng to U.S. Naval Attaché, Berlin, 17 Oct. 1925; BuEng to JAG, 3 Nov. 1925; and Wilbur to Sargent, 19 Nov. 1925, box 2191, SecNav Gencorr 1916–26 (for further material, examine the 26817 correspondence series). See also Hans Maurer, *Die Echolotungen des "Meteor,"* Band 2, *Wissenschaftliche Ergebnisse der Deutschen Atlantischen Expedition auf dem Forschungs und Vermessungsschiff Meteor, 1925–1927.*

Hayes continued to refine the SDF for the navy at the NRL, and interested scientific groups and shipping concerns turned to Submarine Signal for commercial depth measuring devices. In the case of the famous German *Meteor* Expedition of 1925–27, for example, Prof. Alfred Merz of the University of Berlin approached Atlas Werke for the Submarine Signal Fathometer when obtaining a version from the NRL became impossible.

4. Lillie to Rose, 3 Dec. 1925, NAS Committee on Oceanography, 1928, NAS. Lillie succeeded C. O. Whitman as director of the Marine Biological Laboratory and assumed the chairmanship of the University of Chicago Zoology Department in 1910 at the age of forty. See Cochrane, *National Academy of Sciences,* 369–71.

5. "Temple Bill: Short Synopsis of the Origins of the Committee on Oceanography," n.d., ca. 1932, NAS Committee on Oceanography, general, 1928–29, NAS.

6. Report of the NRC Division of Geology and Geography for the year 1925–26, 24 Apr. 1926, NRC Division of Geology and Geography, general, 1926–27, NAS.

7. AGU Program and Order of Business for the Annual Meeting, 1926, Sections: Oceanography, American Geophysical Union 1923–33, NRC: Executive Board, NAS.

8. Vaughan to Barrows, 8 Feb. 1927, general, NRC: Foreign Relations: Committee on Pacific Investigations, 1925–29, NAS.

9. Thomas Wayland Vaughan, "International Cooperation in Oceanographic Investigations in the Pacific" (paper presented at the Third Pan-Pacific Science Congress, Tokyo, 1926), Pan-Pacific Science Association, Committee on the Oceanography of the Pacific, general, International Organizations, 1927–39, NRC Division of Foreign Relations, NAS.

10. Annual Report 1926, app. D, "Report of the Committee on Submarine Topography and Structural History of the Caribbean-Gulf Region, 24 Apr. 1926"; and Kellogg to President Coolidge, "Resolution on Gravity at Sea," 10 May 1926, NRC Division of Geology and Geography, NAS.

11. This committee was part of the Division of Geology and Geography.

12. The Coast Guard created the Ice Patrol in 1912 after the RMS *Titanic* disaster. Its purpose was to plot the position and movement of major icebergs along the North Atlantic commercial shipping routes.

13. Annual Report, 1926, app. E, "Report of the Committee on Submarine Configuration and Oceanic Circulation, 24 Apr. 1926," NRC Division of Geology and Geography, NAS.

14. Memorandum for Dr. Kellogg: Bi-Monthly Report, 1 June 1926; Bi-Monthly Reports, 1924–26; and memorandum for members of the Executive Committee, NRC Division of Geology and Geography, 22 Oct. 1926, general, 1926–27, NAS.

15. Thomas Wayland Vaughan, "International Committee on the Oceanography of the Pacific" (paper presented to the Conference Celebrating the Centennial of the Geographic Society, Berlin, 24–26 May 1928), Pan-Pacific Science Association, Committee on the Oceanography of the Pacific, general, International Organizations, 1927–39, NRC Division of Foreign Relations, NAS.

16. Lillie to Rose, 11 May 1927; and Vaughan to White, 26 May 1927, Committee on Oceanography 1927, NAS.

17. Lillie to committee members, 15 June and 8 Sept. 1927, NAS.

18. Vaughan to Lillie 9 Oct. 1927, NAS.

19. Lillie to Rose, 3 Dec. 1925, NAS; and Lillie, "Third Circular Letter to the Committee on Oceanography," 21 Oct. 1927, Committee on Oceanography 1928, NAS.

20. Lillie to White, 22 Nov. 1927; and Lillie to Rose, 15 Dec. 1927, Committee on Oceanography 1927, NAS.

21. Frank Lillie, "History of the Oceanographic Project Now in the Hands of the Committee on Oceanography of the National Academy of Sciences," 6 June 1928, Committee on Oceanography 1928, NAS.

22. White to Lillie, 7 and 13 Dec. 1927; Lillie to Brockett, 8 Dec. 1927; and Lillie to White 9 and 16 Dec. 1927, Committee on Oceanography 1927, NAS.

23. Brierley to Lillie, 8 June 1928, general, 1928–29, Committee on Oceanography, NAS.

24. The following sources provide a good sense of just how the committee was recruited and assembled: Lillie to members of the Committee on Oceanography, 22 Dec. 1927, 15 and 27 Feb. 1928, and 5 Apr. 1928, Committee on Oceanography 1927; Lillie to general Education Board, 26 Mar. 1928; telegram, Lillie to White, 27 Feb. and 26 Mar. 1928; telegram, White to Lillie, 24 Mar. 1928; and minutes of the Committee on Oceanography meeting, 22 Apr. 1928, NAS.

25. Houghton Mifflin published the report in 1931 under the title, *Oceanography: Its Scope, Problems, and Economic Importance.* For more on this, see Lillie to Brockett, 28 July 1928; Lillie to Bigelow, 6 Sept. 1928; Lillie, "Second Report of the Committee on Oceanography to the National Academy of Sciences," 9 Oct. 1928; and "Temple Bill," Committee on Oceanography, 1928–29; and Lillie, to members of the Committee on Oceanography, 7 Sept. 1928 (plus appendices); Committee on Oceanography meetings, 1927–28, NAS.

26. Memorandum for hydrographer, 11 June 1928; Kempff to chief, BuNav, 11 Nov. 1929; Gherardi to Wheat, 26 June 1930; Jensen to asst. hydrographer, 25 June 1931 (note comment on aerial surveys); G. Medina, memorandum for Commander Jensen, 25 June 1931 (note comment on aerial surveys); and HO memorandum for the officer in charge, Division of Chart Construction, Subject: *Hannibal* and *Nokomis* survey season 1931–32, 27 June 1931, box 145, HO Gencorr 1924–45; and Annual Report, 28 Apr. 1928, NRC Division of Geology and Geography, NAS.

27. Kempff to SecNav, 21 Oct. and 16 Nov. 1927; Freeman, memorandum for hydrog-

rapher, Subject: Gravity Determination by Submarine, 16 Aug. 1928; and Roberts to SecNav, 10 Sept. 1928, box 145, HO Gencorr 1924–45; Annual Report, 27 Apr. 1929, NRC Division of Geology and Geography, NAS; and Schlee, *On the Edge,* 334. Note in this documentation Hydro's desire to learn more about the measurement techniques and operation of the Vening-Meinesz device.

For an analysis of these events that suggests the efforts of Bowie, Wright, Collins, and Vening-Meinesz laid the groundwork for a scientific tradition in geophysics see Naomi Oreskes, "Weighing the Earth from a Submarine: The Gravity Measuring Cruise of the U.S.S. *S-21,*" in *The Earth, the Heavens, and the Carnegie Institution of Washington: Historical Perspectives After Ninety Years,* ed. Gregory Good. See also F. A. Vening-Meinesz, "Gravity Expedition of the U.S. Navy," *Nature,* 23 Mar. 1929, 473–75; and F. A. Vening-Meinesz and F. E. Wright, *The Gravity Measuring Cruise of the U.S. Submarine S-21,* esp. chaps. 1 and 2.

28. Naval Board on Oceanography to SecNav, 22 Mar. 1929, Committee on Oceanography, general, 1928–29, NAS; Thomas Wayland Vaughan et al., *International Aspects of Oceanography,* preface.

29. BuEng memorandum, Subject: Echo Sounding Systems Installed on Government Vessels, 9 May 1929, box 833, SecNav Gencorr 1926–40; and John A. Fleming to SecNav, 18 Oct. 1932, with attached copy of the Schofield Board report, 22 Mar. 1929, and a cover letter dated 1 Dec. 1932, box 141, HO Gencorr 1924–45.

30. Naval Board on Oceanography to SecNav, 22 Mar. 1929, Committee on Oceanography, general, 1928–29, NAS.

31. Susan Schlee, "Real Men Won't Winter on Cape Cod—A Bit of Unofficial W.H.O.I. History," *Woods Hole Notes* (Feb., 1973), Administration/Publications, 1973; Mason to Lillie, 14 Nov. 1929; Lillie to White, 23 Nov. 1929; NAS Annual Meeting 1930, Report of the Committee on Oceanography, 28 Apr. 1930, NAS Committee on Oceanography 1930–31; White to Mason, 2 Dec. 1929; and Brockett to Littlehales, 8 Nov. 1929, NAS Committee on Oceanography, general, 1928–29; and minutes of the meeting of the Committee on Oceanography, 12 Oct. 1929, Committee on Oceanography meetings, 1927–29, NAS.

Bigelow, "Developing Viewpoint in Oceanography," 84–86; Cochrane, *National Academy of Sciences,* 497–502; Deacon, "The Woods Hole Oceanographic Institution: An Expanding Influence," 25–31; Schlee, "The R/V Atlantis and her First Oceanographic Institution," 49–56; Burstyn, "Reviving American Oceanography: Frank Lillie, Wickliffe Rose and the Founding of the Woods Hole Oceanographic Institution," 57–66; Haedrich and Emery, "Growth of an Oceanographic Institution," 67–82.

The R/V *Atlantis* was built in Copenhagen as a steel-hulled sailing ketch. It came equipped with a diesel auxiliary engine, but sail was its primary means of propulsion. The two alternatives were considered the safest for exploration at sea.

32. Minutes of the Committee on Oceanography meeting, 27 Apr. 1930; and NAS Annual Meeting 1931, Report of the Committee on Oceanography 1930–31, NAS.

33. "Report of the Committee on Submarine Configuration and Oceanic Circulation," 3 May 1930; and Jahncke to Burgess, 17 Oct. 1930, Annual Meeting of the NAS Division of Geology and Geography, 1930; and Gherardi, memorandum for the chief, BuNav, 19 Sept. 1930, Navy 1924–30, NAS—GOVT: AG and Departments, NAS.

34. "Report of the Committee on Submarine Configuration and Oceanic Circulation."

Chapter 4. Common Practice and Uncommon Business, 1930–40

1. Recall that WHOI was created in 1930 and Bigelow assumed the directorship at that time.

2. Bigelow to Gherardi, 8 Sept. 1932; Gherardi to Bigelow, 9 and 30 Sept. 1932; and Bigelow to Jensen, 20 Sept. 1932, box 2, Administrative and Individuals, 1930–35, OD/WHOI.

3. Gherardi to Bigelow, 10 and 21 Oct. and 8 Nov. 1932, 27 Jan. 1933, and 17 Apr. and 28 May 1935; Bigelow to Gherardi, 20 Apr. and 24 Oct. 1932 (for a sense of the nature and extent of WHOI's working relationship with Hydro), 6, 14, and 19 Jan., 1 and 7 Feb., and 14 Aug. 1933; Iselin to Gherardi, 20 Oct. 1932; Church to Hydro, 8 July 1931; and Campbell to Sec-Nav, 17 Apr. 1935, box 2 Administrative and Individuals, 1930–35, OD/WHOI.

4. Gherardi to Vaughan, 28 July 1931 and 3 and 16 Aug. 1932; Vaughan to Gherardi, 17 May 1932; Moberg to Gherardi, 15 Sept. and 14 Dec. 1932; Gherardi to commissioner, Bureau of Fisheries, 20 Sept. 1932; Moberg to Jensen, 1 Oct. 1932; Gherardi to Moberg, 14 Oct. and 9 Dec. 1932; Gherardi to SecNav, 16 Nov. 1932; memorandum for the hydrographer, 22 Nov. 1932; HO (Gherardi) to USS *Hannibal,* telegram, 25 Nov. 1932; and commander, USS *Hannibal,* to HO, 5 Dec. 1932, box 141, HO Gencorr 1924–45.

5. ComEleven Action, BuNav Info, Hydro Office, telegram, 7 Jan. 1933; Gherardi to Moberg, 12 Jan. and 3 Oct. 1933; Moberg to Gherardi, 17 Jan. and 14 Oct. 1933; Gherardi to Bigelow, 27 Jan. 1933; commander, USS *Hannibal,* to HO, 31 Mar. 1933; Moberg to Vaughan, 18 Apr. 1933; and Vaughan to Gherardi, 2 and 20 Dec. 1933, box 141, HO Gencorr 1924–45.

6. "Aleutian Islands Survey Expedition, Final Report, 19 Jan.–4 Oct. 1933," box 104, SecNav Corr 1927–39; and HO to commander, USS *Patoka,* 18 Apr. 1933, box 141, HO Gencorr 1924–45.

7. "Aleutian Islands Survey Expedition, Final Report"; Gherardi to commander, Aleutian Islands Expedition, 3 May 1933; Soule to HO, 17 Jan. 1934; Robinson to HO, 21 Feb. 1934; Gherardi to Thompson, 2 Apr. 1934, box 141, HO Gencorr 1924–45.

8. Vaughan to Committee on the Oceanography of the Pacific, 23 Oct. 1934; "Studies Japan Current," 11 Oct. 1934, 2 ff; and Vaughan et al., "Report of the Committee on Oceanographic Activities of North America during Apr. 1933 to Apr. 1934," AGU, 1934, International Organizations 1927–39, NRC Division of Foreign Relations, NAS.

Vaughan to SecNav, 30 Apr. 1934; Taussig to Vaughan, 5 and 18 May and 1 June 1934; Vaughan to Taussig, 12 and 24 May 1934; and CNO to commander, *Bushnell,* 18 May 1934, box 833; Gherardi to commander, Aleutian Islands Expedition, 12 Apr. 1934, box 839, SecNav Gencorr 1926–40.

Sellers to CNO, 29 Jan. 1934; commander, Minecraft, Battleforce to commander in chief, U.S. Fleet, 13 Jan. 1934, box 104, SecNav Corr 1927–39.

Commander, Aleutian Islands Survey Expedition to hydrographer, 9 May 1934, box 142; Leahy to commander, USS *Louisville,* 19 Aug. 1936; and R. C. Needham, memorandum for files (detailed information with enclosures on the procedures and equipment for dynamic ocean surveys), 9 June 1930, box 143, HO Gencorr 1924–45.

Raitt and Moulton, *Scripps,* 117.

9. Pan-Pacific Science Association, general, Committee on the Oceanography of the Pacific, International Organizations 1927–34, NRC Division of Foreign Relations, NAS; Vaughan to Barrows, 2 July 1934; enclosure with Vaughan to Gherardi, 29 June 1934; Gherardi to Vaughan, 18 Dec. 1934; Thompson to Leahy, 1 Aug. 1935; and Thompson to HO, 7 Oct. 1935, box 142, HO Gencorr 1924–45.

10. Bowie to Bigelow, 16 and 29 May 1931; and Bigelow to Bowie, 1 June 1931, box 1; Campbell to SecNav, 17 Apr. 1935, box 2, Administrative and Individuals, 1930–35, OD/WHOI.

In 1923, Vening-Meinesz used the Netherlands submarine K-II on an expedition to Indonesia, and in 1926 he was aboard the K-XIII headed for the southwest Pacific once again, only this time via the Caribbean and Panama Canal.

11. Annual Report of the Committee on Submarine Configuration and Oceanic Circulation, 23 Apr. 1932, Annual Meeting of the NAS Division of Geology and Geography, 1932, NAS; Thomas Townsend Brown, "The Navy-Princeton Gravity Expedition to the West Indies in 1932," *Hydrographic Review* 11, no. 2 (1934): 82–89; George W. Littlehales, "The International Scientific Expedition to the West Indies in 1932," *Hydrographic Review* 9, no. 2 (1932): 124–28; and William Bowie, "Weighing the Earth from a Submarine," *Scientific American*, Mar., 1929, 218–21; and *The Navy-Princeton Gravity Expedition to the West Indies in 1932*.

12. Oreskes, "Weighing the Earth from a Submarine."

13. The O-12 displaced 491 tons on the surface and was built between 1916 and 1918 by the Lake Submarine Torpedo Boat Company of Bridgeport, Conn. Simon Lake also participated in the alterations made to the O-12, employing his experience with the *Argonaut's* diver lockout feature to permit the gathering of samples and the use of instruments while submerged in the O-12.

14. Sverdrup to Bigelow, 7 Oct. 1931; and "Rough Map of the Arctic Region Traversed by Wilkins and a Schedule of Gravity Stations Taken by the Expedition, Aug.–Sept. 1931," box 2, Administrative and Individual, 1930–35, OD/WHOI; and Schlee, *On the Edge,* 276–78.

15. Bigelow to Sverdrup, 28 Oct. 1931 and 11 Jan. 1932; and Sverdrup to Bigelow, 17 Dec. 1931, box 2, Administrative and Individuals, 1930–35, OD/WHOI.

16. Field to Bigelow, 11 Oct. 1932 and 13 June 1933; Bigelow to Field, 24 Oct. 1932 and 21 June 1933; Field to Gerhardi, 15 May 1933; Gerhardi to Field, 8 June 1933; and Field to Bowie, 24 June 1933, box 1; Sverdrup to Bigelow, 9 Jan. 1932, 31 July and 17 Oct. 1933; and Bigelow to Sverdrup, 15 Apr., 17 July, 7 Sept., 4 Oct., and 3 Nov. 1933, box 2, Administrative and Individuals, 1930–35, OD/WHOI.

17. Field to Bigelow, 23 June 1933, box 1, OD/WHOI.

18. Allyn Vine interview, 30 Aug. 1993, author's personal collection; and N. H. Heck, "The Role of Earthquakes and the Seismic Method in Submarine Geology," *Proceedings of the American Philosophical Society* 79 (Apr., 1938): 97–108.

19. Bigelow to Field, 18 Nov. 1935, box 1, Administrative and Individuals, 1930–35, OD/WHOI; and Heck to Ewing, 10 June 1935; and J. M. Smook, "U.S. Coast and Geodetic Survey Special Report no. 1, 1935: Deep Sea Electric Bombs," box 151, CAH/Ewing.

20. The foredeep is a deep, elongated oceanic depression fronting a mountainous land area. It is a trench if it has steep sides, and a trough if it has gently sloping sides. For information on the region explored by the expedition see Harry H. Hess, "Geological Interpretation of Data Collected on Cruise of USS *Barracuda* in the West Indies—Preliminary Report," *Transactions of the American Geophysical Union,* 1937, 69–77.

21. "Report of the Special Committee on Geophysical and Geological Study of Oceanic Basins," 1936, box 161, CAH/Ewing.

22. Ewing to Leahy, 10 June 1936; Leahy to Ewing, 29 July 1936; Ewing to Gorry, 8 Oct. 1936; Ewing to superintendent, U.S. Naval Observatory, 14 Oct. 1936; and "Navy Gravity Expedition," n.d., box 109; Maurice Ewing, "Gravity Measurements on the USS *Barracuda,*" *Transactions of the American Geophysical Union,* 1937, 66–69; Ewing to Vening-Meinesz, 27 July 1936; and Field to Ewing, 17 Aug. 1936, box 161, CAH/Ewing; and Field to Bigelow, 20 Jan. 1936, box 3, Office of the Director, Administrative and Individuals, 1936–40, WHOI.

23. Ewing, "Gravity Measurements," 66–69; Hess, "Geological Interpretation," 69–77; Harry H. Hess, "Gravity Anomalies and Island Arc Structure with Particular Reference to the West Indies," *Proceedings of the American Philosophical Society* 79 (Apr., 1938): 71–96; and the itinerary contained in the program for the 1936 Christmas dinner menu composed by the crew of the USS *Barracuda,* box 109, CAH/Ewing.

24. Field to Bigelow, 15 Jan., 19 Feb., and 22 Mar. 1937; Field to Ewing, Heck, and Iselin, 19 Feb. 1937; and Bigelow to Field, 10 Mar. 1936 and 13 Mar. 1937, box 3, Administrative and Individuals, 1936–40, OD/WHOI.

Ewing had a hard time securing sufficient funding for his seismic work. Initially, his grant from the Geological Society of America was not renewed in 1937 and the research would have foundered if not for support from Field and the AGU. Ewing seems to have had some success in getting the GSA to reverse its decision to not support his work from the Penrose Fund (Ewing to Bigelow, 1 Feb. 1937, box 3, Administrative and Individuals, 1936–40, OD/WHOI).

25. J. Lamar Worzel interview, 22 Aug. 1990, NHC Oral Histories, AR/NHC.

26. Maurice Ewing, "A Plan for the Use of the Ship *Atlantis* in Geophysical Studies of the Continental Shelf," 24 July 1935; Field to Bigelow, 24 July, 13 and 30 Aug., and 2 Dec. 1935; and Bigelow to Field, 19 Aug., 4 Sept., and 18 Nov. 1935, box 1, Administrative and Individuals, 1930–35, OD/WHOI. These documents reveal Bigelow's concern for *Atlantis* given the literally explosive nature of Ewing's seismic method.

27. "Cable Connected OBS (1937–1938)," and "Free Floating OBS" from the taped and transcribed recollections of J. Lamar Worzel with introductory remarks dated 10 Dec. 1980, Tharp.

Marie Tharp worked for years at Columbia University's Lamont-Doherty Geological Observatory. Together with geophysicist Bruce Heezen, Tharp was responsible for some of the most intricate and advanced ocean-floor maps available in the 1960s and 1970s. See deck log of *Atlantis,* Cruise 81, 8–21 Sept. 1938, WHOI.

28. Barrows to Bowen, 6 Apr. 1940; Harrison to Colbert, 7 Nov. 1940; Colbert to Harrison, 13 Nov. 1940; Bowman to Jewett, 21 May 1943; L. M. to Bowman, 30 May 1943; and Bowman to Byrd (Draft), 3 June 1943, general, Committee on U.S. Antarctic, Advisory Expedition 1943, NRC: Executive Board, NAS; and G. E. Fogg, *A History of Antarctic Science,* 162–64.

Chapter 5. Research, Relationships, and Policy: 1930–40

1. Michelson graduated from the academy on 31 May 1873 and resigned from the navy on 30 Sept. 1881. During the Great War he received a reserve commission and was eventually promoted to commander on 13 May 1919. By 1892 he was head of the Physics Department at the newly established University of Chicago.

On heterogeneous engineering, see Donald MacKenzie, *Inventing Accuracy: A Historical Sociology of Nuclear Missile Guidance,* 28–29. Note the similar observation on pages 4–5 in Harvey M. Sapolsky, *Science and the Navy: The History of the Office of Naval Research.*

2. Lillie to Brockett, 17 Dec. 1931; Lillie to Bowie et al., 18 Dec. 1931; Brockett to Brierley, 19 Dec. 1931; "Statement of Funds from general Education Board for Work of Academy Committee on Oceanographic Research, 14 Dec. 1931"; and Gardiner, "Review of H. B. Bigelow's *Oceanography,*" 551–55, Committee on Oceanography 1930–31, NAS.

3. Capt. C. S. McDowell, "Naval Research," 1928, Agencies and Departments: Navy, general, 1933–35, NAS. An EDO is an engineering duty officer. These officers were designated

by choice for an engineering career within the navy that would give no opportunity for a seagoing command.

4. David K. van Keuren, "Science, Progressivism, and Military Preparedness: The Case of the Naval Research Laboratory, 1915–1923," *Technology and Culture* 33 (Oct., 1992): 729–30; and Thomas Parke Hughes, *Elmer Sperry: Inventor and Engineer,* 89 ff.

The manner in which the naval bureaus used the NRL might have agreed more closely with Edison's personal intent for the facility. The great Progressive Era inventor envisioned the laboratory as a test facility and machine shop—a place for practical innovation based upon knowledge already available in the public domain through international scientific investigation. The research mandate given the laboratory by the Naval Consulting Board came via a majority report not supported by the "Wizard of Menlo Park."

5. Senior Member of the General Board to SecNav, "Naval Policy Regarding Research and the Naval Research Laboratory," Agencies and Departments: Navy, general, 1933–35, NAS.

6. Wheeler to Robins, 7 July 1933, NAS; and A. Hoyt Taylor, *The First Twenty-Five Years of the Naval Research Laboratory,* 29–30. Briefly cut off from federal service, Wheeler later distinguished himself on the staff of the Federal Communications Commission.

7. Office memorandum no. 5, "The Situation at the *Naval Research Laboratory,* as discussed with Dr. L. P. Wheeler, Monday evening, July 17, 1933, by Isaiah Bowman"; confidential memorandum (no longer classified) by Lynde P. Wheeler, 19 July 1933; and Jewett to McDowell, 19 July 1933, Agencies and Departments: Navy, general, 1933–35, NAS.

8. Schlee, *On the Edge,* pp. 246 ff.; Kevles, *Physicists,* 118–124; Kargon, *Rise of Robert Millikan,* 86–87; Scott, *Naval Consulting Board,* 67–83; and *History of the Bureau of Engineering,* 47–54.

The Mason apparatus focused the sound drawn from the water to ascertain its source. To determine the direction from which the sound came, the operator needed only to seek the maximum output on his earphones by turning a dial. It emerged as one of the navy's first practical instruments for submarine detection.

9. Coolidge to McDowell, 14 July 1933, Agencies and Departments: Navy, general, 1933–35, NAS.

10. Office memorandum no. 10, "Report of Conferences with Captain C. S. McDowell, U.S.N., concerning proposed *Naval Research Committee,* July 19 and 20, 1933 by Isaiah Bowman," Agencies and Departments: Navy, general, 1933–35, NAS.

11. Office memorandum no. 8, "Interview with Admirals Leigh and Standley, 2 P.M., July 20, 1933," Agencies and Departments: Navy, general, 1933–35, NAS.

12. The National Recovery Act became law on 16 June 1933.

13. Lewis E. Auerbach, "Scientists in the New Deal: A Pre-War Episode in the Relations between Science and the Government in the United States," *Minerva* 3 (summer, 1965): 459–61; and Cochrane, *National Academy of Sciences,* 349–51.

14. Office memorandum no. 18, "Conference with Doctors L. P. Wheeler and Harvey C. Hayes, of the Naval Research Laboratory, Sunday evening, July 23, 1933, by Isaiah Bowman," Agencies and Departments: Navy, general 1933–1935, NAS.

15. NRC: Executive Board–Science Advisory Board, Appointments: Members: Executive Orders 6238 and 6725, 1933–34, NAS; Auerbach, "Scientists in the New Deal," 459–61; and Cochrane, *National Academy of Sciences,* 349–51.

16. For details on the NAS-SAB conflict and the extent to which the debate continued over the entire life of the latter agency (1933–35) see Wilson to Merriam, 26 Jan. and 13 Aug. 1935; Compton: "Memorandum of Conversation with Dr. Campbell on the second day of the

February meeting of the Science Advisory Board," 18 Mar. 1935; Merriam to Wilson, 29 Apr. and 29 July 1935, Relationship of SAB to NAS—1935; W. W. Campbell—NAS files, "Conference with Professor Robert A. Millikan in the University Library building, Room 110, Berkeley, Thursday, 21 June, 1934, 4:20 to 5:35 pm"; Campbell, memorandum for files, 1934; Cushing-Wilson correspondence, July–Aug. 1934; Campbell to Compton 28 Aug. (with attached memo) and 29 Oct. 1934; Compton to Campbell, 14 Sept., 17 Oct., and 14 Nov. 1934; Day to Campbell, 25 Oct. 1934; and Millikan to Compton and Bowman, 14 Dec. 1934, Executive Board: Science Advisory Board, Relationship of SAB to NAS—1934, NAS.

17. Cochrane, *National Academy of Sciences,* 351–53.

18. Excerpt from the first report of the SAB, 31 July 1933 to 1 Sept. 1934, Joint with Science Advisory Board—1934, Committee on Naval Research, NRC: Executive Board, NAS.

19. Almy to McDowell, 22 Nov. 1933, NRC: Executive Board–Science Advisory Board, general, Committee on the War and Navy Departments, 1933–35, NAS.

20. McDowell to Bowman (with attached memo entitled "Naval Research Committee"), 25 July 1933; agenda and attached papers for the SAB Executive Committee meeting 9–10 Nov. 1933, NRC: Executive Board–Science Advisory Board, Executive Committee—Meetings; and McDowell to Jewett, 28 Nov. 1933, Agencies and Departments: Navy, general, 1933–35, NAS.

21. Standley to Bowman, 28 Dec. 1933; Barrows, office memorandum no. 66, 15 Jan. 1934; Barrows to Marquart, 27 Jan. 1934; Bowman to Millikan, 24 Feb. and 30 Nov. 1934; and Wilson to Millikan, 3 May 1934, NRC: Executive Board–Science Advisory Board, general, Committee on the War and Navy Departments 1933–35, NAS.

22. Barrows to Millikan, 13 Nov. 1934, NRC: Executive Board–Science Advisory Board, general, Committee on the War and Navy Departments, 1933–35, NAS.

23. Memorandum, Lewis to Barrows, 25 Oct. 1934; and Barrows to Millikan, 26 Dec. 1934, NRC: Executive Board–Science Advisory Board, general, Committee on the War and Navy Departments, 1933–35, NAS; folder 7, Jan.–June 1937 correspondence (on fouling research for the navy), box 1, Office of the Director (Sverdrup),1936–48, SIO; Columbus O'Donnell Iselin, "WHOI History During the War Years, 1941–1950," ca. 1960, 1–2; and Harrison to Jewett, 6 Sept. 1940, box 4, Administrative, Institutions, Ships, Miscellaneous, and Individuals, OD/WHOI.

24. Congress established the NACA in 1915 to "supervise and direct the scientific study of the problems of flight." See James Phinney Baxter III, *Scientists Against Time,* 13 ff.

25. "A Proposal for Research Mobilization for the Navy Department" prepared by Lt. Comdr. R. W. Gruelick, 9 Nov. 1934; and memorandum concerning mobilization of research for the Navy Department, 10 Nov. 1934, Executive Board: Committee on Naval Research, Committee on Naval Research, Joint with the Science Advisory Board—1935 (EB:CNR), NAS.

26. NRC, "Memorandum upon Possibilities for Scientific Services to the Navy Department," final draft, 18 Mar. 1935, EB: CNR, NAS.

27. Bowman to Compton, 19 Feb. 1935, EB: CNR, NAS.

28. Barrows, "Memorandum on Navy Plan," EB: CNR, NAS.

29. Bowman (NRC) and Compton (SAB) to chair, Navy Department Liaison Committee on Naval Research, 18 Mar. 1935; and NRC, "Memorandum upon Possibilities for Scientific Services to the Navy Department," final draft, 18 Mar. 1935, Agencies and Departments: Navy, Committee on Naval Research: Liaison with the NRC 1933–35, NAS.

30. Brockett to Walker, 8 Jan. 1936, Executive Board, Committee on Naval Research, Joint with Science Advisory Board—1935; "Projects: Scientific Personnel for Naval Nominations,"

15 July 1936; and "Projects: Research Conducted at Colleges and Universities—Responses," NAS: Committee on Oceanography, general, 1936–37, NAS.

31. Lillie to Fosdick, 29 Nov. 1937, NAS.

32. Lillie served as NAS president from 1935 to 1939 and as NRC chairman to 30 June 1936, when the chairmanship reverted to Ludvig Hektoen.

33. On the Navy Department Council for Research see the file entitled "NAS: Committee Advisory to Naval Bureau of Engineering—1939," NAS.

34. NRC office memorandum no. 619 (Barrows), 26 Sept. 1939, Executive Board, Committee on Relationships with the War and Navy Departments—1936–39 (EB: CRWND), NAS. Hooper and the NAS/NRC were still discussing scientific personnel, expansion of university relationships in a very fundamental way three weeks into the European war that began on 1 September 1939.

35. Jewett to Day, 18 Sept. 1939, EB: CRWND, NAS. Ross Harrison was the chairman of the NRC from 8 February 1938 through 30 June 1946. Robert Millikan, chairman of the committee, was away on a scientific expedition in the South Pacific.

36. Harrison to Bush, 15 June 1940, executive files, NDRC, Council of National Defense, NAS. In a short congratulatory note to Bush written in mid-June, Ross Harrison confirmed that he was of a common mind with Jewett and welcomed the NDRC as a solution to the difficulties of scientific relations with the military and the mobilization of science for national defense. As far as Harrison was concerned, Bush and company had provided a solution just in time.

37. Bush, *Pieces of the Action*, 31–33.

38. "Organic Documents Relating to the Establishment of the NDRC," June, 1940, Executive Office files, NDRC 1940, Council of National Defense, NAS; and Irvin Stewart, *Organizing Scientific Research for War: The Administrative History of the Office of Research and Development*, 7.

At the president's behest, three others joined the founders to complete the membership of the National Defense Research Committee. Conway P. Coe, an attorney and commissioner of patents, joined as the fifth civilian member. Rear Adm. Harold G. Bowen and Brig. Gen. George V. Strong represented the navy and army.

Interpolation. Interwar Observations

1. "The Oceanographic Activities of the Hydrographic Office and the United States Navy During Apr. 1935 to Apr. 1936," 1936; and OPNAV War Plans Division to OPNAV Ships Movements Division, 4 Aug. 1939, box 833, SecNav Gencorr 1926–40, RG 80, NA.

Chapter 6. Finding a Niche, 1940–41

1. MacKenzie, *Inventing Accuracy*, 28–29.

2. Ibid.; Edward W. Constant III, "Genesis N + 1: The Origins of the Turbojet Revolution" (Ph.D. diss., Northwestern University, 1976); Edwin T. Layton, Jr., "Technology as Knowledge," *Technology and Culture* 15, no. 1 (1974), 31–41; and Derek J. De Solla Price, "Is Technology Historically Independent of Science? A Study in Statistical Historiography," *Technology and Culture* 6, no. 4 (1965): 553–95.

3. The best source on the timing of the UCDWR's origins is the "Completion Report made to the Chief of BuShips covering the operations of the University of California's Division of War Research at the U.S. Navy Electronics Laboratory, San Diego CA," Contract Nobs-2074 (formerly OEMar-30), 26 Apr. 1941–30 June 1946, NOSC/TL.

4. Columbus O'Donnell Iselin, "Memorandum concerning the research facilities available at Woods Hole Oceanographic Institution," reproduced in Iselin, "WHOI History During the War Years"; and Iselin to Briscoe, 22 Aug. 1940, box 4, Administrative, Institutions, Ships, Miscellaneous, and Individuals, OD/WHOI.

Scripps was approached even earlier to do this kind of work. Scripps services that might have special application for wartime—including bottom-fouling research—were solicited as early as 1937, and money for the project increased through 1940. See esp. box 1, Office of the Director (Sverdrup), 1936–48, SIO.

5. Fritz Spiess, *The Meteor Expedition: Scientific Results of the German Atlantic Expedition, 1925–1927*, trans. William J. Emery; and Eric Mills, "'Physische Meereskunde': From Geography to Physical Oceanography in the Institute for Meereskunde, Berlin, 1900–1935," *Historisch-Meerskundliches Jahrbuch* 4 (1997): 45–70.

6. Memorandum to NDRC members from Irvin Stewart, 21 Aug. 1940, Executive, NDRC 1940, Council of National Defense, NAS.

7. Jewett to Iselin, 13 Aug. and n.d. (probably 29 Aug.) 1940; and Iselin to Jewett, n.d. (probably 23 or 24 Aug.), n.d. (probably 27 or 28 Aug.), 6 Sept., and 11 Dec. 1940, box 4, Administrative, Institutions, Ships, Miscellaneous, and Individuals, OD/WHOI. Some of these sources are handwritten; some dates were determined from the context.

8. John Herrick, "Subsurface Warfare," chap. 3; memorandum, Bowen to NAS, 16 Oct. 1940; Jewett to Mason, 3 Dec. 1940 and 13 Feb. 1941; Jewett to Brockett, 3 Dec. 1940; Jewett to Bush, 20 Jan. 1941; Colpitts to Bowen, 27 Jan. 1941; Jewett to Bowen, 11 and 13 Feb. 1941; and Jewett to Colpitts et al., 18 Feb. 1941, organization files, Subcommittee on Submarine Detection (Colpitts Committee)—1940, NAS: Committee Advisory to the Navy Department on Research, NAS. The NAS documentation cited here contains a great deal of information on Bowen's disputation of the report and his attitude toward the Colpitts Committee's analysis of the problem.

9. "Report of the Subcommittee on the Submarine Problem" (Colpitts Report), 28 Jan. 1941 (see esp. sections 9–11), NSB.

Attenuation is the reduction of the intensity of sound as it moves farther away from the source. Reverberation is the continuation of a sound at a given point after direct reception from the source has ceased. It is usually caused by reflection off another surface, scattering due to various kinds of matter often present in air or water, or vibration excited by the original sound (Hackmann, *Seek and Strike,* 251).

10. Iselin, "Memorandum concerning the research."

11. R/V *Atlantis* Logs for Jan.–Feb., 1941; and Iselin to Mac Murray, 14 Jan. 1941, Cruise III, WHOI.

12. Ewing to Iselin, n.d., 1941; Iselin to Ewing, 13 and 17 (telegram) Feb., and 22, 23 (morning and evening letters), and 25 July 1941; and "Improvements in the Bathythermograph [typed and handwritten report]," n.d., 1941, Maurice Ewing file, box 4, Administrative, Institutions, Ships, Miscellaneous, and Individuals, OD/WHOI; Iselin to Ewing, n.d. (probably Feb.), 1941; Iselin to Ewing, 8 Mar. 1941, Columbus Iselin file; and "Sound Transmission in Seawater: A Preliminary Report," 1 Feb. 1941, 1423-AA Atlas Shelf, WHOI.

13. Jewett: "Objectives of Proposed Laboratory for Major Research in Underwater Sound." organization files, NAS: Committee Advisory to the Navy Department on Research, Subcommittee on Submarine Detection (Colpitts Committee)—1941, NAS.

14. Robinson to Chairman, NDRC, 10 Apr. 1941; and memorandum: "Plan for Handling the Problem of a Comprehensive Investigation of Submarine Detection," n.d. (probably 18 Apr.), 1941, box 30, SONRD Gencorr.

15. Bush to NDRC, 21 Apr. 1941; Bowen to BuShips, 7 June 1941; BuShips to Navy Department member, NDRC, 10 June 1941; Jewett to Briscoe, 19 June 1941; "Agenda—Meeting Section C-4," 30 June 1941; "Program of New London and San Diego Laboratories," 2 July 1941; Briscoe, memorandum to Van Keuren, 5 Sept. 1941; and Stephenson to Briscoe, 12 Sept. 1941, box 30; and memorandum to Rear Adm. A. H. van Keuren from Tate (on conference scheduled for) 5 Sept. 1941, box 96, SONRD Gencorr.

16. Jewett: "Objectives of Proposed Laboratory for Major Research in Underwater Sound," 17 Feb. 1941, organization files, NAS: Committee Advisory to the Navy Department on Research, Subcommittee on Submarine Detection (Colpitts Committee)—1941, NAS.

17. SecNav to all bureaus, 26 Oct. 1942, box 2; Tate to Knight, 8 Jan. 1942, box 61; SecNav to all bureaus and offices of the Navy Department, 10 Oct. 1941, box 64; Burwell and Parker to coordinator of research and development, 2 Dec. 1942; and MacKenzie to Furer, 14 Jan. 1943, box 96, SONRD Gencorr.

The laboratories doing work under NDRC contracts reported their results via the NDRC to Admiral Furer at the coordinator's office. Any work initiated by BuShips at the wartime laboratories required a report to the bureau on progress and completion.

18. Service biography of Capt. Lybrand Smith, records of the Department of Naval Personnel, AR/NHC; and memorandum by Lt. James P. Parker, 16 Oct. 1942, box 3, SONRD Gencorr.

19. Completion report to the chief of BuShips covering the operations of the UCDWR at the NEL, San Diego, Calif., NOSC T/L.

20. "Proceedings of an Informal Board of Representatives from United States Navy Radio and Sound Laboratory San Diego and National Defense Research Committee: Oceanographic Program for the Collection of Information on Sound Transmission Conditions in the Pacific Ocean," 20 Aug. 1941, box 96, SONRD Gencorr; "Report on Work in Nov. 1941," UCDWR Oceanographic Division, 16 Dec. 1941, box 672, OSRD/NDRC; and memorandum for Captain Ruble (possibly composed by Revelle), 15 July 1941, folder 8, "U.S. Navy Radio and Sound Laboratory, 1941," box 27, SIO subject files, SIO.

21. K. O. Emery interview, 21 Nov. 1993, NHC Oral Histories, AR/NHC.

22. Walter Munk interview, 13 July 1993, NHC Oral Histories, AR/NHC.

23. Roger Revelle, interview by Robert A. Calvert, 4 July 1976, Oral History Program, Texas A&M (a copy of the original tape is available at the Data Library/Archives, WHOI), Roger Revelle Papers, SIO, 17–23.

On the extent of international interwar scientific work in the Pacific see Gary E. Weir, "Antecedents and Origins: The Pacific Science Congresses and the Origins of International Scientific Cooperation in the Pacific Region, 1920–1943" (paper presented at the Pacific Basin Meeting of the Oceanography Society, Honolulu, July, 1994).

Revelle completed his doctoral work in oceanography at Scripps in 1936 and received his degree from the University of California, Berkeley. The year he graduated he also obtained a reserve commission in the navy. He went on active duty in July, 1941. Walter Munk joined the UCDWR Point Loma laboratory as a junior oceanographer on 17 Nov. 1941 (Shor, *Scripps Institution of Oceanography*, 24–29; 242–251; and Raitt and Moulton, *Scripps*, 135–149).

24. Oceanographic Division, UCDWR, "Report on Work in July 1941," 2 Aug. 1941, "Report on Work in Aug. 1941," 4 Sept. 1941; "Report on Work in Sept. 1941," 9 Oct. 1941; and "Report on Work in Nov. 1941," 16 Dec. 1941, box 672; Adams, "Bi-Weekly Report Covering Period July 13–July 25, 1942," 28 July 1942; and "Bi-Weekly Report Covering Period Aug. 23–Sept. 5, 1942," 5 Sept. 1942, box 673; Adams, "Annual Report of the Division of National Defense Research," University of California, July 1, 1941–June 30, 1942," 30 June 1942,

box 674, OSRD/NDRC; and Ruble to BuShips (with attached report), 14 Aug. 1941, box 62, SONRD Gencorr.

Chapter 7. The Critical Innovation, 1940–41

1. The bottom fouling research went well and Iselin suggested that a textbook on the problem might serve both the navy and civilian science. In a letter to Jewett on 6 September 1940, the WHOI director noted that Commanders Briscoe and Fowler had endorsed his idea, and with several authors working with both the navy and civilians in mind, the projected publication cost of $25,000 might quickly disappear through retail sales. While making this suggestion to the NAS president, Iselin revealed the growing volume and intensity of the work at WHOI. His letter to Jewett also mentioned the possibility of research money for meteorology to address surface visibility problems.

2. Iselin to Jewett, 6 Sept. 1940. See also Jewitt to Iselin, 13 Aug., n.d. (27 or 28 Aug.—determined from context); and handwritten unaddressed note, n.d. (29 Aug.—determined from context), 1940; Iselin to Jewett, n.d. (either 23 or 24 Aug.), 11 Dec., and n.d., 1940; and Iselin to Briscoe, 22 Aug., 3 Sept., 1 Oct., and 2 Dec. 1940, box 4, Administrative, Institutions, Ships, Miscellaneous, and Individuals, OD/WHOI.

3. Iselin, "Memorandum concerning the research facilities."

4. Athelstan Spilhaus, interview by Robert A. Calvert, 11 Feb. 1976, A&M/OH. For Spilhaus's continuing involvement with the BT and his comments on the Vine modifications, see Athelstan Spilhaus, "On Reaching 50: An Early History of the Bathythermograph," *Sea Technology* 28, no. 11 (Nov., 1987): 19–28.

5. Sloat F. Hodgson, "The Development of a Mechanical Bathythermograph," Instrument Development, 1938–69, accession no. 90–3, WHOI.

6. WHOI, "Instructions for the Installation, Operation, and Servicing of Woods Hole Model Bathythermograph and Equipment," 12 June 1941, WHOI; and Worzel interview.

7. Allyn Vine interview, 27 Apr. 1989, NHC Oral Histories, NHC/AR.

8. "Instructions for the Installation, Operation, and Servicing of Woods Hole Model Bathythermograph and Equipment,"; Worzel interview; and Vine interview, 27 Apr. 1989.

9. Vine interview, 27 Apr. 1989.

10. Burhans to Sverdrup, 7 May and 22 Aug. 1940; Sverdrup to Iselin, 27 Mar. and 2 Apr. 1941; and Francis P. Kinnicutt to Sverdrup, 4 Mar. 1942, box 1, Office of the Director (Sverdrup), 1936–48, SIO.
Before the BTs arrived on the West Coast, the navy borrowed Sverdrup's reversing thermometers for temperature work.

11. W. M. Ewing et al., "Bathythermograph," patent application filed 27 May 1944, granted 11 July 1950, WHOI; "A Suggested Plan for the Quantity Manufacture of BT's and Associated Gear," ca. 1940, box 74; and Ewing to Long, 24 June 1963, box 265, CAH/Ewing; Worzel interview; and Vine interview, 27 Apr. 1989. Iselin often welcomed the ability of his institution to build BTs because some vendors proved very slow in production. This was the case with Submarine Signal in 1940–41. Iselin occasionally voiced concern about the situation. For example, see Iselin to Murphy, 1 Aug. and 23 Oct. 1940; and Murphy to Iselin, 1, 7, and 21 Oct. and 23 Nov. 1940, Robert Cushman Murphy Papers, M957, AmPhil.

12. Blueprint Plans for the Woods Hole BT, 1944, WHOI.

13. Although King's methodology, very similar to Ewing's prewar seismic work, did not lend itself to shipboard ASW operations, he did make significant discoveries. King's research suggested that the navy should investigate the lower frequencies with longer wavelengths

that might travel longer distances and excite less reverberation. He proposed a triangulation method using depth charges as sound sources: "Inquiry indicated that this scheme of submarine location has not been tried by either the American or the British Navies." See Robert W. King, "A Low Frequency Sound Ranging Method for Submarine Detection," 10 June 1941; Jewett to Bush, 1 July 1941; Robert W. King, "Submarine Detection by Underwater Explosion," 1 July 1941; and King to Briscoe, 3 July 1941, box 30, SONRD Gencorr.

14. The UCDWR designed and manufactured its own slide rule, which followed the more common straight or linear type rather than the WHOI circular style.

15. Maurice Ewing, "Sound Ranging Experiments at Key West, July 1941," box 752, OSRD/NDRC. On 23 May 1942, Ewing submitted a report based upon sound ranging experiments he had conducted at Key West in July, 1941, that confirmed the analysis of sound refraction offered by he and Iselin in 1940. Thus, the basis for much of this science and its application to underwater sound problems rested on firm scientific ground.

16. Reverberation is vibration initiated by a sound, such as a sonar projector or ship noise, which lasts beyond the duration of the causal event. Thus, in a sonar search, active sonar pinging will initiate vibrations in the immediate area of its operation that last longer than the ping itself. It thus becomes difficult to distinguish between the sounds from sonar, ships, and sound-initiated vibration in the environment.

17. WHOI, UCDWR, BuShips, "Prediction of Sound Ranges from Bathythermograph Observations," June, 1942, box 752, OSRD/NDRC.

18. Columbus Iselin, memorandum for Admiral Furer, 2 Apr. 1942, single unnumbered box, Institution Confidential (declassified) Collection, WHOI.

19. UCDWR, "Workbook for Prediction of Maximum Echo Ranges"; *Bathythermograph Observations,* U.S. Navy Hydrographic Office Publication 606-c, 1956, CH/NHC; and John W. Stimpson interview, 17 Aug. 1994, NHC Oral Histories, AR/NHC.

A WHOI technician, Stimpson perfected a process for reproducing the lacquered BT slide images on photosensitive paper to create a permanent record of the data on a surface roughly the size of a three-by-five-inch file card. Thousands of these cards accumulated by the end of the war to represent one of the largest and most useful sources of oceanographic data available. The National Oceanographic and Atmospheric Administration (NOAA) still uses the collection of BT cards at the WHOI Archive to correct its computer database on the world ocean.

20. E. C. LaFond and B. K. Cooper, "The Mechanical Bathythermograph, An Historical Review," gift from Allyn C. Vine, CH/NHC; and a series of telegrams beginning with BuShips (Code 940) to director, NEL, San Diego, 9 July 1946 (Howard Gould, Allyn Vine, and William Schevill as SBT instructors with the Pacific Fleet), box 1574, BuShips Confidential Centcorr.

21. Maurice Ewing, "Report on Laboratory Tests—Submarine Model Bathythermograph—Serial #5," 17 Mar. 1942; and Maurice Ewing, "Progress Report on the Submarine Model Bathythermograph," 23 Apr. 1942, box 753; "The Use of a Density Meter as an Aid to Submarine Diving, Preliminary Edition," 11 Dec. 1942; Allyn Vine, Dean Bumpus, and Alfred Redfield, "Direct Observations of Factors Affecting Buoyancy of Submarines, Preliminary Report," 11 Aug. 1943; and WHOI Bi-Weekly Report for 9–22 Jan. 1944 dated 26 Jan. 1944, box 752, OSRD/NDRC.

22. Iselin to Bennett, 12 Feb. 1942; Tate to Jones, 19 Mar. 1942; Jones to Tate, 24 Mar. 1942; Furer to BuShips, 20 Mar. 1943; and BuShips to ComSubLant, 5 Oct. and 27 Nov. 1943, box 62, SONRD Gencorr.

23. Vine interview, 27 Apr. 1989.

24. Ibid.

25. Iselin, preliminary draft of letter to Admiral Furer, n.d., 1943; "Instruction Notes on the Use of the Submarine Bathythermograph with Respect to Diving," n.d., ca. 1942, Institution Confidential [declassified], WHOI; and a series of telegrams beginning with BuShips (Code 940) to director, NEL, San Diego, 9 July 1946 (Howard Gould, Allyn Vine, and William Schevill as SBT instructors with the Pacific Fleet), box 1574, BuShips Confidential Centcorr.

26. Gordon Riley, "Reminiscences of an Oceanographer," 53, biographical files, Gordon Riley folder, WHOI.

27. Worzel interview. See also Corput to BuEng, 9 Jan. 1942, box 54; F. A. Everest, et al. (NDRC/UCDWR), "Water Noise Survey, San Francisco," 19 June 1942, box 158; commander, Naval Bases South Pacific Area to BuShips, 8 Apr. 1943; commander, Service Squadron South Pacific to BuShips, 2 Sept. 1943 (Harbor Detection Study enclosed), and BuShips to commander, Service Squadron South Pacific, 5 Oct. 1943, box 159, ser. 1943, BuShips Secret Gencorr, 1943; Lt. Comdr. W. B. Goulett, memorandum for files, 9 Jan. 1942; and NRL to BuShips, 12 Jan. 1942, box 61; Ewing to Bennett, 11 Jan. 1942; and BuShips to ComSubLant, 13 Aug. 1943, box 62, SONRD Gencorr.

28. Bennett to ComSubLant, 12 Jan. 1943, box 62, SONRD Gencorr.

29. Vine interview, 27 Apr. 1989. Both Street, commander of the *Tirante,* and Ramage received the Medal of Honor for their efforts against the Japanese in World War II. See also Mary Sears interview, 22 June 1992, NHC Oral Histories, AR/NHC; Knight to BuShips, 7 Jan. 1943, box 65, SONRD Gencorr; Iselin to Sylvester, 4 Feb. 1942; Sylvester to Iselin, 6 Feb. 1942; and Iselin to Spitzer, 3 Mar. 1942, P–T alpha file; Iselin to Tate, 28 Jan. 1942, John Tate file; Furer to Forrestal, n.d., 1942; and Iselin to Furer, 26 Mar. 1942, F–J alpha file, box 5; Iselin, "Report of Progress—NYBOS Contract no. N1115-38137," 31 Dec. 1943, unfinished business file, box 7, OD/WHOI; and Revelle interview.

Chapter 8. Operational Applications, 1942–43

1. Revelle interview.

2. Iselin to Sylvester, 23 Jan. 1942; WHOI, "Summary of Bathythermograph Observations," 23 Jan. 1942; and Roger Revelle, "Charts of Bottom Sediments Prepared by NDRC," 12 Feb. 1943, box 62; Iselin to Sylvester 23 Jan. 1942; Hutchisson to Sylvester, 26 Mar. 1942; Report of Conference "Oceanographic Information with Respect to Sound Ranging," 3 Apr. 1942; Sylvester: "Memorandum on Underwater Sound Ranging," 4 Apr. 1942; "Joint Navy-NDRC Conference on the Practical Prediction of Assured Echo Ranges," 9 Apr. 1942 (date received in SONRD); Spitzer to Sylvester, 8 Apr. 1942; Iselin to Sylvester, 13 Apr. 1942; Hutchisson to Furer, 11 July 1942; Tate to Furer, 7 Dec. 1942; UCDWR, "Supplement to Sound Ranging Chart of the North Pacific Ocean, Winter Season," Jan., 1943; Jacobs to Tate, 2 Jan. 1943; Colpitts to CNO, 28 Jan. 1943; Colpitts to Furer, 5 Feb. 1943; Furer to director, U.S. Coast and Geodetic Survey, 9 Feb. 1943; Furer to commander in chief, U.S. Fleet, 9 Feb. 1943; NDRC, "Agenda: Conference on Charts of Bottom Sediments," 12 Feb. 1943; Redfield to Revelle, 1 Mar. 1943; Revelle to Redfield, 8 Mar. 1943; NDRC, "Navy Project NS-140: Acoustic Properties of the Sea Bottom" (research summary conclusions); Listening Section, UCDWR, "Outline of Proposed Program of Sound Propagation Measurements in Deep Water," 9 Nov. 1943; and minutes of the Conference on Bottom Sediments, 5 Jan. 1944, box 65, SONRD Gencorr.

Hydrographer to CNO, 28 Jan. 1942; and Bryan to commander in chief, U.S. Fleet,

21 Apr. 1942, box 177; BuShips to commander, Service Force, Pacific Fleet, 4 Feb. 1943; Bu-Ships to commander in chief, U.S. Fleet, 13 Feb. 1943; and UCDWR, "Supplement to Sound-Ranging Chart of the North Pacific Ocean, Winter Season," Jan., 1943, box 553; "Hydro-graphic and Navigational Extracts from Submarine War Patrol Reports," HO Misc. no. 10,564, Hydrographic Office, Oct., 1943, Confidential 1943 (unclassified), box 554; and Bryan, memorandum to CNO, 3 Nov. 1943, Secret (unclassified) 1943, box 678, SecNav Gen-corr 1927–39.

UCDWR, "Preliminary Submarine Supplement to Hydrographic Office Publication no. 123, Asiatic Pilot," vol. 2, "Late Summer Hydrographic Conditions in the Japanese Area," July, 1943, box 678, OSRD/NDRC.

Glennon and Johnson to HO, 5 Oct. 1942, box 137; Allen to Bryan, 31 Aug. 1940; and Allen to editor, *Hydrographic Bulletin,* 16 Oct. 1942, box 140, HO Gencorr.

"Preliminary Chart of Echo Ranging Conditions in the Gulf Sea Frontier, Aug.–Oct. 1943," Historical Chart Collection, WHOI.

Sears interview; Emery interview; and Revelle interview.

On the West Coast, supplement composition eventually moved from the UCDWR alone to a joint effort by Scripps and UCDWR under contract to Hydro. See Fleming to file, 21 Aug. 1943, box 1, Office of the Director (Sverdrup), 1936–48, SIO.

3. Sears interview.

4. Ibid.; and Revelle interview.

5. Revelle, "Joint Meteorological Committee: Subcommittee on Oceanography, min-utes of meeting held on 22 Dec. 1942"; and Revelle to Furer, 13 Jan. 1943, box 62, SONRD Gencorr; Revelle to Sverdrup, 16 July 1943, box 2, Roger Revelle Papers, SIO; "National Roster of Specialized Personnel, Committee on Wartime Requirements for Specialized Per-sonnel, 1942," NAS; Revelle to Iselin, 6 Feb. 1943, box 5, Office of the Director, NDRC and Navy, 1942–45, WHOI; and Sears and Revelle interviews.

6. "Annual Report of the Division of National Defense Research, University of Cali-fornia, July 1, 1941–June 30, 1942," 30 June 1942, box 673, OSRD/NDRC.

Sverdrup's situation seems to have bothered only the navy, which fastidiously enforced the alien rule. Sverdrup served other groups within the U.S. armed forces during the war, in addition to some help given the Norwegians and British. See Sverdrup to Sproul, 15 June 1942, box 1, Office of the Director (Sverdrup), 1936–48, SIO.

7. UCDWR, "Report on Work in January and February 1942," box 672; UCDWR Bi-Weekly Reports dated 6 Nov. 1942 (covering the period 20 Sept.–3 Oct.), 6 Nov. 1942 (for 18–31 Oct.), 19 Nov. 1942 (for 1–14 Nov.), and 11 Jan. 1943 (for 27 Dec. 1942–9 Jan. 1943), box 675, OSRD/NDRC.

8. J. K. Nunan and W. B. Snow, CUDWR, "A Discussion of Fundamental Concepts Pertinent to the Design of Submarine Sonar Gear," 3 Apr. 1944, box 704, OSRD/NDRC.

9. MIT Bi-Weekly Reports covering the period Oct.–Dec., 1943, related to Navy Proj-ect NS-164, box 160, BuShips Secret Gencorr, 1943; commander, Anti-Submarine Develop-ment Detachment Atlantic Fleet (AsDevLant) to commander in chief, U.S. Fleet, 26 Jan. 1944; NRL to commander in chief, U.S. Fleet et al., "Underwater Sound: Expendable Noise-maker Employing Ammonia Vapor," 22 Jan. 1944; UCDWR Bi-Weekly Report on Project NS-164, 12 Jan. 1944; and commander, AsDevLant to commander in chief, U.S. Fleet, 12 Jan. 1944, box 154, BuShips Secret Gencorr, 1944; Conrad to commander in chief, U.S. Fleet (ASW), 12 Feb. 1945; and Tate to Coordinator of Research and Development, n.d., 1945, box 153; UCDWR Monthly Progress Report for Apr., 1945 dated 15 May 1945; and Kurie to Bu-Ships (Code 940), 19 May 1945, box 154, BuShips Secret Gencorr, 1945.

10. MIT, Project DIC 6187, Navy Project to AsDevLant, "Measurements made on Dec. 2, 1943 of the Sound Output of Three Sets of FXR-II Gear," 23 Dec. 1943, box 154, Bu-Ships Secret Gencorr, 1944.

11. Spitzer to Kingdon, 28 Mar. 1944; Commander, AsDevLant to Tenth Fleet, 10 May 1944; Speedletter: AsDevLant to ComOpDevFor, 16 May 1946; Pryor to BuShips, 17 May 1946; MIT, Navy Project NS-164 Bi-Weekly Report covering 26 June–9 July 1944 dated 18 July 1944; and NRL, "Sound Division Interim Report: FXP–Mark 1, Tests Continue," box 155, BuShips Secret Gencorr, 1944; UCDWR to BuShips Code 940, 11 Mar. 1946; and UCDWR to BuShips Code 940 (monthly progress report for Jan., 1946), box 153, BuShips Secret Gencorr, 1945; CNO to BuShips Code 940, 31 July 1946, box 51, BuShips Secret Gencorr, 1946; McCann to Lockwood, 9 May 1944, box 15, ser. 4, file 1, SOC; and USNUSL, "A Catalogue of Underwater Sound Equipments," 15 July 1943, box 690, OSRD/NDRC.

12. Spitzer to Eckart, 18 Feb. 1944, box 154; K. H. Kingdon, "Comments on UCDWR Memorandum M172-a: Data at 45 Kc on Echoes from a Diving Submarine and its Wake," 24 Mar. 1944, box 155; Spitzer to Eckart, 28 July 1944, box 156; Spitzer to Eckart, 9 and 26 Nov. 1943; and Kingdon to Eckart, 1 Dec. 1943, box 160, BuShips Secret Gencorr, 1944; Hammond to BuShips Code 940, 9 Aug. 1944 (surface scar); and NRL to BuShips, "Sonar Wake Detectors," 2 Mar. 1945, box 153, BuShips Secret Gencorr, 1945; and Jeffries Wyman, Wendel Lehman, David Barnes, WHOI, "Laboratory Studies of the Acoustic Properties of Wakes," pt. 1, n.d., Mar., 1944, box 754, OSRD/NDRC.

13. Carl Eckart, "The Theory of Secure Echo-Sounding," 8 Jan. 1944, box 154; Furer to BuShips, 9 June 1944; UCDWR, Bi-Weekly Report no. 16, "NS-221," 15 June 1944; UCDWR, Bi-Weekly Report no. 19, "NS-221," 28 June 1944; and asst. director, UCDWR to ComSubPac, 11 July 1944, box 156; Kittel, memorandum for Comdr. Rawson Bennett et al., 22 Nov. 1944, box 158, BuShips Secret Gencorr, 1944; and Harnwell to BuShips, 28 Feb. 1945, box 153, BuShips Secret Gencorr, 1945.

14. Commander, USS *Spot,* to director, UCDWR, 3 Feb. 1945, box 153, BuShips Secret Gencorr, 1945.

15. UCDWR Bi-Weekly Report covering the period 26 Dec. 1943–8 Jan. 1944 dated 13 Jan. 1944, box 680, BuShips Secret Gencorr, 1945.

16. BuShips to Colpitts, 25 Apr. 1944, box 62; and Revelle to Colpitts, 14 Apr. 1945, box 63, SONRD Gencorr. Note here the effort to combine the technical developments at HUSL with the sonar work at UCDWR—same effort at comprehensiveness.

17. WHOI Bi-Weekly Report covering the period 30 Apr.–13 May 1944 dated 17 May 1944, box 754, OSRD/NDRC. This document noted William Schevill's trip to Scripps to affect the comprehensive coverage desired on the summer supplements to Hydro's navigation publications. He returned in late June or early July, 1944. See "Travel Report Memorandum, Conrad and Burwell to RCA Laboratories, Princeton, N.J.," 1 Apr. 1944, box 4, OSRD/NDRC.

18. Submarine supplement to "Sailing Directions: Japanese Empire Area, Jan.–Mar.," HO Misc. no. 11,381-B, HO, 11 Nov. 1944. Various examples of these supplements appear in boxes 1112, 1115, and 1807, Cominch, 1944; UCDWR, "Shallow Water Sonar: Bi-Weekly Report on Project NS-142," 17 May 1944; and UCDWR, "Detection of Small Objects: Bi-Weekly Reports 4 and 5," 6 and 21 Apr. 1944, respectively, box 155, BuShips Secret Gencorr, 1945.

19. UCDWR, "Shallow Water Sonar," Bi-Weekly Report no. 10, box 156, BuShips Secret Gencorr, 1944.

20. Malcolm Henderson, UCDWR, "Subsight," 20 Feb. 1943, box 676; and Roshon to

Henderson, 12 Aug. 1944, (Subsight and FM sonar tests in the Mediterranean Sea), box 683, OSRD/NDRC.

21. OSRD to Bennett, 4 May 1944, box 155; Spitzer to Eckart, 31 July 1944, box 156; and Aitchison, memorandum to the NRL director, 5 Aug. 1944, box 157, Buships Secret Gencorr, 1944; Headquarters, U.S. Army Air Forces, Pacific Ocean Area, Operational Research Section, "Tests on the Underwater Sound Direction and Range System (USDAR)—Model 500A," 14 Jan. 1945, box 153, BuShips Secret Gencorr, 1945; NRL, "Report on the Pro-Submarine Program of the Sound Division," 5 Apr. 1945, box 63, BuShips Secret Centcorr, 1945; Revelle to Iselin, 25 Apr. 1944, box 7, OD/WHOI; Emery interview; and Hackmann, *Seek and Strike,* 286–91.

22. "Completion Report Made to the Chief of the Bureau of Ships Covering the Operations of the University of California Division of War Research at the U.S. Navy Electronics Laboratory, San Diego Calif., 30 June 1946," NOSC/TL.

UCDWR "Detection of Small Objects," Bi-Weekly Report no. 2, 7 Mar. 1944; Smith to commander in chief, U.S. Fleet (Readiness), 28 Apr. 1944; UCDWR, "Shallow Water Sonar," Bi-Weekly Report no. 6, 6 May 1944; Bennett to commander in chief, U.S. Fleet, 10 May 1944; Colpitts to Bennett, 18 May 1944 (see attached "Mine Location with FM Sonar: Resume of Experimental Results"; and Conrad to Stewart, 22 May 1944, box 155; Lt. D. W. Atchley, Jr., to director, NRL, 21 July 1944, box 156; memorandum to commander, Service Squadron 5, 28 July 1944; BuShips to commander, Service Force, Atlantic Fleet, 16 Aug. 1944; Kappler to Myers, 17 Aug. 1944; Furth to BuShips, 13 Sept. 1944; and Bennett to commander, Service Squadron 5, 6 Dec. 1944, box 157, BuShips Secret Gencorr, 1944; UCDWR, "Shallow Water Sonar," Bi-Weekly Report no. 25, 26 Jan. 1945; Bennett to BuOrd, 7 Feb. 1945; and Harnwell to BuShips, 21 Feb. 1945, box 153; QLA Project Officer to BuShips, 5 July 1945; BuShips to CNO, 24 Sept. 1945; and BuOrd to BuShips, 9(?) Jan. 1945, box 155, BuShips Secret Gencorr, 1945.

Lockwood to Watkins, 1 and 23 June 1945; Watkins to Lockwood, 9 June 1945; and Watkins to Irvin, 26 June 1945, box 15, ser. 4, file 1, SOC. Burwell, memorandum to CRD, Subject: Sonar Tests at New London, 28 Feb. 1944, box 61, SONRD Gencorr.

Melvin O. Kappler, "Observations Made During Ten Days at Sea with FM Sonar on the USS *Tinosa* (SS-283)," 11 Nov. 1944, box 684, OSRD/NDRC.

23. As defined by sociologist Donald MacKenzie in his book on the development of inertial missile guidance: "It is a technical artifact—or, more loosely, any process or program—that is regarded as just performing its function, without any need for, or perhaps any possibility of, awareness of its internal workings on the part of users" (MacKenzie, *Inventing Accuracy,* 26).

24. Emery interview.

WHOI, "Summary of Bathythermograph Observations," 23 Jan. 1942; Iselin to McMurray, 30 Jan. 1942; Iselin to Furer, 30 Jan. 1942; Furer to Iselin, 2 Feb. 1942; Furer, memorandum for files, 11 Sept. 1942; and Hutchisson to Furer, 12 Oct. 1942, box 62; commander, West Coast Sound Training Squadron, to SONRD, 20 Oct. 1942, box 64; Fleming to Knudsen, 4 Mar. 1942; "Joint Navy-NDRC Conference on the Practical Prediction of Assured Echo Ranges," 9 Apr. 1942; Iselin to Fleming, 17 July and 13 Oct. 1942; Fleming to Iselin, 14 Aug. 1942; commander, Gulf Sea Frontier to the NRL, 19 Jan. 1943, box 65; and "Summary of Work Underway at U.S. Radio and Sound Laboratory, San Diego Calif.," 27 Jan. 1943, box 96, SONRD Gencorr.

"Submarine Listening Ranges," preliminary ed., Aug., 1943, box 720; "Minutes of a Conference on Maximum Echo Ranges," 14 Dec. 1943, box 721; WHOI, "Summary of Bathythermograph Observations," 10 Dec. 1941; and WHOI, UCDWR et al., "Preliminary Draft: Prediction of Echo Ranges from Submarine Bathythermograph Operations, Instruction Manual for Bathythermograph Observers," pt. 2, 1 Sept. 1942, box 752; and WHOI Bi-Weekly Report covering the period 7–20 Mar. 1943 dated 24 Mar. 1943, box 753, OSRD/NDRC.

Fleming to Kinnicutt, 30 Jan. 1942, box 5, Office of the Director, NDRC and Navy, 1942–45, WHOI.

Revelle interview, SIO.

Chapter 9. Unfinished Dialogue, 1942–45

1. ASDIC was a sonarlike acoustic submarine detection device named for the committee that investigated the problem in Britain early in the war: Anti-Submarine Detection Investigation Committee.

2. Iselin to Knight, 3 Feb. 1943, box 158, BuShips Secret Gencorr, 1943.

3. W. E. Schevill, "Wartime Oceanography at the Deutsche Seewarte and the Marine observatorium," Technical Report no. 363–45, Sept., 1945, box 35, records of the U.S. Naval Technical Mission in Europe, AR/NHC.

4. Ibid.

5. Michael L. Hadley, *U-boats Against Canada;* David Zimmerman, *The Great Naval Battle of Ottawa;* Marc Milner, *North Atlantic Run: The Royal Canadian Navy and the Battle for the Convoys;* Marc Milner, *The U-boat Hunters: The Royal Canadian Navy and the Offensive Against Germany's Submarines;* and W. Schott, *Early German Oceanographic Institutions, Expeditions, and Oceanographers.*

6. Spitzer to Iselin, 11 Nov. 1943; and Spitzer to Revelle, 11 Nov. 1943, box 160, BuShips Secret Gencorr, 1943; VCNO to commander, Gulf Sea Frontier, 6 July 1942, box 54, BuShips Secret Gencorr, 1941–42.

7. Spitzer, memorandum to Bolt, 29 Apr. 1944, box 155, BuShips Secret Gencorr, 1944.

8. Hussey to commandant, 13th Naval District, 8 (?) July 1944; Furer to commander in chief, U.S. Fleet (Readiness), 13 July 1944; and captain, H.M.E. Establishment, Fairlie, to director Anti-Submarine Warfare, Naval Staff, "Type 149X ASDIC Set for Self protection against Torpedo Attack," 12 June 1944, box 156, BuShips Secret Gencorr, 1944.

NRL Bi-Weekly Progress Report, "Projector for Sonar Countermeasures," 12 Apr. 1945; and Schevill to Revelle, 17 Apr. 1945; box 154, BuShips Secret Gencorr, 1945.

"Echo Detection of Small Targets" (British), Sept., 1943, box 159, BuShips Secret Gencorr, 1943.

Capt. L. B. Olson, United Kingdom Travel Report Memorandum, 16 Oct. 1943, box 4; and BuShips to Coordinator of Research and Development, 15 Nov. 1942, box 61; Iselin to Hutchisson, 10 July 1942, box 62; Hutchisson to Furer, 15 May 1942, box 96; and ONI Report, ser. 1394, Naval Attaché, 28 May 1942, box 98, SONRD Gencorr.

Jewett to Wilson (cover), 19 June 1941, Council of National Defense, Executive, NDRC, Jan.–June 1941, NAS.

Lyman Spitzer and F. E. Croxton, Jr., "Analysis of British Echo Ranging Data," 15 Aug. 1942, box 698; W. B. Snow, "Memorandum for File: Report of Trip to the United Kingdom, Feb. 1945," 28 Feb. 1945, box 709; WHOI Bi-Weekly Report covering the period 3–16 Oct. 1943 dated 20 Oct. 1943; and WHOI Bi-Weekly Report covering the period 25 July–7 Aug. 1943 dated 11 Aug. 1943, box 753, OSRD/NDRC.

British Admiralty, "Sea Conditions Affecting ASDIC Operating," Sept., 1943; and Iselin to Edgell, 25 Apr. 1945, box 6, OD/WHOI; and William E. Schevill, "Report on Trip to the United Kingdom as Special Representative of the office of Scientific Research and Development . . . ," 16 Oct. 1943, Institution Confidential (Declassified), WHOI.

Eric Mills, "Bringing Oceanography into the Canadian University Classroom," *Scienta Canadensis* 18, no. 1 (winter, 1994–95): 3–21.

9. Iselin to Furer, n.d., 1943, box 6, OD/WHOI.

10. Martin W. Johnson, "Underwater Sounds of Biological Origin," 15 Feb. 1943, box 676, OSRD/NDRC.

11. Revelle interview.

Commander in chief, U.S. Fleet, memorandum for Admiral Lee, 20 July 1942; F. A. Everest to Harnwell, 1 July 1942; commander, Atlantic Fleet ASW Unit to commander in chief, U.S. Fleet, 1 July 1942; BuShips to commander in chief, U.S. Fleet, 10 July 1942; Krause to Furer, 13 July 1942; director, USNRSL to commander in chief, U.S. Fleet, 14 July 1942; Everest to Harnwell, 3 Aug. 1942; BuOrd to commander in chief, U.S. Fleet, n.d., 1942; commander, Atlantic Fleet ASW Unit to commander in chief, U.S. Fleet, 3 Aug. 1942; Furer to Iselin, 21 Oct. 1942 (for comments on early attempts by USN sonar operators at sea to distinguish between fish and possible submerged contacts—USS *Nourmahal* [PG-72]); "Report of Meeting: Underwater Background Noise," 1 Apr. 1943 (note esp. the attached essay on "Ambient or Background Water Noise" by Vern Knudsen); and Furer to commandant, 6th Naval District, 31 Mar. 1943, box 61, SONRD Gencorr.

Revelle to Furer, 25 Feb. 1943, box 554, Cominch.

UCDWR Bi-Weekly Report covering the period 24 Jan.–6 Feb. 1943 dated 10 Feb. 1943, box 676; UCDWR memorandum, "Bottom Reverberation," 16 May 1944, box 682; UCDWR, "Reverberation in Echo-Ranging, Navy Project NS-140," 14 Apr. 1943, box 692; "Memorandum on Preliminary Results of the Shallow Water Program" (a synopsis of the A. P. Crary and Dean Bumpus "Report of Cruise #1"), 3 June 1942, box 700; Columbus Iselin, "Notes on the Oceanographic Origins of Reverberation," n.d., Nov., 1941; memorandum for Dr. (Phillip) Morse, "Oceanographic Factors Entering into Harbor Protection by Underwater Sonic Devices," 21 Jan. 1942; Columbus Iselin, "Gulf of Maine Sound Contact, May 27, 1942," 14 July 1942; and Columbus Iselin and Alfred Woodcock, "Preliminary Report on the Prediction of 'Afternoon Effect'," 25 July 1942, box 752; and Jefferies Wyman, WHOI, "Progress Report on Role of Bubbles in Acoustic Properties of Wakes, Navy Project NS-141," 18 Feb. 1943, box 753, OSRD/NDRC.

Lyman Spitzer, "Harvard–Woods Hole Conference on Subsurface Wake Measurements," 26 Mar. 1943, box 158; and Spitzer to Eckart, 9 Nov. 1943, box 160, BuShips Secret Gencorr, 1943.

BuShips to commander, USS *Livermore,* 29 Jan. 1943, box 5, OD/WHOI.

12. Woollard to Iselin, "Report on Conferences at U.S. Radio and Sound Laboratory, San Diego, 3–15 Apr. 1944," n.d., Apr., 1944, box 7, OD/WHOI.

13. Colpitts to Iselin, 19 Apr. 1944; Sverdrup, comments on "A Manual of Ocean Waves, of Surf and of Beaches," by Henry B. Bigelow and Henry C. Stetson, WHOI, 12 July 1944; Sverdrup to Iselin, 14 July 1944; and Sverdrup to Bigelow, 14 July 1944, box 7, OD/WHOI.

BuShips to commander, U.S. Naval Forces Europe, 23 Aug. 1944, box 63, SONRD Gencorr.

The process of moving ideas and research results to the fleet is still the most difficult task to accomplish. Today, *transitioning,* as it is now called, still relies on the personal touch, individual initiative, and contacts between able and compatible professionals. See Melbourne Briscoe interview, 27 June 1996, NHC Oral Histories, AR/NHC.

14. Iselin to BuPers, 8 Feb. 1944; and Iselin to Revelle, 7 Mar. and 27 July 1944, box 5, OD/WHOI.

15. Iselin to Revelle, 9 Nov. 1944, box 5, OD/WHOI. See also Colpitts to BuShips, 22 Feb. 1944, box 65, SONRD Gencorr; and George L. Clark, "Report of Progress—Contract no. N111s-38137," n.d., 1944, box 7, Office of the Director, NDRC and Navy, 1942–45, WHOI.

16. Riley to Iselin, 8 Mar. 1944, box 5, OD/WHOI. For additional comments on the charts, see Riley to Iselin, 10 Apr. 1944, box 5, OD/WHOI.

17. Riley to Iselin, 8 Mar. 1944.

18. Ibid.

19. Ibid.

20. Ewing to Dyk, 22 June 1953; and Maurice Ewing, G. P. Woollard, A. C. Vine, and J. L. Worzel, "Recent Results in Submarine Geophysics," *Bulletin of the Geological Society of America* 57 (Oct., 1946): 909–34, box 4; "Sofar," broadcast by WGY, Schenectady, N.Y., 17 Apr. 1946, box 232, CAH/Ewing.

Robert W. King, "Submarine Detection by Underwater Explosion," June, 1941, box 30; Jewett to Bush, 1 July 1941; R. W. King to Briscoe, 3 July 1941; E. B Stephenson, memorandum for Commander Briscoe, 12 Sept. 1941; memorandum to Rear Adm. A. H. van Keuren from John Tate, n.d. (probably late Sept.–early Oct.), 1941; Iselin to Tate, 8 Dec. 1941; Iselin to Colpitts, 16 Dec. 1941; Tate to Sylvester, 19 Dec. 1941; Furer to Tate, 24 Dec. 1941; and R. J. McCurdy, "Report on Preliminary Tests of a Submarine Model Temperature-Depth Recorder," Dec., 1941, box 61; and WHOI, memorandum for Tate, 19 Aug. 1941, box 96, SONRD Gencorr.

Iselin to Burhans, 21 Apr. 1941, U.S. Navy file; Rumble to Iselin, 8 Apr. 1941; Iselin to commander, USS *Capella,* 9 Apr. 1941; Pryor to Ellis, 2 June 1941; Iselin to Kirkpatrick, 5 and 18 June 1941; Brook to Iselin, 18 Aug. 1941; Iselin to Brook, 25 Aug. 1941; and Bigelow to Robinson, 20 June 1941, BuShips file; Iselin to Tate, 12 Aug. and 1 Oct. 1941; and Iselin, memorandum for Dr. Tate, 19 Aug. and 4 Oct. 1941, box 4, Administrative, Institutions, Ships, Miscellaneous, and Individuals, OD/WHOI.

Maurice Ewing and J. Lamar Worzel, "Long Range Sound Transmission," *Geological Society of America Memoir 27,* 15 Oct. 1948.

21. Furer, memorandum for files, 27 Aug. 1943, box 4, SONRD Gencorr.

22. Newhouse to Iselin, n.d., July, 1945, box 6, OD/WHOI.

23. Recollections of J. Lamar Worzel (cover letter dated 10 Dec. 1980), "Sofar—Sonic Fixing and Ranging," 39–54, CH/NHC.

24. Maurice Ewing, "Long Distance Transmission of Underwater Sound," 12 July 1943, file cabinet 3, drawer D, file "SOFAR Original Proposal," Office Papers of William Maurice Ewing, Creighton A. Burke, and J. Lamar Worzel, InGeo/UT. (Because these cabinets were not marked, I used a simple shorthand system for keeping track of the proper home drawer for a file or document. File drawer "3D" is the third file cabinet of the four, counting from left to right, and the fourth drawer down.) When Worzel and Ewing first established the existence of the sound channel and the SOFAR concept, the latter was not used as the name. Worzel preferred LORASTO (for Long Range Sound Transmission in the Oceans). However, BuShips had its own idea for a name, hence the selection of SOFAR. See Worzel "Sofar," 39–54.

25. Ingram to VCNO Communications, 21 July 1943; Iselin to Ingram, 13 July 1943; and Ingram to Iselin, 2 Aug. 1943, file cabinet 3, drawer D, Office Papers of William Maurice Ewing, Creighton A. Burke, and J. Lamar Worzel, InGeo/UT.

26. Revelle to Colpitts, 14 Sept. 1944; and Special Studies Group, Sonar Analysis

Section, CUDWR, "Proposed Woods Hole–San Diego Cooperative Transmission Program," 12 June 1944 (note the emphasis on WHOI-UCDWR cooperation), box 63, SONRD Gencorr; Oceanographic Division, UCDWR, to Harnwell, 3 Sept. 1942, box 683; CUDWR, memorandum for file, "Maximum Listening Ranges of Underwater Sound Equipment," 13 Mar. 1944, box 704; and George P. Woollard, "A Summary of Factors Governing Sound Ranges," n.d., June, 1944, box 754, OSRD/NDRC.

27. W. Maurice Ewing and J. Lamar Worzel, "Long Range Sound Transmission," interim report no. 1, 1 Mar. 1944–20 Jan. 1945, Clark Laboratory Document Library, WHOI.

28. "Sound in the Sea," Apr., 1945, no box number, Institution Confidential (declassified) Collection, WHOI; Maurice Ewing, "The Role of the Saluda in the Demonstration of SOFAR," ca. 1943–44, box 6; "Explosive Sound in Shallow Water," ca. 1945, box 110, CAH/Ewing; and Naval History Division, *Dictionary of American Naval Fighting Ships,* 6:270–71.

29. Ewing and Worzel, "Long Range Sound Transmission."

30. Worzel interview.

31. Ewing and Worzel, "Long Range Sound Transmission."

32. Lockwood to McCann and Watkins, 2 June 1944, box 15, ser. 4, file 1, SOC.

33. Newhouse to Iselin, n.d., July, 1945, box 6, OD/WHOI; and Ewing and Worzel, "Long Range Sound Transmission."

34. "Mar. 15, 1945 Conference at W. H. O. I.," Sofar general file, file drawer 3D, Office Papers of William Maurice Ewing, Creighton A. Burke, and J. Lamar Worzel, InGeo/UT. 34. BuOrd to BuShips, 29 Aug. 1945, box 155, BuShips Secret Gencorr, 1945.

Chapter 10. Transition, 1945–46

1. Furer to Lockwood, 3 Feb. 1945, box 84, SONRD Gencorr.

2. Ibid.; Lockwood to Furer, 20 Jan. 1945; and T. E. Shea and T. K. Glennan to Lockwood, 15 Feb. 1945, box 84, SONRD Gencorr; and ComSubPac to commander in chief, U.S. Fleet, 24 Feb. 1945, box 63, BuShips Confidential Centcorr, 1945.

3. Bergmann to staff, CUDWR Sonar Analysis Group, 6 Mar. 1946; NRL to chief, ORI, 23 Apr. 1946; and Office of Emergency Management to BuShips, 5 Apr. 1946 (final report of the CUDWR Sonar Analysis Group), box 51, BuShips Secret Gencorr, 1946.

4. HUSL, "Final Report for Contract OEMsr-58 and Contract OEMsr-287, 1941–1946—Applied Acoustics in Subsurface Warfare," 31 Jan. 1946, box 738, OSRD/NDRC.

5. W. V. Houston, "Final Report of the Columbia Division of War Research Summarizing the Work Done by the Programs Analysis Group," 1 Aug. 1944, box 687; "Final Report: Contracts OEMsr-20 and OEMsr-1128," CUDWR New London laboratory, no box (bound volume); "Final Report: Contract OEMsr-1128: A Summary of the Work of the New London Laboratory on Equipment and Methods for Submarine and Subsurface Warfare, 1941–1945," 30 June 1945, no box (bound volume); "Final Report—Contract OEMsr-1128—Field Engineering Group, Columbia University Division of War Research," vol. 1, 30 June 1945; and "Final Report—Contract OEMsr-1130—Activities of the Underwater Sound Reference Laboratories, May 1, 1942 to Nov. 1, 1945," 31 Oct. 1945, box 720, OSRD/NDRC. Completion report made to the Chief of BuShips covering the operations of the UCDWR at the NEL, Contract NObs2074 (OEMar-30), 26 Apr. 1941 to 30 June 1946, no box (bound volume), NOSC/TL.

6. Harnwell to Underhill, 8 Jan. 1945; Underhill to Sproul, 11 Jan. 1945; and Underhill to Sproul, 10 May 1945, CH/NHC.

7. Kurie to Sproul and Sproul to Cochrane, 8 Feb. 1946; and Sproul to Kurie, 7 Mar. 1946, CH/NHC.

8. Cochrane to Sproul, 31 Jan. 1946, CH/NHC.

9. Sverdrup to Sproul, 8 Mar. 1946; Underhill to Sproul, 9 Mar. 1946; and Sproul to Sverdrup, 14 Mar. 1946, CH/NHC.

10. BuShips to ORI, 10 May 1946; asst. SecNav/ORI to Sproul, and part of ORI Contract N6ori111, 15 May 1946, CH/NHC.

11. Eckart to Sproul, 25 May and 17 June 1946, and 28 Jan. 1947; Day to Watson, 28 May 1986; Sproul to Eckart, 7 June and 5 July (basic MPL budget) 1946; Kurie to Sproul, 10 June 1946; NEL "All Hands Memorandum," 8 July 1946; BuShips to Regents, University of California, 25 Nov. 1946; Eckart to Knudsen (Eckart proposal that MPL become a department within the Scripps Institution), 14 Apr. 1948; and Carolyn L. Ebrahimi and Kenneth M. Watson, "Marine Physical Laboratory, A Brief History," n.d., CH/NHC; Shor, *Scripps Institution of Oceanography,* 79–87, esp. page 80.

12. Public relations brochure, "The U.S. Navy Electronics Laboratory," 1947, file N, box 9, OD/WHOI; and completion report made to the Chief of BuShips covering the operations of the UCDWR at the NEL, Contract NObs2074 [OEMar-30], 26 Apr. 1941 to 30 June 1946.

13. Marvin Lasky, "Historical Review of Undersea Warfare Planning and Organization 1945–1960 with Emphasis on the Role of the Office of Naval Research," Submarines/Undersea Warfare Division (Submarine Officers' Conference), box 15; and Styer to Lockwood, 31 Jan. 1945, series 4, file 1, post–1 Jan. 1946 command file, AR/NHC.

14. Iselin to Spitzer, 13 Jan. and 20 June 1947; Spitzer to Iselin, 15 Jan. and 16 June 1947; and Spitzer to Bergmann, 15 Jan. 1947, file S; Bergmann to Iselin, 3 July 1947; and Bergmann to "All our Friends," 27 June 1947, file B; and Iselin to Harnwell, 4 Aug. 1947, file H, box 9, OD/WHOI; BuShips to SONRD, 15 Feb. 1945; Furer to Stewart, 22 Feb. 1945; and BuShips to Iselin, n.d. (probably Apr.) 1945, box 63, SONRD Gencorr.

15. Cochrane to Iselin, 31 Jan. 1946, box 8, OD/WHOI.

16. Ibid.

17. BuShips to BuOrd, n.d., 1946; and BuShips to director, NEL, 21 June 1946, box 8, OD/WHOI.

18. Worzel to Sands, 5 and 7 Feb. 1946; and Worzel to Nicholson, 7 Mar. 1946, box 8, OD/WHOI.

19. Peoples to Sands, 15 Feb. 1946; Sands to BuShips, 25 Feb. 1946; Bennett to director, NEL et al., 13 June 1946; and Berkley to chief, ORI, 9 Aug. 1946; and commander, Mare Island Naval Shipyard to BuShips, SOFAR Project Informal Status Report, 7 Jan. 1947, box 8, OD/WHOI; and Sands to Iselin, 10 May 1947, box 9, OD/WHOI.

Allyn Vine interviews, 27 Apr. and 23 Sept. 1989, NHC Oral Histories, AR/NHC.

Work on continued wartime projects, e.g., "Indirect Measurement Techniques," n.d., and Spitzer to BuShips (Code 940) with attached reports (harbor defense beacons), 24 Jan. 1945, box 153, BuShips Secret Gencorr, 1945; and memorandum for technical director, US-NUSL (sonar, torpedo tracking), 30 Aug. 1945; "Sonar—The Present State," paper no. 5, 1946, box 51, BuShips Secret Gencorr, 1946.

Memorandum for Admiral Bryan (Hydro management of BT data program), 8 Feb. 1945, box 2, Roger Revelle Papers.

NEL and Scripps collection of BT observations on the Pacific Bathythermograph Program, Jan., 1946, box 20, Folsom Papers, SIO.

Conrad to Massa, 22 Apr. 1945; Conrad to BuShips, 23 Apr. 1945; Shankland to Furer, 24 May 1945; Schindler to ORI, 18 June 1945; Dow to ORI, 7 July 1945; van Keuren to BuShips,

19 May 1945; and Burwell, memorandum to director, ORI Research and Development Division, 11 June and 9 July 1945, box 24; memorandum to SecNav, 1 Aug. 1945; and Stephenson to director, NRL, 11 July 1945, box 61; and Revelle to Colpitts (hydrophone development), 14 Apr. 1945, box 63, SONRD Gencorr.

20. Worzel interview.

21. Ibid.; and Worzel Oral Commentary (see sections on SOFAR and the creation of Lamont Geological Observatory), 10 Dec. 1980 (copy from Tharp), CH/NHC.

Interpolation. Wartime Observations

1. Ewing to Iselin, 29 Dec. 1943, box 5, OD/WHOI.

2. Clifford Geertz, *Local Knowledge: Further Essays in Interpretative Anthropology*, 10, 96.

3. Constant, "Genesis N+1"; Layton, "Technology as Knowledge," 31–41; Derek J. De Solla Price, "Is Technology Historically Independent," 553–95; and MacKenzie, *Inventing Accuracy*, 28–29.

4. Harald Sverdrup, Richard Fleming, and Martin Johnson, *The Oceans*.

Chapter 11. Crossroads, 1945–46

1. Iselin to Bigelow, 25 Feb. 1946, box 8, OD/WHOI.

2. Ibid.

3. Iselin to Sears, 4 Mar. 1946, box 8, OD/WHOI.

4. Ibid. See also Joint Task Force One, O13B to O1B and O14K, 11 May 1946, box 8, OD/WHOI; Willard Bascom interview, 13 July 1993, NHC Oral Histories, AR/NHC, and W. A. Shurcliff, *Bombs at Bikini: The Official Report of Operation Crossroads*, esp. chaps. 2, 3, 7, and 8.

5. Engeman talk with Holter, 13 Feb. 1946; Blandy to Westmore, 25 Feb. 1946; Blandy to USGS, 8 Mar. 1946; Iselin to Sverdrup, 28 Mar. 1946; "Organization and Personnel of Oceanographic Section," Apr., 1946; Dow to director, NEL, "History of Oceanographic Section JTF-1"; Revelle, "Bikini Revisited," n.d., 1947; Harry S. Ladd et al., "Drilling on Bikini Atoll," n.d., 1947; Ladd to Johnson, 30 Apr. 1947; Leahy to SecNav, 16 May 1947; Sherman to CNO, 2 June 1947; BuShips to NEL, 13 June 1947; and Revelle to BuShips, 28 June 1947, box 6, SIO subject files, 1903–81; John D. Isaacs, "Surveying by Cameras at Bikini," box 16, Isaacs Papers; Revelle to Bumpus, ca. 4 Mar. 1947; Schultz to Revelle, 18 Mar. 1947; and "Bikini 1951–1953," box 2, Roger Revelle Papers, SIO.

6. Austin to Iselin, 23 Mar. 1946, box 8, OD/WHOI.

7. Ibid.

8. Riley to Iselin, 31 Mar. 1946, box 8, OD/WHOI.

9. Iselin to BuShips Code 940, 25 June 1946, box 8, OD/WHOI.

10. Iselin to Riley, 9 Apr. 1946. The USS *Mentor* (PYC-37), a converted coastal patrol yacht, supported cooperative underwater sound work done by WHOI and the USNUSL from May, 1945, through August, 1946.

11. Ibid.

12. Vine to Iselin, 28 Apr. 1946, box 8, OD/WHOI.

13. Ibid., 9 May 1946.

14. Ibid., 20 May 1946.

15. Ibid., 16 May 1946. The New London Group to which Vine referred was the US-NUSL.

16. Ibid., 28 Apr. 1946.

17. Johnson to Sverdrup, 10 Mar. and 7 Apr. 1946, box 6, SIO subject files, 1903–81, SIO.

18. Operation Crossroads, JTF-1, Bikini Atoll, Marshall Islands, 1946–47, accession no. 92–18, box 1 (of 1); and JTF-1 technical staff, "Organizational Chart of Oceanographic Section," Atomic Bomb Tests, NRC: Navy: Operation Crossroads, Agencies and Departments 1946, NAS.

19. Carr to Bigelow, 11 Mar. 1946; and Bigelow to Carr, 18 Mar. 1946, Henry Bryant Bigelow Papers, MCZ.

"Oceanographic Expedition to Bikini Atoll, Feb. 21–Mar. 28, 1946 Preliminary Draft," 15 Jan. 1946; Iselin: memorandum to Revelle, 20 Mar. 1946; Allyn Vine, "Oceanographic Instrumentation Program," n.d., 1946; and Bumpus to Iselin, 4 July 1946, box 8, OD/WHOI.

"National Academy of Sciences' Nominations for Civilian Scientific Observers for Operation Crossroads," n.d., 1946, Atomic Bomb Tests, NRC: Navy: Operation Crossroads, Agencies and Departments 1946, NAS.

William von Arx (WHOI), "Summary of the Preliminary Current Survey at Bikini Atoll," box 55, CAH/Ewing.

Comdr. Charles Hendricks, "The Urgent Need for a Controlled Atomic Test for Obtaining Underwater Pressure Data for Submarine Warfare," n.d., 1952; and ONR research and liaison officer, Scripps Institution, to CNR, 1 Oct. 1952, box 19, Roger Revelle Papers; Willard Bascom, "Mike," n.d., box 9, Bascom Papers; and "Results of Oceanographic Survey at Enewetok, Nov.–Dec. 1956" (Redwing aftermath?), 5 Aug. 1957, box 16, Isaacs Papers, SIO.

The staff at Scripps also assisted in this very same way at the other major nuclear tests during the 1950s. Operations Greenhouse and Ivy, especially the "Mike" shot during the latter series in 1952, preoccupied Willard Bascom, Walter Munk, John Isaacs, John Knauss, Ted Folsom and a considerable roster of others. The six shot Castle series took place between Mar. and May of 1954 also at the Bikini and Enewetok sites. The thermonuclear "Mike" detonation and the unexpected force and fallout levels presented particular difficulties in wave measurement and team logistics. See Bascom to Van Dorn et al., 30 Sept. 1952, box 13, Folsom Papers, SIO.

Chapter 12. Shaping the Postwar Dialogue, 1946–50

1. Spitzer to Iselin, 22 Jan. 1945, folder 24, Jan.–June 1945 correspondence box, Office of the Director (Sverdrup), 1936–48, SIO.

2. Iselin to Bryan, 3 Jan. 1946, box 7, OD/WHOI.

3. Bowen to HO, 31 Jan. 1946; and Bowen to CNO (OP-43 [Sears]) et al, 24 Apr. 1946, box 7, OD/WHOI.

4. Iselin to Sears, 16 Apr. 1946, box 7, OD/WHOI.

5. Sears to Iselin, n.d. Apr. 1946, box 7, OD/WHOI.

6. The above discussion is also derived from the following: Iselin to Redfield, 10 June 1945, box 6; Pollak to Iselin, 22 Nov. 1945; Oceanographic Meeting called by ORI, 8 Feb. 1946; Iselin to Fleming, 24 Sept. 1946; and Division of Oceanography Quarterly Reports, 30 Nov. 1946, 8 Aug. 1947, 5 Feb. 1947, box 7, OD/WHOI.

7. Iselin to Fleming, 14 Nov. 1946, box 8, OD/WHOI.

8. Iselin to Fleming, 29 Oct. 1946 and 27 Jan. 1947; Fleming to Iselin, 16 Dec. 1946; and Proposed Organization Chart, Oceanographic Division, 15 Sept. 1947, box 8, OD/WHOI.

9. Ewing to Fleming, 10 Nov. 1947; memorandum for Mr. Traynor, 11 Feb. 1948; Ryder to Pollak, 14 Apr. 1948; Fleming to Iselin, 15 Apr. 1948; Fuglister to Fleming, 2 June 1948; Annual Report of the Division of Oceanography, Fiscal Year 1948, n.d., 1949; and Coleman to

Iselin, 22 Dec. 1948, box 9; and Wehe (aboard the USS *Maury*) to Iselin, 29 Sept., 26 Oct., and 18 Nov. 1948, box 10, OD/WHOI.

10. Iselin to Pollak (aboard *Atlantis*), 31 Mar. and 10 May 1948, box 9, OD/WHOI.

11. Iselin to Coleman, 14 Dec. 1948, box 9, OD/WHOI.

12. Fleming to administrative services officer, 16 Feb. 1949; Fleming to HO, 12 Apr. 1949; and Lyman to Iselin, 8 Nov. 1949, box 10; Austin to Iselin, 2 Aug. 1949, notice of Fleming's departure from Hydro for the University of Washington, 1950, HO file; and Hobbs to director, Naval Ordnance Laboratory, 7 July 1950, box 11; and HO, Division of Oceanography, monthly report for Dec., 1949, 17 Jan. 1950, box 12, OD/WHOI.

13. Spitzer to Iselin, 22 Jan. 1945, folder 24, Jan.–June 1945 correspondence box, Office of the Director (Sverdrup), 1936–48, SIO.

Naval Research Reviews, May, 1969, 5–6, box 7, subject files, NSRDC/DTNSRDC, RC-7, NSWC.

Richard H. Fleming, ONR Undersea Warfare Branch, 14 Nov. 1949 (Confidential), accession no. 65-A-4221, box 12, ONR, WNRC.

Richard H. Fleming, memorandum for file, "Project AMOS," 23 Feb. 1951; Undersea Warfare Branch, ONR to CNR, 27 Jan. 1950; Project AMOS, n.d. 1951 (possibly 1952); John Lyman, memorandum for file, "Background of AMOS Surveys and Sonar Charts," 5 Aug. 1953; H. E. Nash, H. W. Marsh, Jr., and Morris Schulkin, "Summary of AMOS Project Program as Presented in Three Papers before the Oceanographic Panel of the Research and Development Board Meeting at USNUSL," 19–20 Feb. 1953, USL Technical Memorandum no. 1100–08–53, accession no. 67-A-7292, box 1, Navoceano.

14. Bowen to division directors, 31 May 1945; and ORI organization chart, 31 May 1945, box 5, SONRD Gencorr.

15. Programs Division, Research and Development Board, "Organization and Functions of the Secretariat," 10 Aug. 1948; "Research and Development Board: History and Functions," Committee on Technical Information—Meetings, Research and Development Board, NRC: National Military Establishment, Agencies and Departments 1948, James Forrestal, SecNav directive, "Research and Development Board," 18 Dec. 1947, Organizations and Functions 1949: Research and Development Board, National Military Establishment, NRC, Agencies and Departments, 1949, NAS.

16. Piggot to Iselin, 28 Feb. and 5 Mar. 1947; and Iselin to Piggot, 21 Feb. and 5 Mar. 1947, box 9, OD/WHOI. Bush to Lee, 1 Nov. 1946; Bush to Joint Research and Development Board, 1 Nov. 1946; CNO to BuAer et al., 6 Nov. 1946; and Lee to Bush, 15 Nov. 1946, Naval Research Advisory Committee, NRC: Navy: ONR; Noyes to Truman, 21 Aug. 1950; Kimball to Steelman, 22 Aug. 1950; and Bronk to Waterman, 9 Sept. 1950, Navy, NAS-NRC Agencies and Departments 1950, NAS. Charter, Joint Research and Development Board, 3 July 1946; preliminary draft directives, "Formation of the Panel on Oceanography of the Committee on Geophysical Sciences of the Joint Research and Development Board," 11 and 27 Mar. 1947; suggested membership, Panel on Oceanography of the Committee on Geophysical Sciences, 25 Mar. 1947, box 25, subject files; and O. W. Helm, "Genesis of the Naval Research Advisory Committee," *Research Reviews* (reprint), Sept., 1957, 23 ff, box 13, Roger Revelle Papers, SIO.

Capt. Robert Conrad, memorandum for chief, Research and Inventions, 21 June 1945, box 5; Capt. M. J. Lawrence, memorandum for Admiral Bowen, 25 Oct. 1945; and SONRD Planning Division to BuShips et al, n.d. [Spring] 1945, box 63, SONRD Gencorr.

17. Gordon Lill interview, 9 May 1995, NHC Oral Histories, AR/NHC.

18. Loeb to Sverdrup, 30 Oct. 1945, box 1, Officer of the Director (Sverdrup), 1936–48, SIO.

19. Lill interview, 9 May 1995; and Gordon Lill, interview by Robert A. Calvert, 15 Mar. 1977, A&M/OH.

20. Lill interview, 9 May 1995.

21. Ibid.; and Lill interview 15 Mar. 1977.

22. Lill interview, 9 May 1995.

23. Arthur Maxwell interview, 23 Sept. 1993, NHC Oral Histories, AR/NHC.

24. Ibid.; and Lill interviews, 15 Mar. 1977 and 9 May 1995.

25. John Knauss interview, 26 June 1995, NHC Oral Histories, AR/NHC.

26. Ibid.

27. Lill interview, 15 Mar. 1977; and Robert Abel interview, 11 Mar. 1996; John Brackett Hersey interviews, 4, 11, and 23 Oct. 1991; and Gordon Hamilton interview, 15 Sept. 1993, NHC Oral Histories, AR/NHC.

S. W. Doroff, "Trip Report, Visit to the CUW Noise Reduction Group, MIT," 4 Apr. 1955 (Confidential); and "Trip Report, 3rd Noise Group Control Meeting," 24 June 1955 (Confidential); and Lt. Comdr. Paolucci, "Trip Report [along the northeast corridor]," 25 July 1955 (Confidential), accession no. 64-A-4211, box 10, rrip report illustrations, ONR, WNRC.

Sverdrup to Dykstra, 26 Dec. 1945, box 1, Office of the Director (Sverdrup), 1936–48, SIO.

Iselin to ORI, 11 June 1946 (Iselin's effort to make WHOI the center for correlating all existing Pacific marine biological data as per a resolution passed by the Pacific Science Conference); Daspit to Iselin, 25 June 1946 (accepting Iselin's 11 June proposal); Iselin to ORI Planning Division, 12 July 1946; Bowen to Iselin, 21 July 1946; Fleming to Iselin, 30 Aug. 1946; Iselin to Fleming, 7 Sept. 1946, box 8; Revelle to Iselin, 30 Jan. 1947; and Iselin to Liebermann, 16 May 1947, box 9, OD/WHOI. House Resolution 5911 (an act to create ONR within the Navy Department), 21 May 1946, general, NAS-NRC: Navy, Agencies and Departments, 1946, NAS.

Sapolsky, *Science and the Navy,* chaps. 1–3.

When Revelle left BuShips for ONR, the bureau needed someone as familiar with oceanography and its applications as Revelle. Although not interested in leaving Woods Hole permanently, Allyn Vine stepped into Revelle's shoes at BuShips until the bureau found a replacement. Revelle suggested that Vine had replaced BuShips's Jeff Holter, but the job description was too much like Revelle's to suggest that Vine moved into Holter's position. Neither Holter nor Vine had sufficient experience to replace Revelle as administrator, but since both Revelle and Holter were leaving, Vine was the nominee (Revelle to Steenland, 18 Feb. 1947, box 9, OD/WHOI).

28. Lee to Assistant SecNav, 19 Dec. 1946, general, NRC: Navy: ONR Agencies and Departments, NAS.

29. Compton to Weaver, 24 Dec. 1946, general, NRC: Navy: ONR Agencies and Departments, NAS.

30. Ibid.; and Weaver to Bronk et al., 20 Dec. 1946, general, NRC: Navy: ONR Agencies and Departments, NAS.

31. Bronk to Jayne, 28 Feb. 1950; and Jayne to Bronk, 22 Feb. 1950, Navy, NAS-NRC Agencies and Departments 1950; organization chart, RDB Committee on Geophysics and Geography, 15 Mar. 1949, RDB organization charts, NRC: National Military Establishment, Agencies and Departments; comments by Planning Division, NRAC, 12 Nov. 1946; chairman to NRAC members, 3 and 16 Jan. 1947, ONR Planning Division memorandum, 14 Jan. 1947; NRAC minutes, 16 Jan. 1947; Jayne to NRAC Members, 24 Jan. 1947; Appointment of

Committee on Geophysics Advisory to ONR, 7–8 Jan. 1948; Gibbs to Newhouse et al., 13 May 1948; Gibbs to Beers et al., 28 Sept. 1948; Report on the Second Meeting of the Committee on Geophysics Advisory to ONR, 12–13 Mar. 1948, Committee on Geophysics Advisory to ONR, Executive Board; draft letter to SecNav James Forrestal, n.d.; and Lee to Bronk, 28 June 1948, general, NRC: Navy: ONR Agencies and Departments; and Alan T. Waterman, "Research and the Navy," 27 Dec. 1947, AAAS: Annual Meeting: Speeches, Institutions, Associations, Individuals, Division NRC: PS, NAS.

Rear Adm. Thorvald A. Solberg, "Office of Naval Research in Anti-Submarine Problems," 16 May 1949, box 7, RC 21–5, Lasky.

Robert J. Mindak, "Management Studies and Their Effect on the Navy," 1 Nov. 1947, post-1 Jan. 1946 command file, AR/NHC. "Report of the Working Group on Major Unsolved Problems in Oceanography," Panel on Oceanography, Committee on the Geophysical Sciences of the Research and Development Board, 16 Jan. 1948, box 10; RDB directive, "Committee on Geophysics and Geography," 9 Nov. 1948; Louis Johnson, "Directive: Research and Development Board," 14 Sept. 1949; and Committee on Geophysics and Geography organizational chart, 19 May 1950, box 11, OD/WHOI.

Coleman to BuShips, 4 Aug. 1947, box 906, BuShips Centcorr.

"Naval Research," 28 May 1948, box 9, General Board Navy Department, "Office of Naval Research: A Statement of Policy and Organization," Dec., 1948, box 22 (ser. 3, file A-11), General Board S/F, NA. Sapolsky, *Science and the Navy,* 60–63; Vannevar Bush, *Science: The Endless Frontier;* and Jeffrey G. Barlow, *Revolt of the Admirals: The Fight for Naval Aviation, 1945–1950.* Barlow's is the best work on the naval and political effects of the "revolt of the admirals" episode.

32. James Probus, "History and Activities of the Committee on Undersea Warfare, 1946–1956," Sept., 1955 (Confidential), NSB. Bronk to Harnwell, 18 July 1946, NAS.

33. Probus, "History and Activities of the Committee." The fourth in this series of responsibilities was added on 15 June 1951.

34. Coleman to Bronk, 17 Oct. 1946; and Bowen to Bronk, 23 Oct. 1946, NAS.

35. Probus, "History and Activities of the Committee." ONR memorandum, Surface and Subsurface Section to Program Branch, 31 Oct. 1946; Meid to Tate, 18 Nov. 1946; Harnwell to Daspit, 11 Nov. 1946; Lucker to Harnwell, 21 Nov. 1946; Mills to ONR, 2 Dec. 1946 (this is Mill's initial proposal to have CUW absorb the Sonar Analysis Group; Tate was against it); "Annual Report of the CUW," 3 July 1947; Zwemer to Bronk, 3 Dec. 1947; Zwemer to Hibbard, 3 Sept. 1948; "Annual Report of the CUW 1949–50," 30 June 1950, NRC Executive Board—Committee on Undersea Warfare; and G. D. Meid to Daspit, 12 Dec. 1946, NRC-Executive Board-Committee on Undersea Warfare 1946, NAS.

"Meeting of the CUW Committee on the Hydrodynamics of Submerged Bodies," 1 Mar., 20 Oct., and 7 Nov. 1949; "Digest of the Minutes, Joint NIA-NRC Conference and report on the Joint NIA-NRC Conference," 17 Nov. 1948; CUW Panel on Underwater Acoustics, Basic Problems of Underwater Acoustics Research; and CUW, "A Survey Report on Antisubmarine Weapons Systems," July 1950 (Confidential), NSB.

John Coleman interview, 26 May 1994, NHC Oral Histories, AR/NHC.

Thompson to Iselin (with enclosure), 14 Aug. 1947, box 9, OD/WHOI.

Hess to Worzel, 21 Mar. 1950; and Worzel to Hess, 14 June 1950, box 250, CAH/Ewing.

36. Lucker to Harnwell, 21 Nov. 1946

37. Program for the Second Symposium on Undersea Warfare, 5–6 May 1947, box 9, OD/WHOI.

39. Probus, "History and Activities of the Committee."

Chapter 13. The Forest and the Trees, 1946–50

1. Columbus O'Donnell Iselin, "Twenty-First Annual Report of the Woods Hole Oceanographic Institution, 1949–1950," n.d., 1950(discusses the effect of the war on the institution and the discipline), Annual Reports, WHOI.

2. Iselin to Ewing n.d. (mid-Feb.), 1946; Iselin to Ewing, n.d. (early Mar.), 1946, box 8, OD/WHOI.

3. Couper to Iselin, 14 Aug. 1946, box 8, OD/WHOI.

4. Iselin to Sears, 24 Sept. 1946, box 8, OD/WHOI.

5. Ibid.

6. Digest of Iselin's statements (notetaker unknown), Second Undersea Warfare Symposium, Sept. 1947, NRC: Executive Board: Committee on Undersea Warfare, 1947, NAS.

7. Ibid.

8. Ibid.

9. Iselin to Sears, 13 Feb. 1946, box 8, OD/WHOI.

10. Sears to Iselin, n.d., 1946, box 8, OD/WHOI.

11. Iselin to Sears, 14 and 8 Mar. 1946; Sears to Iselin, n.d. (one letter probably written in Mar. and another in Mar. or Apr.), 1946; Iselin to Sverdrup, 13 Nov. and 4 Dec. 1946, box 8; Iselin to Revelle, 20 Feb. 1947; Revelle to Iselin, 17 Mar. 1947; Bigelow to Jewett, 21 Aug. 1947; Iselin to Bigelow, 24 Nov. 1947; and Iselin to Jewett, 11 Dec. 1947, box 9; Spilhaus to Eckart et al., 7 Jan. 1949; Iselin to Spilhaus, 13 Jan. 1949; and Iselin to Coleman, 22 Aug. and 19 Dec. 1949, box 11; and Vine to Landsberg, 23 June 1950; box 12, OD/WHOI.

12. Record of the meeting of the WHOI Board of Trustees, 12 Aug. 1948, Woods Hole Oceanographic Institution, Institutions, Associations, and Individuals 1948, NAS. See also "Meeting held at Woods Hole Oceanographic Institution to Talk about Revival of Interest in Oceanography by the Academy," 11 Aug. 1948, ibid.

13. Columbus O'Donnell Iselin, memorandum to Dr. Alfred N. Richards and Dr. Detlev Bronk, 1 Sept. 1948, Woods Hole Oceanographic Institution, Institutions, Associations, and Individuals 1948, NAS.

14. "Report of the President," 14 Nov. 1948, BP:COO, NAS.

15. Richards to Bigelow, 11 Feb. 1949, BP:COO, NAS.

16. Bigelow to Richards, 24 Feb. 1949, BP:COO, NAS.

17. Iselin to Richards, 8 Apr. 1949, BP:COO, NAS.

18. Richards to Ewing, 25 Apr. 1949, BP:COO, NAS.

19. Iselin to Richards, 11 May 1949, BP:COO, NAS.

20. C. O'D. Iselin, "Summary of a Conference on Oceanography sponsored by the National Academy of Sciences, May 20–21, 1949" (Wardman Park), n.d., 1949, BP:COO, NAS.

21. Carl Eckart and Roger Revelle, "Proposed Report to the National Academy of Sciences," (rough draft of *Oceanography 1951*), 29 Aug. 1949, BP:COO, NAS.

22. "Proceedings: Conference on Oceanography Sponsored by the National Academy of Sciences," 20–21 May 1949, Wardman ParkHotel, Washington D.C., Proceedings, NAS: Committee on Oceanography Conference, Organization May 1948; Smith to Richards, 15 Jan. 1949; "Panel to Inspect Oceanographic Installations," press release, Office of Public Information, National Military Establishment National, 27 Apr. 1949; Jewett to Richards, 9 May 1949; Eckart to Iselin, 30 Aug. 1949; Eckart to Redfield, 30 Aug. 1949; Eckart to Bronk, 30 Aug. 1949; and Redfield to Eckart, 19 Sept. 1949, general, NAS: Committee on Oceanography, Organization May 1949; Columbus O'Donnell Iselin, "Twentieth Annual Report of the Woods Hole Oceanographic Institution, 1948–1949," n.d., 1949 (includes

Iselin's argument for better balance in research priorities and projects), Woods Hole Oceanographic Institution, Institutions, Associations, and Individuals 1949; Redfield to Bronk, 10
Jan. 1951; and Redfield to Bronk, with enclosures, 27 Aug. 1951, Committee on Oceanography, NAS Organizations 1951, NAS.

Mallory to BuShips Code 940, 21 Mar. 1946, box 52, BuShips Secret Gencorr, 1946.
Maurice Ewing's handwritten notes on the Wardman Park Conference for Saturday, 21 May
1949; and Richards to Ewing, 25 Apr. 1949, box 54, CAH/Ewing.

23. Study of Undersea Warfare (Low Report), 22 Apr. 1950, post-1 Jan. 1946 command
file, AR/NHC.

24. Harnwell to Bronk, 22 Aug. 1951, Committee on Undersea Warfare 1951, NRC Executive Board, NAS.

25. R. B. Livingstone, memorandum on "Conference to Establish Better Communica-
tion Between Civilian and Military Scientists," Executive Board: Committee on Undersea
Warfare 1951, NAS; and Warren Weaver to chairman, Joint Research and Development
Board, 3 June 1953, box 14, OD/WHOI.

26. Ketchum to Bronk, 8 Feb. 1954; S. D. Cornell to Philip Armstrong et al., 16 Feb.
1954; Bronk to Spilhaus, 3 Mar. 1954, NRC Governing Board 1954; Oceanographic Convocation Program, 22-24 June 1954, Oceanographic Convocation at Woods Hole, Sessions;
Oceanographic Convocation at Woods Hole, NRC Steering Committee 1954; Oceanographic Convocation, 9 Apr. 1954; Redfield to Bronk, 17 Aug. 1954; Lill to Redfield, 20 Aug.
1954; and minutes of the Steering Committee, Oceanographic Convocation, 4 and 25 Mar.
1954, Oceanographic Convocation at Woods Hole, NRC general 1954, NAS.

Chapter 14. Back to Sea with a Flourish, 1946-55

1. Ewing to BuShips, 25 Feb. 1948, box 10, WHOI; and Ewing, "Exploring the Mid-
Atlantic Ridge," *National Geographic* 94, no. 3 (Sept., 1948): 275-94.

2. Coordinator of Undersea Warfare to CNO, "Submarine Conference of 25 Feb. 1948,"
9 Mar. 1948, box 7, ser. 2, file 2, SOC.

3. Ibid.; and Iselin to Hendricks, 31 Mar. 1947, box 9, WHOI.

4. Martin J. Pollak, "*Atlantis* Cruise 151 to the Mediterranean Area, Technical Report no.
2: Echo Soundings and Bathymetric Chart of the Aegean Sea," Nov., 1948, Report Collection, WHOI.

5. David M. Owen, "*Atlantis* Cruise 151 to the Mediterranean Area, Technical Report no.
3: Radar Scope and Coastline Photography," Nov., 1948, Report Collection, WHOI.

6. Abel interview.

7. Martin J. Pollak, "*Atlantis* Cruise 151 to the Mediterranean Area, Technical Report no.
1: Navigational Control in Reconnaissance Surveying," July, 1948; "Cruise Reference Log:
Atlantis 151, with Director's orders to the master," 6 Dec. 1947; and "Preliminary Report of
the Scientific Results of *Atlantis* Cruise 151 to the Mediterranean Area," Sept., 1948, Report
Collection, WHOI. For more information on the *Atlantis 151* cruise see the following WHOI
blue cover reports filed by Martin Pollak, Dean Bumpus, and the cruise staff: reference nos.
49-2, 49-5, 49-8, 49-24, 49-31, and 49-39, Report Collection, WHOI.

8. Ewing to BuOrd, 18 Apr. 1948. See also Ewing to Ryder, 22 Apr. 1948; Geyer to Ryder, 9 Mar. 1948 (refers to loan of geophone equipment to WHOI from the Humble Oil and
Refining Company), box 10, OD/WHOI; and Ewing, "Exploring the Mid-Atlantic Ridge,"
275-94.

9. Iselin to Smith, 17 Feb. 1948; Hersey to Iselin, 18 Apr. 1948; Ryder to Hersey, 27 Apr.

1948; Iselin to Montgomery, 9 June 1948; Iselin to Grosvenor, 28 June 1948; Iselin to Bigelow, 30 June 1948; and Marley to Iselin, 29 Dec. 1948 (Allyn Vine was a regular part-time consultant in BuShips Code 942 temporarily taking the position resigned by Revelle), box 10, OD/WHOI.

10. The above discussion is derived mainly from the following sources: Bergmann to Tucker, 13 Dec. 1946 (for work with the U-3008), box 8, OD/WHOI; BuShips to CNO, 30 July 1945, box 10, ser. 3, file 5; William Schevill and Allyn Vine, "Submerged Performance Tests on German Type XXI Submarines," 17 Mar. 1947, box 12, ser. 3, file 1; and WHOI to BuShips, 28 Mar. 1947, box 12, ser. 3, file 3, SOC; Rössler, *Geschichte des deutschen Uboot-baus*, text and appendix charts specific to particular U-boats used for test depth statistics; and Weir, *Forged in War: The Naval-Industrial Complex and American Submarine Construction, 1940–1961*, 130 (for the Balkon array and the GHG).

11. Columbus Iselin, "The Present Status of Long Range Listening," 15 June 1948, NSB.

12. Pollak to Iselin, 30 Apr. 1948, box 10, OD/WHOI.

13. Schevill and Vine, "Submerged Performance Tests on German Type XXI Submarines."

14. Iselin to Hersey, 2 Mar. 1949, box 11, OD/WHOI.

15. WHOI Nineteenth Annual Report, 1947–1948, Hersey to Iselin, 15 Feb. 1949 (note the experimentation with the new ten-thousand-foot depth BT and the presence of Worzel and Gordon Hamilton from Lamont); Hersey to Iselin, 26 Feb. 1949; and Hersey to Iselin, 29 Mar. 1949, box 11, OD/WHOI.

BuShips to CNO, 30 July 1945, box 10, ser. 3, file 5; WHOI to BuShips, 28 Mar. 1947; and Coordinator of Undersea Warfare to CNO, "Submarine Conference of 25 Feb. 1948," 9 Mar. 1948, SOC.

BuShips Code 420 to Code 300, 30 Mar. 1946; and Code 940 to Code 100, 1 Apr. 1946, box 51; USNUSL report no. 59, "Proposal for Extending the Maximum Range of Sonar Passive Ranging" (good reference on the state of passive ranging ca. 1946), 16 Sept. 1946; USNUSL to director, NEL, 6 Sept. 1946; Clement to Hull, 17 Dec. 1946; and L. Batchelder and E. Horton, "Low Frequency Listening Gear for Submarines," 27 Nov. 1946, box 52, BuShips Secret Gencorr, 1946.

Shor, *Scripps Institution of Oceanography*, 26

16. Von Arx's GEK turned out to be a flawed instrument for current measurements. It responded just as well and at the same time to currents and fluctuations in the Earth's magnetic field.

17. F. C. Fuglister and L. V. Worthington, "Hydrography of the Western Atlantic: Some Results of a Multiple Ship Survey of the Gulf Stream, Technical Report no. 18," submitted to the Oceanographic Division, HO under contract no. N6onr-27701 with the ONR, Feb., 1951, WHOI reference no. 51–9; Ryder to Nutt, 6 Mar. 1950; Ford to Ryder, 27 Mar., 17 Apr., and 2 and 16 May 1950; Ryder to Ford, 25 Apr. 1950; and Ryder to Hart, 12 July 1950, box 12, OD/WHOI; Munk to Iselin, 22 and 28 Nov. 1949; LaFond to Munk, 22 Nov. 1949; and Columbus Iselin (?), "A Plan for a Proposed Five Ship Cruise," n.d. (probably 1949), Operation Cabot file, Institution Activities files, 1950–80 (file cabinet); and Cruise Reference Log for *Atlantis* Cruise 163, cruise books, WHOI.

18. "Proposed University of California Mid-Pacific Expedition," 10 Oct. 1949, box 6, SIO subject files, 1903–81, SIO.

19. Hamilton, "We've Only Scratched the Pacific's Bottom," 10 Oct. 1949, box 6, SIO subject files, 1903–81, SIO.

20. "The Mid-Pacific Expedition of the University of California and the U.S. Navy

Electronics Laboratory," Nov., 1950, box 6, SIO subject files, 1903–81, SIO; University of California, Scripps Institution of Oceanography, Progress Report no. 17, July–Sept. 1950, accession no. 16734, box 1, ONR, WNRC; and Shor, *Scripps Institution of Oceanography,* 90–92, 382–94.

21. Lill to Worzel, 8 June 1954, file drawer 2C, office files of Maurice Ewing, InGeo/UT; Ewing to Sverdrup, 12 Feb. 1946; Ewing, "Proposed Measurements at Sea," 30 Apr. 1946; Ewing and Field to Mathes, 1 May 1946; and Ewing to Field, 1 May 1946, box 8; Iselin to Revelle, 17 Feb. and 13 June 1947; Ewing to Iselin, 16 June 1947; Berkley to Iselin, 27 Aug. 1947; WHOI to ONR Geophysics Branch, 19 Jan. 1949; and Ryder to ONR, 26 Jan. 1949, box 9, OD/WHOI.

22. See also the files on the individual gravity activities aboard the USS *Archerfish, Balao, Conger, Corsair, Diablo,* and *Toro,* 1952–56, file drawer 2D, Office Papers of William Maurice Ewing, Creighton A. Burke, and J. Lamar Worzel, InGeo/UT.

23. Stephan to Worzel, 15 May 1954; Worzel to ComSubPac, 1 June 1954; Lill to Worzel, 8 June 1954; Grenfell to Watkins, 9 Nov. 1954; Watkins to Worzel, 13 and 20 Apr. 1955; Worzel to Watkins, 14 and 28 Apr. 1955; G. Lynn Shurbet, Progress Reports no. 19 and 20, 1 Oct. 1954–31 Jan. 1956, N6-onr-271, Task Order VIII, 14 and 26 Feb. 1956; Worzel to Tuve, 28 Feb. 1956; Lynch to Worzel, 7 May 1956; Worzel to Lynch, 31 May 1956 and 10 Oct. 1957; Worzel, "Gravity Measurements at Sea," n.d., 1956; Worzel to CNR, 18 Feb. 1957; and Grenfell to Worzel, 28 June 1957, file drawer 2C, Office Papers of Ewing, Burke, and Worzel, InGeo/UT.

24. Ewing to CNR, 21 Nov. 1960, file drawer 2C, Office Papers of Ewing, Burke, and Worzel, InGeo/UT.

25. "Program Abstracts from the Annual Meeting of the AGU, 30 Apr.–2 May 1951," AGU 1951, NRC Executive Board, NAS; Mackenzie, *Inventing Accuracy,* 193–94, 291–92, 354–64; and Weir, *Forged in War,* chaps. 9–11. Gravity affected the problem of determining the vertical in the basic equation of inertial guidance.

26. Iselin to Revelle, 21 May 1946, box 9, OD/WHOI.

27. Hendricks to Iselin, n.d., 1947; Hendricks to Iselin, 24 July 1947; Iselin to Hendricks, 29 July 1947; and Hendricks to Arctic Institute of North America, 15 Oct. 1947, box 9, OD/WHOI.

28. Iselin to ONR Boston Branch Office, 30 Dec. 1947; and Johns to Iselin, 23 Dec. 1947, box 9, OD/WHOI.

29. Rear Adm. Edward H. Smith, USCG, "Scientific Problems of Naval Warfare in the Arctic," ca. 1948, Institution Confidential (declassified) Collection, WHOI.

30. Smith to Iselin, 26 Oct. 1948; and "A Suggested Method for Implementing Research in Oceanography at the Arctic Research Station, Point Barrow, Alaska," 19–20 Nov. 1948, box 10; "Informal Digest of Information on Ice Research Laboratory, Point Barrow, Alaska," n.d., 1949, box 11; Smith to ONR (proposal), 12 Jan. 1954; and Smith to ONR Code 416, 2 Nov. 1954, box 17, OD/WHOI.

31. Metcalf, Operation Highjump Log; Metcalf, Operation Nanook II Log; Metcalf, U.S. Navy Arctic Operation Log; Metcalf, Point Barrow Log; and Metcalf, *Edisto* Arctic Cruise Log, the William G. Metcalf Historical Manuscript Collection, WHOI.

32. "Chronological History of the Arctic Research Laboratory and Arctic Research Laboratory Advisory Board," Dec., 1948, NAS.

33. ARL Advisory Board, "Minutes of the Second Meeting," 22 Apr. 1948; and Jordan to Miller, "ONR Meeting," 27 Apr. 1948, NAS.

34. Iselin to Revelle, 19 May 1946, box 8, OD/WHOI.

35. "First Report Arctic Research Laboratory Advisory Board," 18 Oct. 1948, NAS.

36. "Scientific Activities 1950–1995" file drawer, Institution History file cabinet; William G. Metcalf and John F. Holmes, "Report on the Oceanographic Program at Point Barrow, Alaska, Feb.–Mar. 1949," June, 1949, Point Barrow folder, the William G. Metcalf Historical Manuscript Collection, WHOI.

37. Eugene C. LaFond, Oceanographic Studies Section, NEL, San Diego, "Summary Report of the Scientific Results of an Oceanographic Cruise to the Bering and Chukchi Seas during the Summer of 1949," CH/NHC; and LaFond, "Arctic Oceanography by Submarines," *U.S. Naval Institute Proceedings* 86, no. 9 (Sept., 1960): 90–96.

38. Lill interview, 9 May 1995; and Waldo Lyon interview, 26 Apr. 1994, NHC Oral Histories, AR/NHC.

39. Lyon interview.

40. Ibid.; and Lill interview, 9 May 1995.

41. T-3, also known in its time as Fletcher Island, was discovered in July, 1950. Ice islands, roughly similar to tabular icebergs in the Southern Hemisphere, are formed when pieces of the shelf ice break away from the Arctic ice pack. They usually drift clockwise around the Arctic Ocean with the prevailing currents. T-3 was 10 miles long and 5 miles wide with a drift rate of usually 2.5 miles per day. In rare circumstances it moved as fast as 8.7 miles per day. The air force kept up its Arctic research on T-3 for at least a few months each year from 1952 until the ice island drifted out of the Arctic Ocean in 1983. It broke up and melted in the Davis Strait between Baffin Island and Greenland's west coast (Dale E. Ingmanson and William J. Wallace, *Oceanography: An Introduction,* 400).

42. Material on Operation Skijump, including Iselin's heavily used "Note on Arctic Oceanography," is found in the ONR Washington file, box 14, OD/WHOI; and the Operation Skijump folder, Institution Confidential (unclassified) Collection, WHOI. See also Smith, "Scientific Problems of Naval Warfare in the Arctic"; Lyon interview; and L. V. Worthington, "Oceanographic Results of Project Skijump I and Skijump II in the Polar Sea, 1951–1952," *Transactions of the American Geophysical Union* 32, no. 4 (Aug., 1953): 543–49.

43. Iselin, "Note on Arctic Oceanography," and other materials from the ONR Washington file; and Operation Skijump folder.

44. "U.S. Program for the IGY, Supplemental Estimate for FY 1955," Mar., 1954, NAS-NRC Agencies and Departments 1954, NAS.

45. "Proposed United States Program for the International Geophysical Year, 1957–1958," Aug., 1955, International Geophysical Year, NRC Central file: IR: IGY 1955, NAS.

46. For basic information on the formation of NSF see "Analysis of legislation now pending in Congress with regard to a postwar program for scientific research," Aug., 1945; "Comparison of Amended National Science Foundation Bills," 18 Oct. 1945; "Initial and Supplementary Statements of Frank B. Jewett . . . Concerning a Postwar Program for Scientific Research," 19 Oct. 1945; and "Memorandum Summarizing Salient Points of Testimony before the Kilgore Committee," 15 Oct. 1945, Hearings: Testimony, Bills: National Science Foundation, Congress, NAS. See also George J. Dufek, *Operation Deepfreeze,* 34–45; and J. Merton England, *A Patron for Pure Science: The National Science Foundation's Formative Years, 1945–1947,* 297–304.

47. Dufek, *Operation Deepfreeze,* 34–45.

Chapter 15. A Closer Relationship, 1950–58

1. Maxwell interview; Worzel interview; Lill interview, 9 May 1995; and Vine interviews, 27 Apr. and 23 Sept. 1989.

2. Elizabeth Bunce interview, 24 Nov. 1993, NHC Oral Histories, AR/NHC.

3. Maxwell interview.

4. Lill interview, 9 May 1995.

5. Ibid.; Maxwell interview; Gordon Hamilton interview; Worzel interview; and Vine interviews, 27 Apr. and 23 Sept. 1989,

6. Comdr. Charles Bishop, "Trip Report," 2 June and 21 Dec. 1956, accession no. 64-A-4211, box 10, ONR, RG 298, WNRC.

7. Bronk to Burke, 16 Dec. 1955; and Burke to Bronk, 23 Dec. 1955, NAS.

8. "Exerpts from Minutes of an Informal Meeting of Member of the Committee on Undersea Warfare, Harvard University," 5 Jan. 1956 (with attached list of suggested participants for Nobska), NAS.

9. Wood to Coleman (with enclosures), 8 Feb. 1956; Lincoln to NAS (ONR contract for the Nobdka Project), 26 Apr. 1956; Bronk to Strauss, 9 May 1956; Strauss to Bronk, 10 May 1956; "Project Nobska Tentative Schedule," 29 May 1956; "A Word about the National Academy of Sciences and the Origins of Project Nobska," 29 May 1956 (with attached "Tentative Group Assignments for the Nobska Faculty" dated 6 July 1956); "Project Nobska: A Tentative Roster of Scientific, Administrative, and Service Personnel," 8 June 1956; "Project Nobska: Activities of Special Interest, 2–9 July 1956," 2 July 1956; Bronk to Burke, 7 July 1956; Burke to Bronk, 11 July and 6 Dec. 1956; "Project Nobska: Partial Listing of Activities of Special Interest Beginning 13 July 1956"; "Project Nobska: Partial Listing of Activities of Special Interest Beginning 23 July 1956"; Wood to Coleman, 7 Sept. and 6 Nov. 1956; Ramage to Bronk, 31 Dec. 1956; George Wood, memorandum to the Committee on Undersea Warfare, 8 Feb. 1957; and CUW Annual Reports, 26 May 1956, 17 Apr. 1957, 7 May 1957, "Physical Sciences—CUW/Project Nobska," NAS.

10. Compiled from a list of basic Nobska preparatory queries in Columbus Iselin, "Project Nobska: A Preliminary Assessment," n.d., 1956, NSB. See also "Project Nobska: Final Report," vol. 1, "Assumptions, Conclusions, and Recommendations," 1 Dec. 1956 (sanitized by the ONR 3 Jan. 1991); and "Review of Project Nobska: Final Report," 5–9 Aug. 1957, CH/NHC; "Project Nobska: Final Report," vol. 2, "Technical and System Studies," 1 Mar. 1957, box 17, 00 files—1956, AR/NHC; and "Program for the Symposium on Aspects of Deep Sea Research, sponsored by ONR and CUW, 29 Feb.–1 Mar. 1956," with attached handwritten notes on the meeting in Maurice Ewing's hand with copies of some paper presentations, box 70, CAH/Ewing.

Because ocean surveillance was absolutely central to the navy's involvement in oceanography, the next chapter will explore this subject in much greater detail than the simple contextual references necessary to the theme of this chapter. For many in the navy, underwater sound was the primary reason oceanography had value to the fleet.

11. "Project Nobska: Final Report," vols. 1 and 2.

12. Ibid., vol. 2.

13. Ibid., vol. 1.

14. Ibid., vols. 1 and 2.

15. Director, OEG, to OP-09, 17 Oct. 1956, box 5, 00 Files–1956, AR/NHC.

16. Director, OEG, office memorandum, 11 Oct. 1956, box 1, 00 Files–1956, AR/NHC.

17. Combs to CNO, 11 Nov. 1956, box 1; BuShips to CNO, "Comments on Project Nobska," 22 Oct. 1956; Combs to CNO, 2 Nov. 1956; Combs, memorandum for VCNO, 7 Nov. 1956; VCNO to DCNO (Fleet Operations and Readiness), 13 Nov. 1956; and Combs memorandum for VCNO, 20 Nov. 1956, box 5; "Project Nobska: Final Report," vol 2, 00 Files–1956, AR/NHC; Director, OEG, to OP-09 (evaluation of "Project Nobska: Final Report," vol. 1),

17 Oct. 1956, (microfilm copy; second copy in oo files—1956, box 5, ooFiles–1956, AR/NHC), CNA.

18. Elliot A. Kearsley, "Survey of Current Research and Development on Long-Range Active Sonar," n.d., 1956, NSB.

19. Comdr. Charles Bishop, "Trip Report," 19 June and 21 Dec. 1956. See also "Deep Water Propagation Committee Meeting, 17 June 1954," 25 Aug. 1954, accession no. 65-A-4221, box 1; CNO to CNR, 18 Oct. 1955 (this is the Burke letter requesting the Nobska Summer Study); CNO to BuShips et al, 14 Nov. 1955; Iselin to Mumma, n.d., 1957; and memorandum by Comdr. D. L. Lassell, "AEC Clearances," 17 Aug. 1957, accession no. 64-A-4211, box 10; Bennett to Strauss, 8 June 1956; Bennett to commander, Operational Development Force, 11 June 1956; CNR to commander, Destroyer Force Atlantic Fleet, 11 June 1956; CNR to CNO et al, 12 June 1956; Ramage to ONR, 13 June 1956; Bennett to chief of staff, USAF, 21 June 1956; Mumma to CNR, 26 June 1956; Welch to Houston, 16 July 1956; "FY58 Plans: Nobska Recommendations—coordination with CNO," 30 Nov. 1956; "The Fifty-One Nobska Recommendations" (excerpted by OP 316), n.d., (probably autumn) 1956; and Code 100 to Code 466, 27 May 1957, accession no. 65-A-4221, box 9; ONR to BuAer, 30 July 1954; accession no. 65-A-4221, box 10; "Deep Water Propagation Committee Meeting," 14 Dec. 1955; and "Fourth Formal Report of the Shallow Water Propagation Committee," 18 Jan. 1956, accession no. 65-A-4284, box 1, ONR, WNRC.

20. "Review of Project Nobska: Final Report, 5–9 Aug. 1957" (sanitized by the ONR 3 Jan. 1991), CH/NHC

21. Ibid., 5, emphasis in original.

22. Ibid.; and "Project Nobska: Final Report," vol. 1; Iselin, "Project Nobska: A Preliminary Assessment"; and "Project Nobska: Final Report Preliminary Edition: Assumptions, Conclusions, Recommendations," 25 Sept. 1956, NSB.

23. Coleman, draft proposal for NASCO, attached to Campbell note, 5 Nov. 1956; and Coleman, final proposal for NASCO, 23 Nov. 1956, Committee on Oceanography—Proposed, NRC Earth Sciences 1956, NAS.

24. Bennett to Bronk, 9 Aug. 1956, NAS-NRC Agencies and Departments—Navy, NAS.

25. Thurston to Coleman, 21 Aug. 1956, Committee on Oceanography—Proposed, NRC Earth Sciences 1956, NAS.

26. Weiss to Campbell, 13 Nov. 1956, Committee on Oceanography—Proposed, NRC Earth Sciences 1956, NAS.

27. Farley to Bronk, 9 Aug. 1956; Cornell to Farley, 14 Aug. 1956; Thurston to Bronk, 27 Aug. and 4 Dec. 1956; Hess to Bronk, 27 Aug. 1956; Fields to Bronk, 26 Sept. 1956; Cornell to Fields, 28 Sept. 1956; Berkner to Cornell, 24 Oct. 1956; Coleman, draft proposal for NASCO; Campbell to Weiss, 8 Nov. 1956; Cornell to Thurston, 19 Nov. 1956; Maynard to Campbell, 20 Nov. 1956; Coleman, final proposal for NASCO; Campbell to Bascom, 24 Nov. 1956; and Thurston to Cornell, 26 Nov. 1956, Committee on Oceanography—Proposed, NRC Earth Sciences 1956; Cornell to Bennett, 14 Aug. 1956, NAS-NRC Agencies and Departments—Navy;

Thurston to Bronk, 7 Jan. 1957; Welpley to Bascom, 7 Jan. 1957; Isaacs to Bronk, 4 Feb. and 4 Mar. 1957; Thurston to Isaacs, 8 Mar. 1957; Hess to Bronk, 8 Apr. 1957; Cornell to Vetter, 14 Oct. 1957; Vetter to Cornell, 16 Oct. 1957; Revelle to Bronk with attachment, 7 Nov. 1957; Iselin to Brown, 2 and 13 Dec. 1957; Brown to Iselin, 4 and 10 Dec. 1957; and Brown to Colbert, 27 Dec. 1957, Committee on Oceanography, General NRC Earth Sciences 1957, NAS.

Coleman interview.

Chapter 16. Listening, 1946–61

1. Aarons to Iselin, 9 Jan., 27 Aug., and 24 Dec. 1948; Fye to Iselin, 31 Jan. 1948; Iselin to Bumpus and Pollak, 10 Mar. 1948; Fye to BuOrd, 7 Oct. 1948; and Iselin to Aarons, 28 Dec. 1948, box 10; Task Assignments under NOrd-9500 dated 1 Sept. 1945 and 2 Jan. 1946; George P. Woollard, "A Manual for SOFAR Observers," 1 July 1947; Aarons to Iselin, 29 Jan. 1949, "Research Program for Summer 1949," 29 Jan. 1949, and memorandum," Suggestions for work which might logically be performed under an extension of Contract NOrd 9500," 9 Feb. 1949; Hersey to Iselin, 29 Mar. 1949; and Fye to Iselin, 8 and 16 Feb. and 17 Nov. 1949, box 11; Fye to Coles, 18 Jan. 1950; and Aarons to Ryder, 15 Mar. and 24 Apr. 1950, box 12, OD/WHOI.

Bunce interview.

2. OSRD/NDRC Division Six, Section 6.1, "Acoustical Treatment for Submarines, MIT," 15 Aug. 1945, box 155, BuShips Secret Gencorr, 1945; Lee E. Holt (USNUSL), "The German Use of Sonic Listening," 12 Nov. 1947, box 10, series 3, file 6, SOC; and Weir, *Forged in War,* chaps. 3 and 5.

3. "Distribution of Temperature and Salinity in the Western North Atlantic Ocean," 19 Sept. 1947; "Pacific Ocean Biology," 19 Sept. 1947; "SOFAR in the Atlantic Ocean," 19 Sept. 1947; "SOFAR in the Pacific Ocean," 19 Sept. 1947; "Underwater Acoustics," 22 Sept. 1947; "Diving Characteristics of Submarines," 22 Sept. 1947; "Current Measuring Techniques," 22 Sept. 1947; Iselin to Revelle, 16 May 1946; George P. Woollard, "Quarterly Progress Report, 4th Quarter 1946 — Projects B-16 and B-17," n.d., 1946; and Woollard to director, WHOI, 16 July, 8 Aug. (includes breakdown of personnel assigned to B-16 and B-17), 18 Sept., and 20 Nov. 1946 (regarding B-16 and B-17), box 8, OD/WHOI.

4. Woollard to SOFAR Project Officer, Mare Island Naval Shipyard, 20 Dec. 1946; Woollard to Sands, 14 Jan. 1947; BuShips SOFAR Project Officer to file (summation of conference at NEL 13–17 Jan. 1947), n.d., 1947; Bennett, NEL, to all SOFAR Project personnel, 17 Jan. 1947; Timmerman, Mare Island Naval Shipyard, to director, NEL, 24 and 31 Jan. and 7, 15, 21, and 28 Feb. 1947; Holtsmark, SOFAR Project Group II, NEL, to BuShips SOFAR Project Officer, 31 Jan., 24 Feb, and 12 and 20 Apr. 1947; memorandum by Joseph Cryden, NEL, 6 Feb. 1947; SOFAR Project Group III, c/o USCG Operating Base, Monterey, Calif., to director, NEL, 8 and 22 Feb., 1, 3, 8, 14, and 31 Mar., and 4, 12, and 19 Apr. 1947; Iselin to Fleming, 4 Apr. 1947; Sands to BuPers, 13 Feb. 1947; Sands to Iselin 14 Feb. and 1 Mar. 1947; Sands to BuShips Code 940, 14 Feb. 1947; Iselin to Sands, 18 Feb. 1947; Hersey to director, WHOI, 5 Mar. and 7 July 1947; BuOrd to BuShips (Code 940), 10 and 18 Mar. 1947; Tucker to director, WHOI, 18 Mar. 1947; Cole to Iselin, 18 Apr. 1947; "SOFAR Project, Informal Progress Report," 4, 11, 18, 25, and 27 July, and 1 Aug. 1947; Iselin to Walker, 22 Aug. 1947; Hydrographer to commander in chief, Atlantic Fleet et al., 6 Oct. 1947; Iselin to BuShips, 24 Dec. 1947; and Iselin to BuOrd, 31 Dec. 1947, box 9, OD/WHOI.

5. CNO to BuShips, 20 June 1946; William J. Horvath, memorandum for Capt. R. S. Benson, 27 May 1946; memoranda for director, OEG, by Glen D. Camp, "Reduction of Sonic Target Strength of Submarines," 21 May 1946, "Silent Ranging by Spectral Distortion," 21 May 1946, and "Selective Listening Based Upon the Discreet Lines in Ship Spectra, 22 May 1946, box 51; BuShips to asst. chief, Air Staff-4, Guided Missile Branch, 3 and 19 Dec. 1946; BuShips Code 940 to NEL, 17 Dec. 1946; BuShips to deputy chief of the air staff for research and development, 3 Dec. 1946, box 52 , BuShips Secret Gencorr, 1946; and Hamilton interview.

6. Hersey, informal notebook memoir (written with pencil in longhand), n.d., ca. 1960, box 5, John Brackett Hersey Papers, WHOI.

7. Ibid.

8. Project Hartwell, MIT, "A Report on Security of Overseas Transport," 21 Sep 1950, post–1 Jan. 1946 command file, AR/NHC.

9. Ibid.

10. Ibid.

11. Interview with Capt. Joseph Kelly by David K. Allison and John Pitts, 9 Nov. 1984, Navy Laboratories Oral History Collection, DTRC (these records are now resident at AR/NHC).

12. Frederick V. Hunt, "New Concepts for Acoustic Detection at Very Long Ranges" (Bombshell Report), Fifth Undersea Symposium, 15–16 May 1950, NSB.

13. Sonar capable of detecting sounds generated in a frequency below 1,000 cycles, but more often than not in the 100–200 cycle range.

14. Project Hartwell, MIT, "A Report on Security of Overseas Transport"; Momsen to Distribution List with attachment, 26 Jan. 1951, box 7, ser. 2, file 5, SOC; "The Hartwell Project," by Dr. Jerold R. Zacharias, MIT, 9–10 May 1951, NSB; "ASW Surveillance, Phase I," vol. 2, app. A, "History of ASW Surveillance," 28 June 1968, Edward Dahlrymple file cabinet 1, drawer 1, Bldg W-1/Rm 1667, TRW Underwater Surveillance Office Archive, McLean, Va.; and Weir, *Forged in War,* 133–38.

15. Project Hartwell, MIT, "Report on Security of Overseas Transport." See also Gaylord P. Harnwell, "Underwater Detection," 25 Apr. 1947; and memorandum from the Special Panel on Low Frequency Sonar to the CUW, NRC, 20 Oct. 1950, NSB.

16. "Report on Project Kayo Conference, New London, Connecticut, 10–11 Jan. 1950," box 7, RC-21–5, Lasky (these records are now resident at the AR/NHC).

Memorandum from the Special Panel on Low Frequency Sonar to the CUW, NRC. Iselin to ONR, 10 June 1949; Iselin to A. A. Brown, OEG, 29 July 1949; and Brown to Iselin, 27 July 1949, box 11, OD/WHOI; WHOI Audograph Soundwriter Disk Collection 1949–50; and sound recordings of the USS *Amberjack,* 5 Mar. 1949, WHOI. Benson to ComSubLant, "SS vs SS Operations," 22 July 1949, and Kayo Report no. 3, 15 Oct. 1949, box 20, series 5 file A 16–3(3), Project Kayo Incoming Correspondence, SOC.

Hamilton and Worzel interviews.

"Project Jezebel: Final Report on Developmental Contract NObsr-57093," 1 Jan. 1961 (covering the period 1 Nov. 1951 to 1 Jan. 1961); and "ASW Surveillance, Phase 1," vol. 2, app. A, TRW Underwater Surveillance Office Archive, McLean, Va.

17. Hamilton and Worzel interviews. For background on the WHOI technique of using explosives as sound sources see, John Brackett Hersey, Essays III (handwritten text in a computation book), box 9 (unprocessed), John Brackett Hersey Papers, WHOI.

18. USS *San Pablo* file, 1949–50, box 139, CAH/Ewing.

19. Hamilton interview.

20. Worzel to Hamilton, 24 Jan. 1950; Ewing to Hamilton, 28 Mar. 1950; and Hamilton to Ewing, 12 Apr. 1950, file cabinet 3, drawer D, Ewing office files, InGeo/UT.

Procurement justification forms for contract N6onr27124, 5 May and 28 Dec. 1950; Pegram to ONR, 29 May 1950 (with attached contract renewal application for N6 onr 27124 and NR 350 013); contract clearance memorandum between ONR and Columbia University, 8 June 1950; Hamilton to Smith, 12 Sept. 1950; Code 418 (Smith) to Code 100 (regarding contract N6onr27124), 13 Sept. 1950; Smith to Hamilton, 6 Nov. 1950; Edward T. Miller, "San

Pablo Sofar Cruise—New London to Trinidad and Trinidad to Newfoundland Legs," 8 Nov. 1950; and "Proposal for Contract Renewal: N6onr271, Task Órder XIII, for the year May 1, 1951–Apr. 30, 1952," accession no. 62-A-3303, box 3 (of 5), ONR, WNRC.

Memorandum from the Special Panel on Low Frequency Sonar to the CUW, NRC. *Dictionary of American Naval Fighting Ships,* 6:302–303.

21. "Project Jezebel: Final Report"; "ASW Surveillance, Phase 1," vol. 2, app. A.

22. Kelly interview.

23. Ibid.

24. Memorandum from the Special Panel on Low Frequency Sonar to the CUW, NRC; Harnwell, "Underwater Detection"; "Report on Project Kayo Conference"; Hunt, "New Concepts for Acoustic Detection" (Bombshell Report); Ryder to commander, Branch Office of Naval Research, 12 May 1950; and Hart to Ryder, 19 May 1950, box 12, OD/WHOI; Project Hartwell, MIT: "Report on Security of Overseas Transport"; "Project Jezebel: Final Report"; and "ASW Surveillance, Phase 1," vol. 2, app. A.

25. "Project Jezebel: Final Report"; "ASW Surveillance, Phase 1," vol. 2, app. A.

Physical Sciences—Committee on Undersea Warfare; and Wood to Ide, 2 Nov. 1956, NAS.

Marvin Lasky, "Historical Review of Undersea Warfare Planning and Organization 1945–1960 with Emphasis on the Role of the Office of Naval Research," *U.S. Navy Journal of Underwater Acoustics* 26, no. 2 (Apr., 1976): 327–55.

Coleman to Iselin, 24 Aug. 1950; Iselin to Coleman, 5 Sept. 1950; Iselin to Eckart, et al, 22 Sept. 1950; Iselin to Harnwell, 11 Oct. 1950; Coleman to Spitzer, 26 Oct. 1949; Spitzer to Coleman, 7 Nov. 1949; Iselin to Coleman, 14 Nov. 1949; and Hunt to Harnwell, 22 Nov. 1950, drawer B, Iselin Collection, 1944–51; Smith to BuShips, n.d. (ca. 1951), Institution Confidential (not classified) Collection; Hersey: memorandum to Smith, 14 Mar. 1951, box 13; and Smith to ONR, 15 July 1955, file Nonr-1790(00) (Hersey), box 15, OD/WHOI.

26. "Subtask Progress Reports: Research Analysis Group, Brown University," 31 Dec. 1950, accession no. 62-A-3303, box 3, ONR, WNRC; CNO, memorandum on "Current and Projected Program of Investigation in Project Kayo," 6 Mar. 1950, ser. 5, file A16–3(3), box 20, Project Kayo Incoming Correspondence, SOC.

27. Capt. William L. Pryor (BuShips), "Detection of Submarines from the Surface and Below" (paper presented at the Sixth Undersea Symposium, sponsored by the CUW, 9–10 May 1951), NSB.

28. CNR to CNO, 16 June 1949; R. H. Fleming, "Rough Draft Notes on the ASW Program," ONR (Undersea Warfare Branch), 14 Nov. 1949; Project Card: "Detection by Acoustic Methods," ONR Code 466 (Undersea Warfare Branch); and Solberg to BuShips et al, 5 Jan. 1950, accession no. 65-A-4221, box 12; Jordan to CNO et al, 2 Feb. 1950, accession no. 65-A-4221, box 3; and Kock to Murphy, 15 Feb. 1951, accession no. 64-A-4211, box 2, ONR, WNRC; "Project Jezebel: Final Report"; "ASW Surveillance, Phase 1," vol. 2, app. A; and Lasky, "Historical Review of Undersea Warfare Planning and Organization 1945–1960," 327–55.

Chapter 17. A Closer Look, 1955–60

1. The bulk of this chapter is entirely dependent on a series of oral histories done by the author with those involved with *Trieste* from its earliest days. The Naval Undersea Museum in Keyport, Washington, sponsored the series. These oral histories are used in combination as sources throughout this narrative, especially for the details surrounding the Project Nekton deep dive in January, 1960. The interviews include Gordon Lill, 15 Mar. 1977 (by Robert A. Calvert) and 9 May 1995; Allyn Vine, 27 Apr. and 23 Sept. 1989, J. Lamar Worzel, 22 Aug.

1990; Arthur Maxwell, 23 Sept. 1993; Rear Adm. J. B. Mooney, 22 May and 6 July 1995; Capt. Donald Keach, 20 June 1995; John Knauss, 26 June 1995; Capt. George Martin, 29 June 1995; Eugene LaFond, 17 Aug. 1995; Andreas Rechnitzer, 9 Aug. 1995; Capt. Don Walsh, 1 Aug. 1995; Chief Jon Michel, 16 Aug. 1995; and Capt. Charles B. Bishop, 14 Aug. 1995, NHC Oral Histories, AR/NHC.

See also:

Robert Dietz, scientific liaison officer, ONR Branch Office, London, "The Bathyscaphe and Deep Sea Research" (paper presented at the Symposium on Deep Sea Research, 29 Feb.–1 Mar. 1956, Washington D.C.), Committee on Oceanography, NRC Earth Sciences 1957; Bronk to Harrison, 18 July 1957; Brown to Bronk, 9 Aug. 1957; Bronk to Brown, 17 Sept. 1957; and memorandum by Richard C. Vetter, 20 Dec. 1957, NAS.

Arthur Maxwell to participants in Project Bathyscaphe, 22 May 1957, box 17, OD/WHOI.

Robert S. Dietz, "Bathyscaph *Trieste,*" n.d. (late 1957, early 1958); "Data on the big dive," n.d., 1960; and Maxwell to Distribution List, 7 Mar. 1960, box 1, Robert Dietz Papers; Arthur Maxwell et al., "A Preliminary Report on the 1957 Investigations with the Bathyscaph *Trieste,*" n.d., 1957, box 9, Roger Revelle Papers; K. O. Emery to Bascom et al., n.d. 1957; and Lill to Distribution List, 11 Feb. 1957, John Isaacs Papers, SIO.

(C) ONR to commander, Navy Purchasing Office, London, 13 Feb. 1959 (C), box 1, accession no. 65-A-4215; (C) Comdr. Charles Bishop, "Report of Foreign Travel," 16 May 1957 (C), box 9, accession no. 64-A-4211; (C) Comdr. Charles Bishop, "Visit to USNUSL and Hudson, 21–22 Mar.," 26 Mar. 1957 (C); and (C) Mommsen to Walsh, 12 Aug. 1959 (C), box 10; ONR, RG 298, WNRC. (Documents marked "(C)" are still classified *confidential* and are not available to the general public.)

Robert Dietz, "Earth, Sea, and Sky: Life and Times of a Journeyman Geologist," *Review of Earth and Planetary Sciences* 22 (1994): 1–32; Eugene C. LaFond, "Dive Eighty-Four," *Sea Frontier* 8, no. 2 (May, 1962): 94–102; George W. Martin, "*Trieste*—The First Ten Years," *U.S. Naval Institute Proceedings* 90 (Aug., 1964): 52–64; Jacques Piccard and Robert Dietz, *Seven Miles Down;* and R. Frank Busby, *Manned Submersibles,* 90–91; 124–25; 192–93.

2. Piccard and Dietz, *Seven Miles Down;* and Kirby to Knudsen (regarding the proposed University of California *Trieste* development contract), 26 Apr. 1956, box 17, SIO subject files, SIO.

3. Rechnitzer interview.

4. Dietz to Vetter, 13 Feb. 1958, box 9, Roger Revelle Papers, SIO.

5. Piccard and Dietz, *Seven Miles Down.*

6. Here begins a description of the events relating to project Nekton. The oral history inverviews are the primary sources here.

7. Maxwell interview.

8. Rechnitzer interview, 9 Aug. 1995.

9. Walsh interview, 1 Aug. 1995.

10. Worzel interview; Keach interview, 20 June 1995; and Martin interview, 29 June 1995.

11. Mooney interviews, 22 May and 6 July 1995.

Chapter 18. Coming Full Circle

1. These documents were used in support of this section. Because this is a concluding, synoptic, and analytical section, most of the narrative and the conclusions drawn emerged

from the accumulated events and ideas revealed in this study for the entire period from 1946 to 1965. Any new material or ideas introduced here came from the following:

Maxwell and Coleman interviews; and Lill interview, 9 May 1995.

CNO to Distribution List, "Long Range Program for Oceanographic Research" (TENOC), 1 Jan. 1959, CH/NHC. Laurence M. Gould, "Antarctic Prospect," 1–28, NRC central file—1957—IGY—U.S. National Committee Antarctic Program; "Report on the Current Status of the International Program in Oceanography for the International Geophysical Year," May, 1956; "Preview of the International Geophysical Year," 1957; and American IGY Budget, 1957, NRC central file—IGY; "U.S. National Committee: U.S. Program, Western Hemisphere—Proposed," July, 1956, NAS-NRC: IR: IGY 1956; "U.S. National Committee, Summary Report, 1 July 1957–31 Dec. 1958," NAS-NRC: IR: IGY 1965; Bronk to Burke, 8 May 1958; Burke to Bronk, 16 May 1958; Vetter to Cornell, 6 May 1958; Gates to Bronk, 29 Jan. 1959; Cornell to Bronk, 6 Mar. 1958 and 4 Feb. 1959; "Oceanography 1960–1970," NASCO Report, n.d., 1959, Committee on Oceanography, NRC Earth Sciences, NAS.

CNO to Distribution List, 1 Jan. 1959 (regarding Project TENOC), Institution History/Miscellaneous History, 1956–95 file, drawer (no number); Iselin to Aarons, 14 July 1958, box 18, OD/WHOI.

"Project TENOC, The Next Ten Years in Oceanography" (a survey of the growth potential at existing institutions), 1958; and "Information and Comments Pertinent to Reconsideration of the TENOC Plan," 25 Nov. 1960, box 26, Isaacs Papers; "U.S. Navy Project TENOC," box 27, SIO subject files, SIO.

Ships Characteristics Board memorandum no. 108–57, 21 May 1957 (preliminary characteristics for oceanographic research ship, SCB project no. 185, box 10, accession no. 64-A-3979, Insurv.

Barlow, *Revolt of the Admirals,* and E. John Long, *Ocean Sciences* (Annapolis, Md.: U.S. Naval Institute, 1964), see esp. the chapter by Steven Anastasion.

2. Bronk to Burke, 8 May 1958.

3. CNR to CNO, "Project TENOC," 1958, SIO.

4. Ibid.

5. CNO to Distribution List, 1 Jan. 1959.

6. Ibid.

7. Long, *Ocean Sciences,* 188.

8. Ibid., 185.

BIBLIOGRAPHY

Archival Sources

Data Library and Archives, Woods Hole Oceanographic Institution, Woods Hole, Massachusetts

Archives of the Scripps Institution of Oceanography, University of California—San Diego, La Jolla, California

Center for American History, University of Texas, Austin
  W. Maurice Ewing Papers

Marie Tharp, Oceanographic Cartographer, South Nyack, New York
  Personal Collection

National Archives, Washington, D.C.
  Navy Hydrographic Office General Correspondence, 1907–24, RG 37
  Navy Hydrographic Office General Correspondence, 1925–45, RG 37
  Naval Consulting Board Correspondence Files, 1915–23, RG 80
  Office of Naval Research, General Correspondence of the Coordinator of Research and Development, 1941–45, RG 298
  Records of the National Academy of Sciences—Paris Report Series, RG 189
  General Correspondence of the Secretary of the Navy, 1916–26, RG 80
  General Correspondence of the Secretary of the Navy 1926–40, RG 80
  Secretary of the Navy Confidential [unclassified] Correspondence, 1927–39, RG 80
  Records of the CNO–Headquarters COMINCH Confidential [unclassified], RG 38
  BuShips Central Correspondence, 1947, RG 19
  General Board Studies, 1947–49, RG 45
  General Board Subject Files, 1950–51, RG 45

National Archives 2, College Park, Maryland
  Office of Scientific Research and Development/National Defense Research Committee Records, RG 227
  Naval Oceanographic Office, RG 456 [National Imaging and Mapping Agency]

Washington National Records Center, Suitland, Maryland
  BuShips Confidential [declassified] Central Correspondence, RG 19
  BuShips Secret [declassified] General Correspondence, RG 19
  Records of the Office of Naval Research, RG 298
  Records of the Board of Inspection and Survey, RG 181

Archive of the National Academy of Sciences, Washington, D.C.

Navy Laboratories Archive, David Taylor Research Center, Carderock, Maryland

Operational Archives, U.S. Naval Historical Center, Washington, D.C.
  Submarine Officers Conference, Submarines/Undersea Warfare Division Files
  Naval Historical Center Oral History Collection

Contemporary History Branch Collection, U.S. Naval Historical Center, Washington D.C.
Cushing Memorial Library, Texas A&M University, Oceanography Collection (oral history)
American Philosophical Society Library, Philadelphia, Pennsylvania
Institute for Geophysics, Pickles Research Campus Warehouse, University of Texas, Austin
 Office Files of W. Maurice Ewing, Creighton A. Burke, and J. Lamar Worzel
Special Collections, Library of the Museum of Comparative Zoology, Harvard University,
 Cambridge, Massachusetts
 Henry Bryant Bigelow Papers
Naval Surface Warfare Center, White Oak, Maryland
Naval Laboratories Archives
Technical Library, Naval Ocean Systems Center, San Diego, California
 "Completion Report made to the Chief of BuShips covering the operations of the Uni-
  versity of California Division of War Research at the U.S. Navy Electronics Labo-
  ratory, San Diego CA," Contract Nobs-2074 (formerly OEMar-30), 26 April
  1941–30 June 1946
Archive of the Center for Naval Analysis, Alexandria, Virginia

Books

Allard, Dean C. *Spencer Fullerton Baird and the U.S. Fish Commission: A Study in the History of American Science.* New York: Arno Press, 1978.

Barlow, Jeffrey G. *Revolt of the Admirals: The Fight for Naval Aviation, 1945–1950.* Washington, D.C.: Naval Historical Center, 1994.

Baxter, James Phinney III. *Scientists Against Time.* Cambridge, Mass.: MIT Press, 1968

Bigelow, Henry Bryant. *Oceanography: Its Scope, Problems, and Economic Importance.* Boston: Houghton Mifflin, 1931.

Bruce, Robert V. *The Launching of Modern American Science, 1846–1876.* New York: Alfred A. Knopf, 1987.

Burstyn, Harold L. "Reviving American Oceanography: Frank Lillie, Wickliffe Rose and the Founding of the Woods Hole Oceanographic Institution." In *Oceanography: The Past,* ed. Mary Sears and Daniel Merriman. New York: Springer, 1980.

Busby, R. Frank. *Manned Submersibles.* Washington, D.C.: Oceanographer of the Navy, 1976.

Bush, Vannevar. *Pieces of the Action.* New York: William Morrow, 1970.

———. *Science, the Endless Frontier.* Washington, D.C.: GPO, 1945.

Cochrane, Rexmond C. *The National Academy of Sciences: The First Hundred Years.* Washington, D.C.: National Academy of Sciences, 1978.

Davis, George T. *A Navy Second to None: The Development of Modern American Naval Policy.* New York: Harcourt, Brace, 1940.

Dufek, George J. *Operation Deepfreeze.* New York: Harcourt, Brace, 1957.

Dupree, A. Hunter. *Science and the Federal Government: A History of Politics and Activities to 1940.* Cambridge, Mass.: The Belknap Press of the Harvard University Press, 1957.

England, J. Merton. *A Patron for Pure Science: The National Science Foundation's Formative Years 1945–1957.* Washington, D.C.: National Science Foundation, 1982.

Ewing, Maurice, and J. Lamar Worzel. *Long Range Sound Transmission.* Geological Society of America Memoir 27. New York: Geological Society of America, 1948.

Fogg, G. E. *A History of Antarctic Science.* Cambridge, U.K.: Cambridge University Press, 1992.

Geertz, Clifford. *Local Knowledge: Further Essays in Interpretative Anthropology*. New York: Basic Books, 1983.

Hackmann, Willem. *Seek and Strike: Sonar, Anti-submarine Warfare, and the Royal Navy, 1914–54*. London: HMSO, 1984.

Hadley, Michael L. *U-boats Against Canada*. Kingston, Ont.: McGill-Queen's University Press, 1985.

Haedrich, Richard L., and Emery, Kenneth O. "Growth of an Oceanographic Institution." In *Oceanography: The Past*, ed. Mary Sears and Daniel Merriman. New York: Springer, 1980.

Hersey, John Brackett, ed. *Deep Sea Photography*. Baltimore: Johns Hopkins University Press, 1967.

History of the Bureau of Engineering, Navy Department, During the World War. Washington, D.C.: Navy Department, 1922.

Hughes, Thomas Parke. *Elmer Sperry: Inventor and Engineer*. Baltimore: Johns Hopkins University Press, 1971.

Ingmanson, Dale E., and William J. Wallace. *Oceanography: An Introduction*. Belmont, Calif.: Wadsworth, 1989.

Kargon, Robert H. *The Rise of Robert Millikan: Portrait of a Life in American Science*. Ithaca, N.Y.: Cornell University Press, 1982.

Kaufman, Robert Gordon. *Arms Control in the Pre-Nuclear Era: The United States and Naval Limitation Between the Two World Wars*. New York: Columbia University Press, 1990.

Kevles, Daniel J. *The Physicists*. New York: Alfred A. Knopf, 1978.

Ingmanson, Dale E., and William J. Wallace, *Oceanography: An Introduction*. Belmont, Calif.: Wadsworth, 1989.

Linklater, Eric. *The Voyage of the Challenger*. London: Cardinal Books, 1974.

Long, E. John. *Ocean Sciences*. Annapolis, Md.: U.S. Naval Institute, 1964.

Maurer, Hans. *Die Echolotungen des "Meteor." Band 2. Wissenschaftliche Ergebnisse der Deutschen Atlantischen Expedition auf dem Forschungs und Vermessungsschiff Meteor, 1925–1927*. Berlin und Leipzig: Verlag von Walter de Gruyter, 1933.

MacKenzie, Donald. *Inventing Accuracy: A Historical Sociology of Nuclear Missile Guidance*. Cambridge, Mass.: MIT Press, 1993.

Mills, Eric. "Problems of Deep Sea Biology." In *The Sea*. Vol. 8, *Deep Sea Biology*, ed. G. T. Rowe. New York: John Wiley, 1983.

Milner, Marc. *North Atlantic Run: The Royal Canadian Navy and the Battle for the Convoys*. Annapolis, Md.: Naval Institute Press, 1985.

———. *The U-boat Hunters: The Royal Canadian Navy and the Offensive Against Germany's Submarines*. Annapolis, Md.: Naval Institute Press, 1994.

Naval History Division, *Dictionary of American Naval Fighting Ships*. Vols. 1–8. Washington, D.C.: Naval History Division, 1959–81.

The Navy-Princeton Gravity Expedition to the West Indies in 1932. Washington, D.C.: U.S. Naval Hydrographic Office, 1933.

Nelson, Stewart B. *Oceanographic Ships Fore and Aft*. Washington D.C.: Office of the Oceanographer of the Navy, 1982.

Oreskes, Naomi. "Weighing the Earth from a Submarine: The Gravity Measuring Cruise of the U.S.S. S-21." In *The Earth, the Heavens, and the Carnegie Institution of Washington: Historical Perspectives After Ninety Years*, ed. Gregory Good. Washington, D.C.: Carnegie Institution, 1993.

Piccard, Jacques, and Robert Dietz. *Seven Miles Down*. New York: G. P Putnam's Sons, 1961.

Pinsel, Marc I. *150 Years of Service on the Seas: A Pictorial History of the U.S. Naval Oceanographic Office from 1830 to 1980.* Vol. 1, *1830–1946.* Washington, D.C.: U.S. Naval Oceanographic Office, 1981.

Raitt, Helen, and Beatrice Moulton. *Scripps Institution of Oceanography: The First Fifty Years.* Los Angeles: Ward Ritchie Press, 1967.

Rössler, Eberhard. *Geschichte des deutschen Ubootbaus.* Munich: J. F. Lehmanns Verlag, 1975.

Sapolsky, Harvey M. *Science and the Navy: The History of the Office of Naval Research.* Princeton, N.J.: Princeton University Press, 1990.

Schlee, Susan. *On the Edge of an Unfamiliar World: A History of Oceanography.* New York: E. P. Dutton, 1973.

———. "The R/V *Atlantis* and Her First Oceanographic Institution." In *Oceanography: The Past,* ed. Mary Sears and Daniel Merriman. New York: Springer, 1980.

Schott, W. *Early German Oceanographic Institutions, Expeditions, and Oceanographers.* Hamburg: DHI, 1987.

Scott, Lloyd N. *Naval Consulting Board of the United States.* Washington, D.C.: Navy Department, 1920.

Shor, Elizabeth Nobel. *Scripps Institution of Oceanography: Probing the Oceans, 1936 to 1976.* San Diego: Tofua Press, 1978.

Shurcliff, W. A. *Bombs at Bikini: The Official Report of Operation Crossroads.* New York: William H. Wise, 1947.

Spiess, Fritz. *The Meteor Expedition: Scientific Results of the German Atlantic Expedition, 1925–1927.* Trans. William J. Emery. New Delhi: Amerind, 1985.

Stewart, Irvin. *Organizing Scientific Research for War: The Administrative History of the Office of Scientific Research and Development.* Boston: Little, Brown, 1948.

Taylor, A. Hoyt. *The First Twenty-Five Years of the Naval Research Laboratory.* Washington, D.C.: Navy Department, 1952.

Vaughan, Thomas Wayland et. al. *International Aspects of Oceanography.* Washington D.C.: NAS, 1937.

Vening-Meinesz, F. A., and F. E. Wright. *The Gravity Measuring Cruise of the U.S. Submarine S-21.* Washington, D.C.: Publications of the U.S. Naval Observatory, 1930.

Weir, Gary E. *Forged in War: The Naval-Industrial Complex and American Submarine Construction, 1940–1961.* Washington, D.C.: Naval Historical Center, 1993.

Williams, Frances Leigh. *Matthew Fontaine Maury: Scientist of the Sea.* New Brunswick, N.J.: Rutgers University Press, 1963.

Zimmerman, David. *The Great Naval Battle of Ottowa.* Toronto: University of Toronto Press, 1989.

Articles, Dissertations, and Papers

Auerbach, Lewis E. "Scientists in the New Deal: A Pre-War Episode in the Relations between Science and the Government in the United States." *Minerva* 3 (summer, 1965): 459–61.

Bigelow, Henry B. "A Developing Viewpoint in Oceanography." *Science,* 24 January 1930, 84–89.

Bowie, William. "Weighing the Earth from a Submarine." *Scientific American,* March, 1929, 218–21.

Brown, Thomas Townsend. "The Navy-Princeton Gravity Expedition to the West Indies in 1932." *Hydrographic Review* 11, no. 2 (1934): 82–89.

Constant, Edward W. III. "Genesis N + 1: The Origins of the Turbojet Revolution." Ph.D. diss., Northwestern University, 1976.

De Solla Price, Derek J. "Is Technology Historically Independent of Science? A Study in Statistical Historiography." *Technology and Culture* 6, no. 4 (1965): 553–95.

Dietz, Robert S. "Earth, Sea, and Sky: Life and Times of a Journeyman Geologist." *Annual Review of Earth and Planetary Sciences* 22, (1994), 1–32.

Ewing, Maurice. "Gravity Measurements on the USS *Barracuda*." *Transactions of the American Geophysical Union,* 1937, 66–69.

———. "Exploring the Mid-Atlantic Ridge," *National Geographic* 94, no. 3 (September, 1948): 275–94.

Heck, N. H. "The Role of Earthquakes and the Seismic Method in Submarine Geology." *Proceedings of the American Philosophical Society* 79 I (April, 1938): 97–108.

Hess, Harry H. "Geological Interpretation of Data Collected on Cruise of USS *Barracuda* in the West Indies—Preliminary Report." *Transactions of the American Geophysical Union,* 1937, 69–77.

———. "Gravity Anomalies and Island Arc Structure with Particular Reference to the West Indies." *Proceedings of the American Philosophical Society* 79 (April, 1938): 71–96.

Hobbs, William H. "A Proposed New 'Challenger' Exploring Expedition," *Science,* 7 March 1924, 237–38.

LaFond, Eugene C. "Dive Eighty-Four." *Sea Frontiers* 8, no. 2 (May, 1962): 94–102.

———. "Arctic Oceanography by Submarines." *U.S. Naval Institute Proceedings* 86, no. 9 (September, 1960): 90–96.

Lasky, Marvin. "A Historical Review of Underwater Acoustic Technology 1916–1939 with Emphasis on Undersea Warfare." *U.S. Navy Journal of Underwater Acoustics* 24, no. 4 (October, 1974): 559–601.

———. "Historical Review of Undersea Warfare Planning and Organization 1945–1960 with Emphasis on the Role of the Office of Naval Research." *U.S. Navy Journal of Underwater Acoustics* 26, no. 2 (April, 1976): 327–55.

Layton, Edwin T., Jr. "Technology as Knowledge." *Technology and Culture* 15, no. 1 (1974): 31–41.

Littlehales, George W. "In Relation to the Extent of Knowledge Concerning the Oceanography of the Pacific." *Proceedings of the National Academy of Sciences* 2 (1916): 419–21.

———. "The International Scientific Expedition to the West Indies in 1932." *Hydrographic Review* 9, no. 2 (1932): 124–28.

Martin, George W. "*Trieste*—The First Ten Years." *U.S. Naval Institute Proceedings* 90 (August, 1964): 52–64.

Mills, Eric. "From Marine Ecology to Biological Oceanography." *Helgoländer Meeresuntersuchungen* 49 (1995): 29–44.

———. "Bringing Oceanography into the Canadian University Classroom." *Scienta Canadensis* 18, no. 1 (winter, 1994–95): 3–21.

———. "'Physische Meereskunde': From Geography to Physical Oceanography in the Institute for Meereskunde, Berlin, 1900–1935." *Historisch-Meereskundliches Jahrbuch* 4 (1997): 45–70.

Schlee, Susan. "Real Men Won't Winter on Cape Cod—A Bit of Unofficial W.H.O.I. History." *Woods Hole Notes,* February, 1973.

Spilhaus, Athelstan. "On Reaching 50: An Early History of the Bathythermograph." *Sea Technology* 28, no. 11 (November, 1987): 19–28.

van Keuren, David K. "Science, Progressivism, and Military Preparedness: The Case of the

Naval Research Laboratory, 1915–1923." *Technology and Culture* 33 (October, 1992): 724, 729–30.

Vening-Meinesz, F. A. "Gravity Expedition of the U.S. Navy." *Nature,* 23 March 1929, 473–75.

Weir, Gary E. "Antecedents and Origins: The Pacific Science Congresses and the Origins of International Scientific Cooperation in the Pacific Region, 1920–1943." Paper presented at the Pacific Basin Meeting of the Oceanography Society, Honolulu, June, 1994.

Worthington, L. V. "Oceanographic Results of Project Skijump I and Skijump II in the Polar Sea, 1951–1952." *Transactions of the American Geophysical Union* 32, no. 4 (August, 1953): 543–49.

INDEX

GARY E. WEIR heads the Contemporary History Branch of the Naval Historical Center in Washington, D.C. His other books include *Forged in War: The Naval-Industrial Complex and American Submarine Construction, 1940–1961*.